MyWorkBook

Christine Verity

Developmental Mathematics: College Mathematics and Introductory Algebra

Eighth Edition

Marvin L. Bittinger
Indiana University Purdue University Indianapolis

Judith A. Beecher
Indiana University Purdue University Indianapolis

Addison-Wesley
is an imprint of

Reproduced by Pearson Addison-Wesley from electronic files supplied by the author.

Publishing as Pearson Addison-Wesley, 75 Arlington Street, Boston, MA 02116.

ISBN-13: 978-0-321-73090-9
ISBN-10: 0-321-73090-9

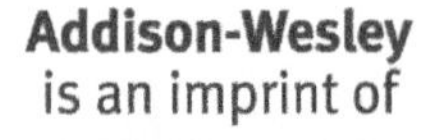

www.pearsonhighered.com

19 2023

Table of Contents

Name: Date:
Instructor: Section:

Chapter 1 WHOLE NUMBERS

1.1 Standard Notation; Order

Learning Objectives
A Give the meaning of digits in standard notation.
B Convert from standard notation to expanded notation.
C Convert between standard notation and word names.
D Use < or > for □ to write a true sentence in a situation like 6 □ 10.

Key Terms
Use the vocabulary terms listed below to complete each statement in Exercises 1–7.

natural	**whole**	**digit**	**standard**
expanded	**equation**	**inequality**	

1. The number 5 thousands + 4 hundreds + 9 tens + 2 ones is expressed in ____________________ notation.

2. The set 0, 1, 2, 3, 4,... is the set of ____________________ numbers.

3. A sentence like $2 + 3 = 5$ is an ________________ .

4. A ____________________ is a number 0, 1, 2, 3, 4, 5, 6, 7, 8, or 9 that names a place-value location.

5. The set 1, 2, 3, 4,... is the set of ____________________ numbers.

6. The number 36,205 is expressed in ____________________ notation.

7. A sentence like $2 < 3$ is an ________________ .

GUIDED EXAMPLES AND PRACTICE

Objective A Give the meaning of digits in standard notation.

Review these examples for Objective A:

1. What does the digit 7 mean in 4,678,952?

   4,6[7]8,952 7 ten thousands

2. In 816,304,259, which digit tells the number of hundreds?

   816,304,[2]59
   The digit 2 tells the number of hundreds.

Practice these exercises:

1. What does the 2 mean in 516,204?

2. In 124,806,357, which digit tells the number of ten thousands?

Objective B Convert from standard notation to expanded notation.

Review this example for Objective B:

3. Write expanded notation for 12,309.

12,309 = 1 ten thousand + 2 thousands + 3 hundreds + 0 tens + 9 ones, or 1 ten thousand + 2 thousands + 3 hundreds + 9 ones

Practice this exercise:

3. Write expanded notation for 2087.

Objective C Convert between standard notation and word names.

Review these examples for Objective C:

4. Write the word name for 36,760,235.

The first period denotes millions. There are thirty-six millions. The second period denotes thousands. There are seven hundred sixty thousands. The last period denotes ones. There are two hundred thirty-five ones.
Thus, a word name for 36,760,235 is thirty-six million, seven hundred sixty thousand, two hundred thirty-five.

5. Write standard notation for eighty-six million, one hundred twenty-three thousand, seven hundred sixty-one.

The number named in the millions period is 86, the number named in the thousands period is 123, and the number named in the ones period is 761. We write each of these numbers in order, separating them with commas. Thus, standard notation is 86,123,761.

Practice these exercises:

4. Write a word name for 5,487,203.

5. Write standard notation for four hundred sixty-five thousand, eight hundred thirteen.

Objective D Use $<$ or $>$ for $\square$ to write a true sentence in a situation like 6 $\square$ 10.

Review this example for Objective D:

6. *Use $<$ or $>$ for $\square$ to write a true sentence:* 23 $\square$ 16.

Since 23 is to the right of 16 on a number line, $23 > 16$.

Practice this exercise:

6. *Use $<$ or $>$ for $\square$ to write a true sentence:* 33 $\square$ 36.

Name: Date:
Instructor: Section:

ADDITIONAL EXERCISES

Objective A Give the meaning of digits in standard notation.

For extra help, see Examples 1–7 on pages 2–3 of your text and the Section 1.1 lecture video.

What does the digit 8 mean in each number?

1. 231,708

2. 897,435

In the number 5,643,970, what digit names the number of:

3. hundreds?

4. ten thousands?

Objective B Convert from standard notation to expanded notation.

For extra help, see Examples 8–10 on page 4 of your text and the Section 1.1 lecture video.

Write expanded notation for each number.

5. 1813

6. 7406

7. 85,126

8. 653, 497

9. 4,306,749

Objective C Convert between standard notation and word names.

For extra help, see Examples 11–15 on page 5 of your text and the Section 1.1 lecture video.

Write a word name.

10. 3452

11. 77,422

Write standard notation.

12. One thousand, two hundred sixty-three

13. Four hundred fourteen thousand, nine hundred sixty-three

14. Three billion

Objective D Use < or > for $\square$ to write a true sentence in a situation like 6 $\square$ 10.

For extra help, see Examples 16–17 on page 6 of your text and the Section 1.1 lecture video.

Use < or > for $\square$ to write a true sentence. Draw the number line if necessary.

15. 51 $\square$ 37

16. 134 $\square$ 143

17. 216 $\square$ 202

18. 1098 $\square$ 1046

Name: Date:
Instructor: Section:

Chapter 1 WHOLE NUMBERS

1.2 Addition and Subtraction

Learning Objectives
A Add whole numbers.
B Use addition in finding perimeter.
C Subtract whole numbers.

Key Terms
Use the vocabulary terms listed below to complete each statement in Exercises 1–4.

difference	**subtrahend**
commutative law of addition	**associative law of addition**

1. The statement $2 + 5 = 5 + 2$ illustrates the ________________ .

2. In a subtraction sentence, the number being subtracted is the ________________ .

3. The statement $(2 + 5) + 3 = 2 + (5 + 3)$ illustrates the ________________ .

4. The ________________ $a - b$ is that unique number c for which $a = c + b$.

GUIDED EXAMPLES AND PRACTICE

Objective A Add whole numbers.

Review this example for Objective A:

1. Add: 8429 + 4098.

$$\begin{array}{r} {}^{1\ 1} \\ 8\,4\,2\,9 \\ +\ 4\,0\,9\,8 \\ \hline 1\,2,5\,2\,7 \end{array}$$

We add ones. We get 17, so we have 1 ten + 7 ones. Write 7 in the ones column and 1 above the tens.
We add tens. We get 12 tens, so we have 1 hundred and 2 tens. Write 2 in the tens column and 1 above the hundreds.
We add hundreds. We get 5.
We add thousands. We get 12.

Practice this exercise:

1. Add: 27,609 + 38,415.

Objective B Use addition in finding perimeter.

Review this example for Objective B:

2. Find the perimeter.

2 mi

14 mi

$$\text{Perimeter} = 14 \text{ mi} + 2 \text{ mi} + 14 \text{ mi} + 2 \text{ mi} = 32 \text{ mi}$$

Practice this exercise:

2. Find the perimeter.

56 in.

22 in.

Objective C Subtract whole numbers.

Review this example for Objective C:

3. Subtract: 8045 – 2897.

$$\begin{array}{r} {}^{7}\,{}^{9}\,{}^{13}\,{}^{15} \\ \not{8}\,\not{0}\,\not{4}\,\not{5} \\ -\ 2\,8\,9\,7 \\ \hline 5\,1\,4\,8 \end{array}$$

We cannot subtract 7 ones from from 5 ones.
We borrow 1 ten to get 15 ones.
We cannot subtract 9 tens from 3 tens.
We borrow 1 hundred to get 13 tens.
We have 79 hundreds.

Practice this exercise:

3. Subtract: 6401 – 3629.

ADDITIONAL EXERCISES

Objective A Add whole numbers.

For extra help, see Examples 1–2 on pages 11–12 of your text and the Section 1.2 lecture video.

Add.

1. $\begin{array}{r} 95 \\ +\ 86 \\ \hline \end{array}$

2. $5346 + 784$

3. $\begin{array}{r} 19 \\ 36 \\ 78 \\ +\ 54 \\ \hline \end{array}$

4. $\begin{array}{r} 13,473 \\ 4,519 \\ +\ \ 7,395 \\ \hline \end{array}$

Name: Date:
Instructor: Section:

Objective B Use addition in finding perimeter.
For extra help, see Examples 3–4 on page 13 of your text and the Section 1.2 lecture video.
Find the perimeter of each figure.

5.

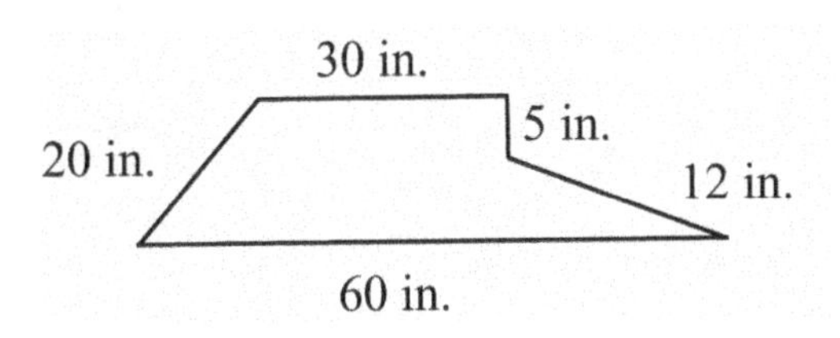

6.

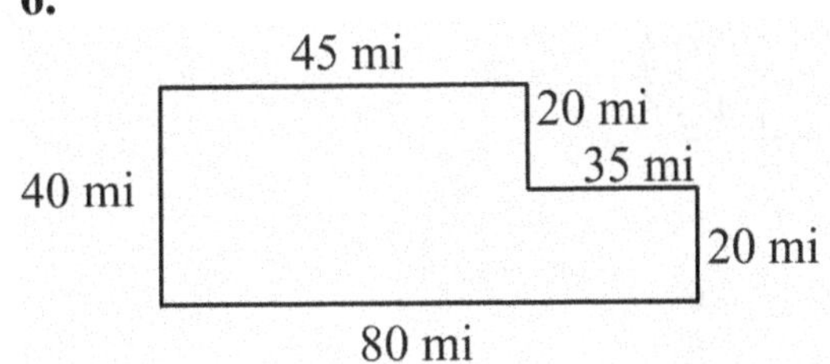

7.

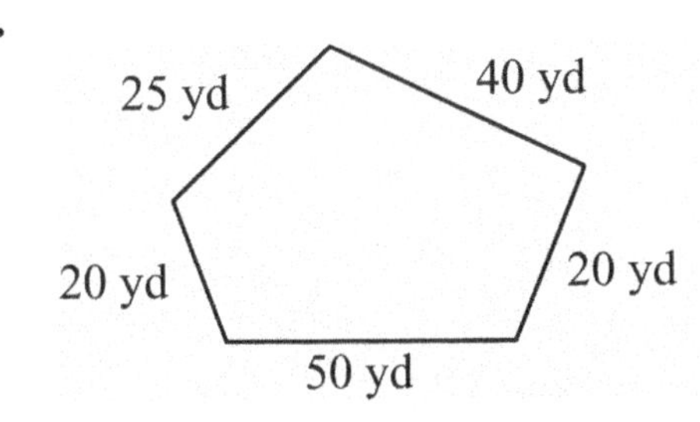

8.

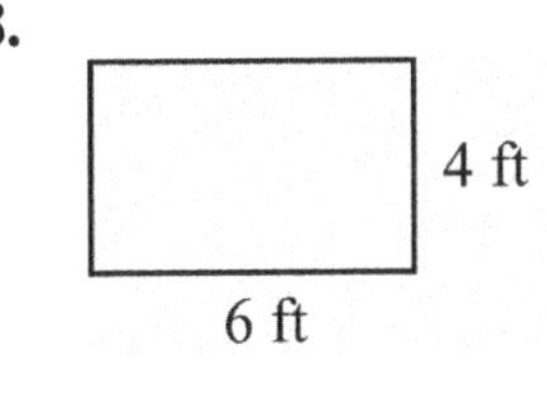

Objective C Subtract whole numbers.
For extra help, see Examples 5–11 on pages 14–16 of your text and the Section 1.2 lecture video.
Subtract.

9. $\begin{array}{r} 764 \\ -\,531 \\ \hline \end{array}$

10. $524 - 139$

11. $\begin{array}{r} 9352 \\ -6418 \\ \hline \end{array}$

12. $\begin{array}{r} 32,156 \\ -\ \ 3,492 \\ \hline \end{array}$

13. $\begin{array}{r} 6005 \\ -1456 \\ \hline \end{array}$

Name: Date:
Instructor: Section:

Chapter 1 WHOLE NUMBERS

1.3 Multiplication and Division; Rounding and Estimating

Learning Objectives
A Multiply whole numbers.
B Use multiplication in finding area.
C Divide whole numbers.
D Round to the nearest ten, hundred, or thousand.
E Estimate sums, differences, products, and quotients by rounding.

Key Terms
Use the vocabulary terms listed below to complete each statement in Exercises 1–7.

one **zero** **not defined** **that same number**
associative law of multiplication **commutative law of multiplication**
distributive law

1. Zero divided by any nonzero number is ________________ .

2. Any number divided by 1 is ________________ .

3. The sentence $2\cdot(3+5)=(2\cdot3)+(2\cdot5)$ illustrates the ________________ .

4. The sentence $2\cdot3=3\cdot2$ illustrates the ________________ .

5. Division by 0 is ________________ .

6. Any nonzero number divided by itself is ________________ .

7. The sentence $2\cdot(3\cdot5)=(2\cdot3)\cdot5$ illustrates the ________________ .

GUIDED EXAMPLES AND PRACTICE

Objective A Multiply whole numbers.

Review this example for Objective A:
1. Multiply: 37×415.

$$\begin{array}{r} {\scriptstyle 1} \\ {\scriptstyle 1\ 3} \\ 4\ 1\ 5 \\ \times\ \ 3\ 7 \\ \hline 2\ 9\ 0\ 5 \\ 1\ 2\ 4\ 5\ 0 \\ \hline 1\ 5,3\ 5\ 5 \end{array}$$

← Multiplying 415 by 7
← Multiplying 415 by 30

Practice this exercise:
1. Multiply: 238×764.

Objective B Use multiplication in finding area.

Review this example for Objective B:

2. Find the area.

2 mi

14 mi

$A = l \cdot w = 14 \text{ mi} \cdot 2 \text{ mi} = 28 \text{ sq mi}$

Practice this exercise:

2. Find the area.

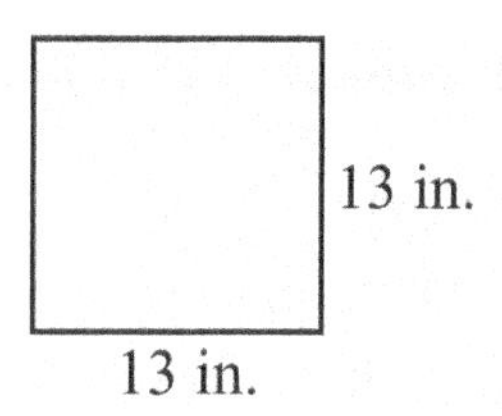

Objective C Divide whole numbers.

Review this example for Objective C:

3. Divide: $8973 \div 36$.

```
     249
36)8973
   72↓     Bring down the 7
   177
   144↓    Bring down the 3
    333
    324
      9    The remainder is 9.
```

The answer is 249 R 9.

Practice this exercise:

3. Divide: $8519 \div 27$.

Objective D Round to the nearest ten, hundred, or thousand.

Review this example for Objective D:

4. Round 8365 to the nearest hundred.

a) Locate the digit in the hundreds place, 3.

8365
↑

b) Consider the next digit to the right, 6.

8365
↑

c) Since that digit, 6, is 5 or higher, round 3 hundreds up to 4 hundreds.

d) Change all digits to the right of the hundreds digit to zeros.

8400 ← This is the answer.

Practice this exercise:

4. Round 27,459 to the nearest thousand.

Name: Date:
Instructor: Section:

Objective E Estimate sums, differences, products, and quotients by rounding.

Review these examples for Objective E:

5. Estimate this sum by first rounding to the nearest ten: 84 + 35 + 49 + 22.

$$\begin{array}{r} 84 \\ 35 \\ 49 \\ +22 \\ \hline \end{array} \qquad \begin{array}{r} 80 \\ 40 \\ 50 \\ +20 \\ \hline 190 \end{array} \leftarrow \text{Estimated answer}$$

6. Estimate this difference by first rounding to the nearest hundred: 7546 – 3271.

$$\begin{array}{r} 7546 \\ -3271 \\ \hline \end{array} \qquad \begin{array}{r} 7500 \\ -3300 \\ \hline 4200 \end{array} \leftarrow \text{Estimated answer}$$

7. Estimate this product by first rounding to the nearest thousand: 4532×8291.

$$\begin{array}{r} 4532 \\ \times 8291 \\ \hline \end{array} \qquad \begin{array}{r} 5000 \\ \times 8000 \\ \hline 40{,}000{,}000 \end{array} \leftarrow \text{Estimated answer}$$

8. Estimate this quotient by first rounding to the nearest thousand: $41{,}624 \div 5803$.

$$5803\overline{)41{,}624} \qquad \begin{array}{r} 7 \\ 6000\overline{)42{,}000} \\ 42{,}000 \\ \hline 0 \end{array}$$

Practice these exercises:

5. Estimate this sum by first rounding to the nearest thousand.

$$\begin{array}{r} 2764 \\ 9076 \\ +4528 \\ \hline \end{array}$$

6. Estimate this difference by first rounding to the nearest hundred.

$$\begin{array}{r} 6328 \\ -4291 \\ \hline \end{array}$$

7. Estimate this product by first rounding to the nearest ten.

$$\begin{array}{r} 23 \\ \times 77 \\ \hline \end{array}$$

8. Estimate this quotient by first rounding to the nearest hundred: $3472 \div 670$.

ADDITIONAL EXERCISES

Objective A Multiply whole numbers.

For extra help, see Examples 1–4 on pages 21–23 of your text and the Section 1.3 lecture video.

Multiply.

1. $\begin{array}{r} 43 \\ \times \ \ 6 \\ \hline \end{array}$

2. $43(62)$

3. $\begin{array}{r} 293 \\ \times 547 \\ \hline \end{array}$

4. $\begin{array}{r} 8042 \\ \times 7633 \\ \hline \end{array}$

Objective B Use multiplication in finding area.

For extra help, see Example 5 on page 23 of your text and the Section 1.3 lecture video.

Find the area of each figure.

5.

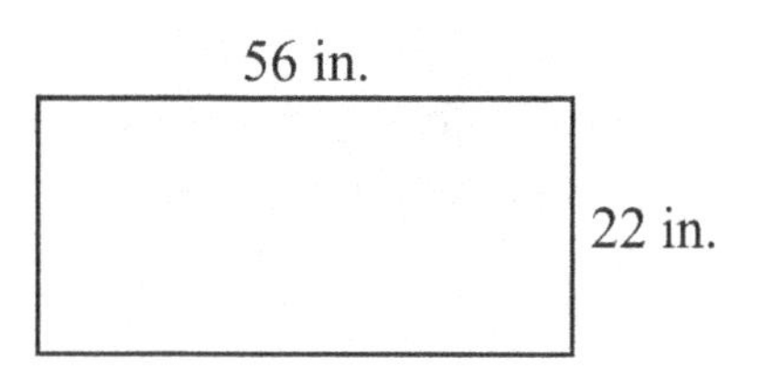

6.

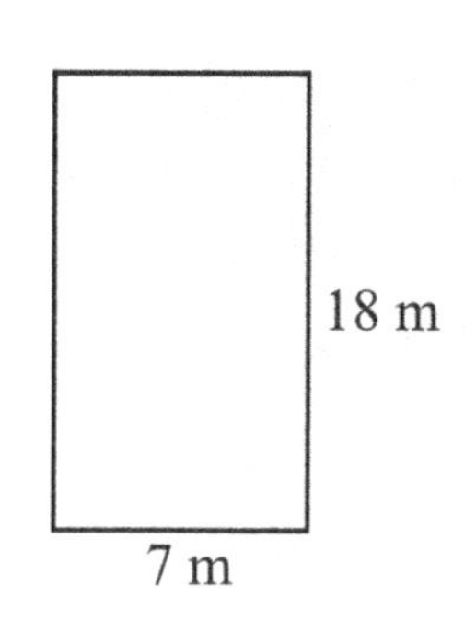

7.

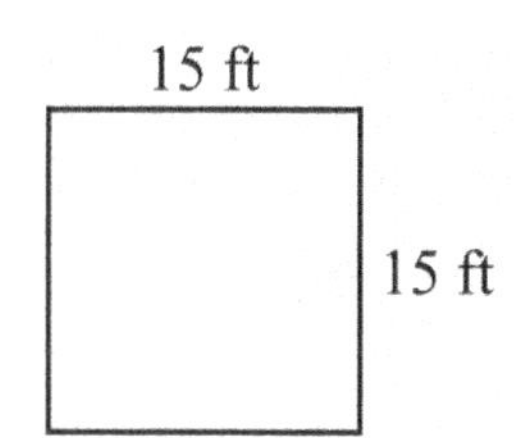

Objective C Divide whole numbers.

For extra help, see Examples 6–10 on pages 25–29 of your text and the Section 1.3 lecture video.

Divide, if possible. If not possible, write "not defined."

8. $\frac{42}{1}$

9. $45 \div 0$

Name: Date:
Instructor: Section:

Divide.

10. $738 \div 6$

11. $5\overline{)948}$

12. $24\overline{)2592}$

Objective D Round to the nearest ten, hundred, or thousand.
For extra help, see Examples 11–17 on pages 30–32 of your text and the Section 1.3 lecture video.

13. Round 763 to the nearest ten.

14. Round 3063 to the nearest hundred.

15. Round 65,812 to the nearest thousand.

Objective E Estimate sums, differences, products, and quotients by rounding.
For extra help, see Examples 18–23 on pages 32–34 of your text and the Section 1.3 lecture video.

16. Estimate this sum by first rounding to the nearest ten: 51 + 67 + 29 + 43.

17. Estimate this difference by first rounding to the nearest hundred: 6783 – 4246.

18. Estimate this product by first rounding to the nearest hundred: 870×209.

19. Estimate this quotient by first rounding to the nearest ten: $278 \div 72$.

20. Movie rentals cost $4.95 each. The total movie rental sales for Tuesday was $905.85. Estimate the number of movies rented for Tuesday.

Name: Date:
Instructor: Section:

Chapter 1 WHOLE NUMBERS

1.4 Solving Equations

Learning Objectives
A Solve simple equations by trial.
B Solve equations like $x+28=54$, $28 \cdot x=168$, and $98 \cdot 2=y$.

Key Terms
Use the vocabulary terms listed below to complete each statement in Exercises 1–3.

equation **solution of an equation** **variable**

1. A(n) ________________ is a replacement for the variable that makes the equation true.

2. A(n) ________________ is a sentence with =.

3. A(n) ________________ can represent any number.

GUIDED EXAMPLES AND PRACTICE

Objective A Solve simple equations by trial.

Review this example for Objective A:

1. Solve $x+5=12$ by trial.

We replace x with several numbers.

If we replace x with 5, we get a false equation: $5+5=12$.
If we replace x with 6, we get a false equation: $6+5=12$.
If we replace x with 7, we get a true equation: $7+5=12$.

No other replacement makes the equation true, so the solution is 7.

Practice this exercise:

1. Solve $x-2=6$ by trial.

Objective B Solve equations like $x+28=54$, $28 \cdot x=168$, and $98 \cdot 2=y$.

Review these examples for Objective B:

2. Solve $t+15=32$.

$$t+15=32$$
$$t+15-15=32-15 \quad \text{Subtracting 15 on both sides}$$
$$t+0=17$$
$$t=17$$

The solution is 17.

Practice these exercises:

2. Solve $y+8=9$.

3. Solve $x = 156 \times 18$.

To solve the equation, we carry out the calculation.

$$\begin{array}{r} 156 \\ \times\ 18 \\ \hline 1248 \\ 1560 \\ \hline 2808 \end{array} \qquad \begin{aligned} x &= 156 \times 18 \\ x &= 2808 \end{aligned}$$

The solution is 2808.

3. Solve $46 \times 61 = n$.

4. Solve $16 \cdot n = 416$.

$$16 \cdot n = 416$$

$$\frac{16 \cdot n}{16} = \frac{416}{16} \quad \text{Dividing by 16 on both sides}$$

$$n = 26$$

The solution is 26.

4. Solve $24 \cdot y = 912$.

ADDITIONAL EXERCISES

Objective A Solve simple equations by trial.

For extra help, see Examples 1–3 on page 42 of your text and the Section 1.4 lecture video.

Solve by trial.

1. $y - 5 = 8$

2. $x + 3 = 7$

3. $48 \div 6 = t$

4. $x \cdot 13 = 26$

Objective B Solve equations like $x + 28 = 54$, $28 \cdot x = 168$, and $98 \cdot 2 = y$.

For extra help, see Examples 4–11 on pages 43–45 of your text and the Section 1.4 lecture video.

Solve. Be sure to check.

5. $108 \div 9 = w$

6. $7 + x = 42$

7. $150 = 6 \cdot y$

8. $8001 - 1469 = x$

Name: Date:
Instructor: Section:

Chapter 1 WHOLE NUMBERS

1.5 Applications and Problem Solving

Learning Objectives
A Solve applied problems involving addition, subtraction, multiplication, or division of whole numbers.

Key Terms
Use the vocabulary terms listed below to complete each statement in Exercises 1–2.

translate the problem to an equation **familiarize yourself with the situation**

1. The first step in solving an applied problem is to ________________ .

2. The second step in solving an applied problem is to ________________ .

GUIDED EXAMPLES AND PRACTICE

Objective A Solve applied problems involving addition, subtraction, multiplication, or division of whole numbers.

Review this example for Objective A:

1. Margaret borrows \$8820 to buy a car. The loan is to be paid off in 36 equal monthly payments. How much is each payment (excluding interest)?

   1. *Familiarize*. Visualize a rectangular array of dollar bills with 36 rows. How many dollars are in each row?
   Let p = the amount of each payment.
   2. *Translate*. We translate to an equation.

Amount of loan	divided by	Number of payments	is	Amount of each payment
↓	↓	↓	↓	↓
8820	÷	36	=	p

Practice this exercise:

1. Rex is driving from Las Vegas to Chicago, a distance of 1749 miles. He travels 1399 miles to Des Moines. How much farther must he travel?

3. *Solve*. We carry out the division.

$$\begin{array}{r} 245 \\ 36\overline{)8820} \\ \underline{72} \\ 162 \\ \underline{144} \\ 180 \\ \underline{180} \\ 0 \end{array}$$

4. *Check*. We can repeat the calculation. We can also multiply the number of payments by the amount of each payment: $36 \cdot 245 = 8820$. The answer checks.

5. *State*. Each payment is $245.

ADDITIONAL EXERCISES

Objective A Solve applied problems involving addition, subtraction, multiplication, or division of whole numbers.

For extra help, see Examples 1–8 on pages 50–57 of your text and the Section 1.5 lecture video.

Solve.

1. In her job as a telemarketer, Jody contacted 952 customers in September, 1058 customers in October, 857 customers in November, and 1314 customers in December. What was the total number of customers contacted?

2. Donna's car gets 26 miles to the gallon in city driving. How many gallons will she use in 1092 miles of city driving?

3. Hudson Manufacturing buys 315 office chairs at $160 each for use at its new corporate headquarters. Find the total cost of the purchase.

4. In 2004, there were 773,200 jobs in the home healthcare industry. It is estimated that there will be 1,310,300 jobs in this industry in 2014. Find the increase in jobs from 2004 to 2014.

Name: Date:
Instructor: Section:

Chapter 1 WHOLE NUMBERS

1.6 Exponential Notation and Order of Operations

Learning Objectives
- A Write exponential notation for products such as $4 \cdot 4 \cdot 4$.
- B Evaluate exponential notation.
- C Simplify expressions using the rules for order of operations.
- D Remove parentheses within parentheses.

Key Terms
Use the vocabulary terms listed below to complete each statement in Exercises 1–3.

average **base** **exponent**

1. In the expression 5^3, 3 is the ________________ .

2. In the expression 5^3, 5 is the ________________ .

3. The ________________ of a set of numbers is the sum of the numbers divided by the number of addends.

GUIDED EXAMPLES AND PRACTICE

Objective A Write exponential notation for products such as $4 \cdot 4 \cdot 4$.

Review this example for Objective A:

1. Write exponential notation for $6 \cdot 6 \cdot 6 \cdot 6$.

Exponential notation is 6^4.

Practice this exercise:

1. Write exponential notation for $2 \cdot 2 \cdot 2 \cdot 2 \cdot 2$.

Objective B Evaluate exponential notation.

Review this example for Objective B:

2. Evaluate: 3^4.

$3^4 = 3 \cdot 3 \cdot 3 \cdot 3 = 81$

Practice this exercise:

2. Evaluate 5^3.

Objective C   Simplify expressions using the rules for order of operations.

Review these examples for Objective C:

3. Simplify: $64 \div 4^2 \cdot 3 + (12 - 7)$.

$64 \div 4^2 \cdot 3 + (12 - 7)$
$= 64 \div 4^2 \cdot 3 + (5)$ Subtracting inside parentheses
$= 64 \div 16 \cdot 3 + 5$ Evaluating exponential expressions
$= 4 \cdot 3 + 5$ Doing all multiplications,
$= 12 + 5$ divisions, and additions in
$= 17$ order from left to right

4. Find the average of 12, 32, 15, and 29.

The number of addends is 4, so we divide the sum of the numbers by 4.
The average is given by
$\frac{12 + 32 + 15 + 29}{4} = \frac{88}{4} = 22$.

Practice these exercises:

3. Simplify: $9 + (19 - 9)^2 \div 5 \cdot 2$.

4. Find the average of 43, 26, 35, and 16.

Objective D   Remove parentheses within parentheses.

Review this example for Objective D:

5. Simplify: $7 + \{15 - [2 \times (6 - 4)]\}$.

$7 + \{15 - [2 \times (6 - 4)]\}$
$= 7 + \{15 - [2 \times 2]\}$ Doing the calculations in the innermost parentheses first
$= 7 + \{15 - 4\}$ Again, doing the calculations in the innermost brackets
$= 7 + 11$ Subtracting inside the braces
$= 18$ Adding

Practice this exercise:

5. Simplify
$25 + \{3 \times [18 - (2 + 6)]\}$.

ADDITIONAL EXERCISES

Objective A   Write exponential notation for products such as $4 \cdot 4 \cdot 4$.

For extra help, see Examples 1–2 on page 67 of your text and the Section 1.6 lecture video.

Write exponential notation.

1. $8 \cdot 8 \cdot 8 \cdot 8 \cdot 8$

2. $4 \cdot 4 \cdot 4$

3. $5 \cdot 5 \cdot 5 \cdot 5$

4. $6 \cdot 6 \cdot 6 \cdot 6 \cdot 6 \cdot 6 \cdot 6$

Name: Date:
Instructor: Section:

Objective B Evaluate exponential notation.
For extra help, see Examples 3–4 on page 68 of your text and the Section 1.6 lecture video.
Evaluate.

5. 2^7

6. 3^5

7. 6^3

8. 10^2

Objective C Simplify expressions using the rules for order of operations.
For extra help, see Examples 5–12 on pages 68–71 of your text and the Section 1.6 lecture video.
Simplify.

9. $5 \cdot 9 + 30$

10. $100 - 5 \cdot 5 - 2$

11. $4^3 - 5 \times 3 - (4 + 2 \cdot 7)$

12. $15(6-4)^2 - 3(2+1)^2$

13. Find the average of 67, 70, 39, and 56.

Objective D Remove parentheses within parentheses.
For extra help, see Examples 13–14 on page 72 of your text and the Section 1.6 lecture video.
Evaluate.

14. $[28-(2+6)\div 2]-[24\div(5+1)]$

15. $8\times\{(31-9)\cdot[(17+23)\div 4-(6-3)]\}$

16. $9\times 20-\{50\div[12-(4+3)]\}$

17. $64\div 4-[3\times(10-4\cdot 2)]$

18. $[56\times(5-3)\div 8]+[6\times(7-1)]$

Name: Date:
Instructor: Section:

Chapter 1 WHOLE NUMBERS

1.7 Factorizations

Learning Objectives

A Determine whether one number is a factor of another, and find the factors of a number.

B Find some multiples of a number, and determine whether a number is divisible by another.

C Given a number from 1 to 100, tell whether it is prime, composite, or neither.

D Find a prime factorization of a composite number.

Key Terms

Use the vocabulary terms listed below to complete each statement in Exercises 1–4.

multiple **divisible** **prime** **composite**

1. A number is a ________________ of another if it is the product of the number and some natural number.

2. A ________________ number has exactly two different factors.

3. A number a is ________________ by another number b if there is a number c such that $b \cdot c = a$.

4. A ________________ number has more than two different factors.

GUIDED EXAMPLES AND PRACTICE

Objective A Determine whether one number is a factor of another, and find the factors of a number.

Review these examples for Objective A:

1. Determine whether 7 is a factor of 381.

```
   54
7)381
  35
   31
   28
    3
```

The remainder is not 0, so 7 is not a factor of 198.

Practice these exercises:

1. Determine whether 6 is a factor of 378.

2. List all the factors of 36.

$36 = 1 \cdot 36 \qquad 36 = 4 \cdot 9$
$36 = 2 \cdot 18 \qquad 36 = 6 \cdot 6$
$36 = 3 \cdot 12$

Factors: 1, 2, 3, 4, 6, 9, 12, 18, 36

2. List all the factors of 20.

Objective B Find some multiples of a number, and determine whether a number is divisible by another.

Review these examples for Objective B:

3. Multiply by 1, 2, 3, and so on to find ten multiples of 8.

$1 \cdot 8 = 8 \qquad 6 \cdot 8 = 48$
$2 \cdot 8 = 16 \qquad 7 \cdot 8 = 56$
$3 \cdot 8 = 24 \qquad 8 \cdot 8 = 64$
$4 \cdot 8 = 32 \qquad 9 \cdot 8 = 72$
$5 \cdot 8 = 40 \qquad 10 \cdot 8 = 80$

4. Determine whether 86 is divisible by 2 and by 4.

```
   43         21
2)86       4)86
  8          8
   6          6
   6          4
   0          2
```

Since the remainder is 0 when 86 is divided by 2, 86 is divisible by 2. When 86 is divided by 4, the remainder is not 0, so 86 is not divisible by 4.

Practice these exercises:

3. Multiply by 1, 2, 3, and so on to find ten multiples of 13.

4. Determine whether 188 is divisible by 8.

Objective C Given a number from 1 to 100, tell whether it is prime, composite, or neither.

Review this example for Objective C:

5. Classify each of the numbers 1, 19, and 24 as prime, composite, or neither.

1 does not have two *different* factors. It is neither prime nor composite.

19 has only the factors 1 and 19. It is prime.

24 has more than two different factors. It is composite.

Practice this exercise:

5. Classify 57 as prime, composite, or neither.

Name: Date:
Instructor: Section:

Objective D Find a prime factorization of a composite number.

Review this example for Objective D:

6. Find the prime factorization of 60.

We can use a string of divisions or a factor tree.

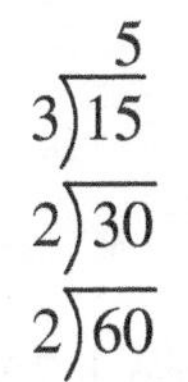

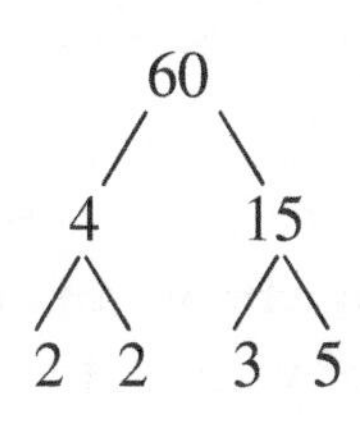

$60 = 2 \cdot 2 \cdot 3 \cdot 5$

Practice this exercise:

6. Find the prime factorization of 63.

ADDITIONAL EXERCISES

Objective A Determine whether one number is a factor of another, and find the factors of a number.

For extra help, see Examples 1–2 on pages 76–77 of your text and the Section 1.7 lecture video.

Determine whether the second number is a factor of the first.

1. 571; 3

List all the factors of each number.

2. 32

3. 60

4. 28

5. 50

Objective B Find some multiples of a number, and determine whether a number is divisible by another.

For extra help, see Examples 3–5 on page 77–78 of your text and the Section 1.7 lecture video.

Multiply by 1, 2, 3, and so on, to find ten multiples of each number.

6. 5

7. 9

8. Determine whether 221 is divisible by 7.

9. Determine whether 1096 is divisible by 4.

Objective C Given a number from 1 to 100, tell whether it is prime, composite, or neither.

For extra help, see Example 6 on page 78 of your text and the Section 1.7 lecture video.

Classify each number as prime, composite, or neither.

10. 15

11. 38

12. 37

13. 1

Objective D Find a prime factorization of a composite number.

For extra help, see Examples 7–10 on pages 79–80 of your text and the Section 1.7 lecture video.

Find the prime factorization of each number.

14. 48

15. 54

16. 270

17. 66

Name: Date:
Instructor: Section:

Chapter 1 WHOLE NUMBERS

1.8 Divisibility

Learning Objectives
A Determine whether a number is divisible by 2, 3, 4, 5, 6, 8, 9, or 10.

Key Terms
Use the vocabulary terms listed below to complete each statement in Exercises 1–4.

divisible **divisible by 3** **divisible by 5** **even**

1. A number is divisible by two if it is ________________ .

2. If a number has a ones digit of 0 or 5, then it is ________________ .

3. A number is ________________ if the sum of its digits is divisible by three.

4. A number a is ________________ by another number b if there is a number c such that $a = b \cdot c$.

GUIDED EXAMPLES AND PRACTICE

Objective A Determine whether a number is divisible by 2, 3, 4, 5, 6, 8, 9, or 10.

Review this example for Objective A:

1. Determine whether 56,340 is divisible by 2, 3, 5, 6, 9, or 10.

The ones digit, 0, is even so 56,340 is divisible by 2.
$5 + 6 + 3 + 4 + 0 = 18$ and 18 is divisible by 3, so 56,340 is divisible by 3.
The ones digit is 0, so 56,340 is divisible by 5.
The ones digit is even and the sum of the digits, 18, is divisible by 3, so 56,340 is divisible by 6.
The sum of the digits, 18, is divisible by 9, so 56,340 is divisible by 9.
The ones digit is 0, so 56,340 is divisible by 10.

Practice this exercise:

1. Determine whether 18,225 is divisible by 2, 3, 5, 6, 9, or 10.

ADDITIONAL EXERCISES

Objective A Determine whether a number is divisible by 2, 3, 4, 5, 6, 8, 9, or 10.

For extra help, see Examples 1–26 on pages 83–86 of your text and the Section 1.8 lecture video.

Test each number for divisibility by 2, 3, 5, 6, 9, and 10.

1. 732

2. 1845

Consider the following numbers. Use the tests for divisibility.

57 171 95 487 48 7317 5001

3. Which of the above are divisible by 3?

4. Which of the above are divisible by 5?

Name: Date:
Instructor: Section:

Chapter 1 WHOLE NUMBERS

1.9 Least Common Multiples

Learning Objectives
A Find the least common multiple, or LCM, of two or more numbers.

Key Terms
Use the vocabulary terms listed below to complete each statement in Exercises 1–3.

least common multiple **factorization** **prime factorization** **multiple**

1. We find the ________________ of a number when we write it as a product of primes.

2. $3 \cdot 4$ is an example of a ________________ of 12.

3. The ________________ of numbers is the smallest number that is a ________________ of both numbers.

GUIDED EXAMPLES AND PRACTICE

Objective A Find the least common multiple, or LCM, of two or more numbers.

Review these examples for Objective A:

1. Find the LCM of 15 and 18 using a list of multiples.

1. 18 is the larger number, but it is not a multiple of 15.
2. Check multiples of 18:
$2 \cdot 18 = 36$ Not a multiple of 15
$3 \cdot 18 = 54$ Not a multiple of 15
$4 \cdot 18 = 72$ Not a multiple of 15
$5 \cdot 18 = 90$ A multiple of both 15 and 18
The LCM is 90.

2. Find the LCM of 9 and 21 using prime factorizations.

We write the prime factorization of each number in exponential notation.
$9 = 3 \cdot 3 = 3^2$ 3^2 is the greatest power of 3
$21 = 3 \cdot 7 = 3^1 \cdot 7^1$ 7^1 is the greatest power of 7
LCM = $= 3^2 \cdot 7^1 = 63$

Practice these exercises:

1. Find the LCM of 12 and 16 using a list of multiples.

2. Find the LCM of 8 and 20 using prime factorizations.

ADDITIONAL EXERCISES

Objective A Find the least common multiple, or LCM, of two or more numbers.

For extra help, see Examples 1–9 on page 89–92 of your text and the Section 1.9 lecture video.

Find the LCM of each set of numbers.

1. 10, 35

2. 18, 45

3. 9, 12, 30

4. 8, 12, 25

Name: Date:
Instructor: Section:

Chapter 2 FRACTION NOTATION

2.1 Fraction Notation and Simplifying

Learning Objectives

A Identify the numerator and the denominator of a fraction and write fraction notation for part of an object.
B Simplify fraction notation like *n*/*n* to 1, 0/*n* to 0, and *n*/1 to *n*.
C Multiply a fraction by a fraction, and multiply a fraction by a whole number.
D Multiply a number by 1 to find fraction notation with a specified denominator.
E Simplify fraction notation.

Key Terms

Use the vocabulary terms listed below to complete each statement in Exercises 1–5.

numerator **denominator** **canceling** **equivalent** **simplest**

1. In a fraction, the __________________ tell us the number of units into which we are partitioning an object.

2. __________________ is a shortcut for removing a factor of 1, which must be used with caution.

3. Two fractions are __________________ if they name the same number.

4. The __________________ of a fraction tells us the number of equal parts we are considering.

5. A fraction is in __________________ form if the numerator and the denominator are the smallest whole numbers possible.

GUIDED EXAMPLES AND PRACTICE

Objective A Identify the numerator and the denominator of a fraction and write fraction notation for part of an object.

Review these examples for Objective A:

1. Identify the numerator and denominator: $\frac{7}{12}$.

   The top number, 7, is the numerator; the bottom number, 12, is the denominator.

Practice these exercises:

1. Identify the numerator and denominator: $\frac{5}{6}$.

2. What part is shaded?

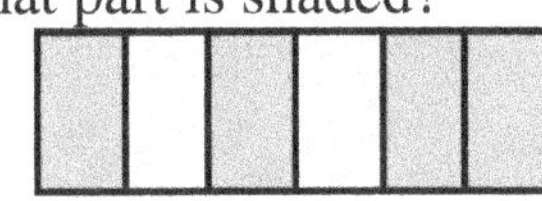

The object is divided into 6 parts of the same size. 4 of them are shaded. $\frac{4}{6}$ of the object is shaded.

2. What part is shaded?

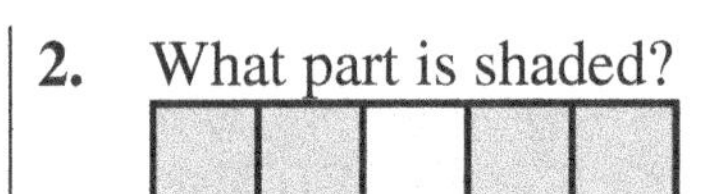

Objective B Simplify fraction notation like *n*/*n* to 1, 0/*n* to 0, and *n*/1 to *n*.

Review this example for Objective B:

3. Simplify: $\frac{6}{6}, \frac{0}{10}$, and $\frac{3}{1}$.

$\frac{6}{6}=1,\ \frac{0}{10}=0,\ \frac{3}{1}=3$

Practice this exercise:

3. Simplify: $\frac{5}{1}, \frac{12}{12}$, and $\frac{0}{2}$.

Objective C Multiply a fraction by a fraction, and multiply a fraction by a whole number.

Review this example for Objective C:

4. Multiply: $\frac{3}{4}\cdot\frac{5}{2}$.

$\frac{3}{4}\cdot\frac{5}{2}=\frac{3\cdot 5}{4\cdot 2}=\frac{15}{8}$

Practice this exercise:

4. Multiply: $\frac{5}{8}\times\frac{7}{6}$.

Objective D Multiply a number by 1 to find fraction notation with a specified denominator.

Review this example for Objective D:

5. Find a number equivalent to $\frac{2}{3}$ with a denominator of 12.

Since $12\div 3=4$, we multiply by $\frac{4}{4}$:

$\frac{2}{3}=\frac{2}{3}\cdot\frac{4}{4}=\frac{2\cdot 4}{3\cdot 4}=\frac{8}{12}$

Practice this exercise:

5. Find a number equivalent to $\frac{3}{4}$ with a denominator of 20.

Objective E Simplify fraction notation.

Review this example for Objective B:

6. Simplify: $-\frac{16}{36}$.

$-\frac{16}{36}=-\frac{4\cdot 4}{9\cdot 4}=-\frac{4}{9}\cdot\frac{4}{4}=-\frac{4}{9}\cdot 1=-\frac{4}{9}$

Removing a factor equal to 1: $\frac{4}{4}=1$

Practice this exercise:

6. Simplify: $\frac{9}{24}$.

Name: Date:
Instructor: Section:

ADDITIONAL EXERCISES

Objective A Identify the numerator and the denominator of a fraction and write fraction notation for part of an object.

For extra help, see Examples 1– on pages 104–106 of your text and the Section 2.1 lecture video.

Identify the numerator and denominator.

1. $\frac{2}{9}$

2. $\frac{15}{4}$

What part of the object or set of objects is shaded?

3.

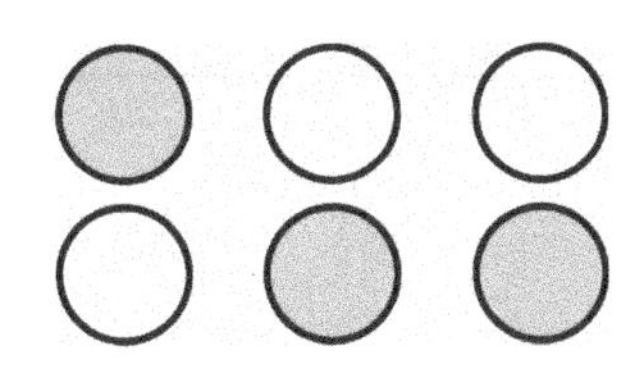

4.

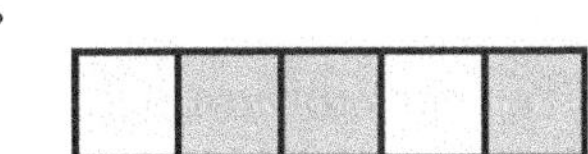

Objective B Simplify fraction notation like *n*/*n* to 1, 0/*n* to 0, and *n*/1 to *n*.

For extra help, see Examples 7–15 on pages 107–108 of your text and the Section 2.1 lecture video.

Simplify.

5. $\frac{0}{19}$

6. $\frac{9}{1}$

7. $\frac{7}{0}$

8. $\frac{7-3}{6-5}$

Objective C Multiply a fraction by a fraction, and multiply a fraction by a whole number.

For extra help, see Examples 16–20 on page 109 of your text and the Section 2.1 lecture video.

Multiply.

9. $\frac{1}{3}\times\frac{1}{7}$

10. $\frac{6}{7}\cdot\frac{5}{11}$

11. $\frac{3}{4}\cdot\frac{3}{8}$

12. $\frac{13}{17}\cdot\frac{5}{9}$

Objective D Multiply a number by 1 to find fraction notation with a specified denominator.

For extra help, see Examples 21–22 on page 110 of your text and the Section 2.1 lecture video.

Find an equivalent expression for each number, using the denominator indicated. Use multiplication by 1.

13. $\frac{2}{3} = \frac{?}{12}$

14. $\frac{7}{15} = \frac{?}{45}$

15. $\frac{11}{21} = \frac{?}{147}$

16. $\frac{10}{3} = \frac{?}{21}$

Objective E Simplify fraction notation.

For extra help, see Examples 23–27 on pages 111–112 of your text and the Section 2.1 lecture video.

Simplify.

17. $\frac{6}{18}$

18. $\frac{56}{7}$

19. $\frac{250}{325}$

20. $\frac{600}{400}$

Name: Date:
Instructor: Section:

Chapter 2 FRACTION NOTATION

2.2 Multiplication and Division

Learning Objectives

A Multiply and simplify using fraction notation.
B Find the reciprocal of a number.
C Divide and simplify using fraction notation.
D Solve equations of the type $a \cdot x = b$ and $x \cdot a = b$, where a and b may be fractions.

Key Terms

Use the terms listed below to complete each statement in Exercises 1–4. Terms may be used more than once.

one **zero** **reciprocal**

1. ________________ has no reciprocal.

2. We find the________________ of a fraction by interchanging the numerator and the denominator.

3. If the product of two numbers is __________, the numbers are reciprocals of each other.

4. We simplify a fraction by removing a factor equal to ________________.

GUIDED EXAMPLES AND PRACTICE

Objective A Multiply and simplify using fraction notation.

Review this example for Objective A:

1. Multiply and simplify: $\frac{3}{4} \cdot \frac{2}{9}$.

$$\frac{3}{4} \cdot \frac{2}{9} = \frac{3 \cdot 2}{4 \cdot 9} = \frac{3 \cdot 2 \cdot 1}{2 \cdot 2 \cdot 3 \cdot 3}$$

$$= \frac{3 \cdot 2}{3 \cdot 2} \cdot \frac{1}{2 \cdot 3} = 1 \cdot \frac{1}{2 \cdot 3}$$

Removing a factor equal to 1: $\frac{3 \cdot 2}{3 \cdot 2} = 1$

$$= \frac{1}{2 \cdot 3} = \frac{1}{6}$$

Practice this exercise:

1. Multiply and simplify: $\frac{5}{6} \cdot \frac{4}{15}$.

Objective B Find the reciprocal of a number.

Review this example for Objective B:

2. Find the reciprocals of $\frac{5}{9}$, 4, and $\frac{1}{6}$.

The reciprocal of $\frac{5}{9}$ is $\frac{9}{5}$. Note that $\frac{5}{9}\cdot\frac{9}{5}=1$.

The reciprocal of 4 is $\frac{1}{4}$. Note that $4\cdot\frac{1}{4}=\frac{4}{4}=1$.

The reciprocal of $\frac{1}{6}$ is 6. Note that $\frac{1}{6}\cdot 6=\frac{6}{6}=1$.

Practice this exercise:

2. Find the reciprocal of 13.

Objective C Divide and simplify using fraction notation.

Review these examples for Objective C:

3. Divide and simplify: $\frac{5}{4}\div\frac{25}{16}$.

$$\frac{5}{4}\div\frac{25}{16}=\frac{5}{4}\cdot\frac{16}{25}$$ Multiplying by the reciprocal of the divisor

$$=\frac{5\cdot 4\cdot 4}{4\cdot 5\cdot 5}$$ Factoring and identifying common factors

$$=\frac{5\cdot 4}{5\cdot 4}\cdot\frac{4}{5}$$ Removing a factor equal to 1: $\frac{5\cdot 4}{5\cdot 4}=1$

$$=\frac{4}{5}$$

Practice these exercises:

3. Divide and simplify: $\frac{2}{3}\div\frac{8}{9}$.

Objective D Solve equations of the type $a\cdot x=b$ and $x\cdot a=b$, where a and b may be fractions.

Review this example for Objective B:

4. Solve: $54=\frac{9}{2}y$.

We multiply by the reciprocal of $\frac{9}{2}$, or $\frac{2}{9}$, on both sides of the equation to get y alone.

$$54=\frac{9}{2}y$$

$$\frac{2}{9}\cdot 54=\frac{2}{9}\left(\frac{9}{2}y\right)$$

$$\frac{2\cdot 6\cdot \not{9}}{\not{9}\cdot 1}=y$$

$$12=y$$ Removing a factor equal to 1: $\frac{9}{9}=1$

Practice this exercise:

4. Solve: $\frac{5}{6}x=30$.

Name: Date:
Instructor: Section:

ADDITIONAL EXERCISES

Objective A Multiply and simplify using fraction notation.

For extra help, see Examples 1–3 on page 116–117 of your text and the Section 2.2 lecture video.

Multiply and simplify.

1. $\frac{8}{7}\cdot\frac{3}{10}$

2. $\frac{12}{5}\cdot\frac{9}{16}$

3. $\frac{48}{7}\cdot\frac{9}{16}$

Objective B Find the reciprocal of a number.

For extra help, see Examples 4–6 on page 117 of your text and the Section 2.2 lecture video.

Find the reciprocal.

4. $\frac{6}{7}$

5. $\frac{1}{9}$

6. $\frac{12}{11}$

7. 19

Objective C Divide and simplify using fraction notation.

For extra help, see Examples 7–9 on pages 118–119 of your text and the Section 2.2 lecture video.

Divide. Don't forget to simplify when possible.

8. $\frac{4}{3}\div\frac{4}{5}$

9. $\frac{3}{5}\div\frac{7}{10}$

10. $\frac{7}{24}\div\frac{21}{5}$

11. $\frac{56}{55}\div\frac{14}{15}$

Objective D Solve equations of the type $a \cdot x = b$ and $x \cdot a = b$, where a and b may be fractions.

For extra help, see Examples 10–11 on page 119 of your text and the Section 2.2 lecture video.

Solve

12. $\frac{5}{8}x = 40$

13. $7a = \frac{14}{3}$

14. $\frac{3}{5}x = \frac{9}{10}$

15. $\frac{4}{45} = t \cdot \frac{8}{15}$

Name: Date:
Instructor: Section:

Chapter 2 FRACTION NOTATION

2.3 Addition and Subtraction; Order

Learning Objectives
A Add using fraction notation.
B Subtract using fraction notation.
C Use < or > with fraction notation to write a true sentence.
D Solve equations of the type $x+a=b$ and $a+x=b$, where a and b may be fractions.

Key Terms
Use the vocabulary terms listed below to complete each statement in Exercises 1–3.

least common denominator **like denominators** **different denominators**

1. We find the LCD before adding or subtracting fraction which have ________________.

2. The ________________ of two fractions is the LCM of the two denominators.

3. When adding fractions with ________________, we add the numerators, keep the denominator, and simplify, if possible.

GUIDED EXAMPLES AND PRACTICE

Objective A Add using fraction notation.

Review these examples for Objective A:

1. Add and simplify: $\frac{3}{8}+\frac{7}{8}$.

Since the denominators are the same, we add the numerators and keep the denominators.

$\frac{3}{8}+\frac{7}{8}=\frac{3+7}{8}=\frac{10}{8}$

$=\frac{2}{2}\cdot\frac{5}{2}$ Simplify by removing a factor of 1: $\frac{2}{2}=1$

$=\frac{5}{2}$

Practice these exercises:

1. Add and simplify: $\frac{1}{12}+\frac{7}{12}$.

2. Add and simplify: $\frac{2}{9}+\frac{1}{6}$.

Since $9=3\cdot 3$ and $6=2\cdot 3$, the LCM of 9 and 6 is $2\cdot 3\cdot 3$, or 18. Thus the LCD is 18.

$$\frac{2}{9}+\frac{1}{6}=\frac{2}{9}\cdot\frac{2}{2}+\frac{1}{6}\cdot\frac{3}{3}$$ We multiply by 1, using $\frac{2}{2}$ and $\frac{3}{3}$.

$$=\frac{4}{18}+\frac{3}{18}=\frac{7}{18}$$

2. Add and simplify: $\frac{3}{4}+\frac{3}{10}$.

Objective B Subtract using fraction notation.

Review this example for Objective B:

3. Subtract and simplify, if possible: $\frac{7}{8}-\frac{3}{8}$.

Since the denominators are the same, we subtract the numerators and keep the denominators.

$$\frac{7}{8}-\frac{3}{8}=\frac{7-3}{8}=\frac{4}{8}$$

$$=\frac{4}{4}\cdot\frac{1}{2}$$ Simplify by removing a factor of 1: $\frac{4}{4}=1$

$$=\frac{1}{2}$$

4. Subtract and simplify, if possible: $\frac{2}{3}-\frac{1}{4}$.

The LCM of 3 and 4 is 12, so the LCD is 12.

$$\frac{2}{3}-\frac{1}{4}=\frac{2}{3}\cdot\frac{4}{4}-\frac{1}{4}\cdot\frac{3}{3}$$ We multiply by 1, using $\frac{4}{4}$ and $\frac{3}{3}$.

$$=\frac{8}{12}-\frac{3}{12}=\frac{5}{12}$$

Practice this exercise:

3. Subtract and simplify: $\frac{7}{9}-\frac{4}{9}$

4. Subtract and simplify: $\frac{4}{5}-\frac{3}{8}$

Objective C Use < or > with fraction notation to write a true sentence.

Review this example for Objective C:

5. Use < or > for □ to form a true sentence: $\frac{3}{5}\,\square\,\frac{5}{8}$.

$\frac{3}{5}\cdot\frac{8}{8}=\frac{24}{40}$; $\frac{5}{8}\cdot\frac{5}{5}=\frac{25}{40}$

Since $24<25$, it follows that $\frac{3}{5}<\frac{5}{8}$.

Practice this exercise:

5. Use < or > for □ to form a true sentence: $\frac{2}{3}\,\square\,\frac{5}{9}$.

Name: Date:
Instructor: Section:

Objective D Solve equations of the type $x+a=b$ and $a+x=b$, where a and b may be fractions.

Review this example for Objective D:

2. Solve: $x+\frac{1}{3}=\frac{4}{5}$.

$$x+\frac{1}{3}=\frac{4}{5}$$

$$x+\frac{1}{3}-\frac{1}{3}=\frac{4}{5}-\frac{1}{3}$$ Using the addition principle: adding $-\frac{1}{3}$ to, or subtracting $\frac{1}{3}$ from, both sides

$$x+0=\frac{4}{5}\cdot\frac{3}{3}-\frac{1}{3}\cdot\frac{5}{5}$$ Multiplying by 1 to obtain the LCD, 15

$$x=\frac{12}{15}-\frac{5}{15}=\frac{7}{15}$$

The solution is $\frac{7}{15}$.

Practice this exercise:

2. Solve: $x+\frac{5}{6}=\frac{7}{8}$.

ADDITIONAL EXERCISES

Objective A Add using fraction notation.

For extra help, see Examples 1–8 on pages 122–124 of your text and the Section 2.3 lecture video.

Add, and if possible, simplify.

1. $\frac{1}{7}+\frac{5}{7}$

2. $\frac{3}{4}+\frac{5}{8}$

3. $\frac{4}{9}+\frac{0}{7}$

4. $\frac{5}{6}+\frac{3}{10}$

5. $\frac{2}{3}+\frac{5}{12}+\frac{1}{6}$

Objective B Subtract using fraction notation.

For extra help, see Examples 9–13 on page 125–126 of your text and the Section 2.3 lecture video.

Subtract, and if possible, simplify.

6. $\frac{7}{9}-\frac{2}{9}$

7. $\frac{15}{16}-\frac{7}{16}$

8. $\frac{5}{6}-\frac{3}{8}$

9. $\frac{1}{3}-\frac{1}{8}$

10. $\frac{3}{4}-\frac{7}{12}$

Objective C Use < or > with fraction notation to write a true sentence.
For extra help, see Examples 14–15 on page 127 of your text and the Section 2.3 lecture video.

Use < or > for $\square$ to form a true sentence.

11. $\frac{5}{4}\square\frac{3}{2}$

12. $\frac{4}{15}\square\frac{3}{10}$

13. $\frac{7}{10}\square\frac{5}{9}$

14. $\frac{9}{16}\square\frac{7}{12}$

Objective D Solve equations of the type $x+a=b$ and $a+x=b$, where a and b may be fractions.
For extra help, see Example 16 on page 127 of your text and the Section 2.3 lecture video.

Solve.

15. $t+\frac{1}{3}=\frac{5}{6}$

16. $\frac{1}{5}+x=\frac{7}{4}$

17. $x+\frac{2}{5}=\frac{7}{10}$

18. $n+\frac{3}{10}=\frac{3}{8}$

Name: Date:
Instructor: Section:

Chapter 2 FRACTION NOTATION

2.4 Mixed Numerals

Learning Objectives
A Convert between mixed numerals and fraction notation.
B Add using mixed numerals.
C Subtract using mixed numerals.
D Multiply using mixed numerals.
E Divide using mixed numerals.

Key Terms
Use the vocabulary terms listed below to complete each statement in Exercises 1–3. Terms may be used more than once.

whole numbers **mixed numeral(s)** **fraction notation** **fractions**

1. To divide using ________________, first write ________________ and divide. Then convert the answer to a mixed numeral, if appropriate.

2. ________________ consist of a whole number part and a fraction less than 1.

3. To add mixed numerals, first add the ________________; then add the ________________, and if possible, simplify the fraction part.

GUIDED EXAMPLES AND PRACTICE

Objective A Convert between mixed numerals and fraction notation.

Review these examples for Objective A:

1. Convert $5\frac{3}{8}$ to fraction notation.

   $5\frac{3}{8}=\frac{43}{8}$ $\quad 5\cdot 8=40,\ \ 40+3=43$

2. Convert $\frac{13}{3}$ to a mixed numeral.

   $3\overline{)13}$ quotient 4, -12, remainder 1 $\quad \frac{13}{3}=4\frac{1}{3}$

Practice these exercises:

1. Convert $3\frac{4}{5}$ to fraction notation.

2. Convert $\frac{11}{6}$ to a mixed numeral.

Objective B Add using mixed numerals.

Review this example for Objective B:

3. Add: $3\frac{5}{8}+4\frac{1}{2}$.

$$\begin{array}{rcr} 3\frac{5}{8} & = & 3\frac{5}{8} \\ +\,4\frac{1}{2}\cdot\frac{4}{4} & = & +\,4\frac{4}{8} \\ \hline & & 7\frac{9}{8} \end{array}$$

$$7\frac{9}{8}=7+\frac{9}{8}=7+1\frac{1}{8}=8\frac{1}{8}$$

Practice this exercise:

3. Add: $5\frac{2}{3}+1\frac{3}{4}$.

Objective C Subtract using mixed numerals.

Review this example for Objective C:

4. Subtract: $6\frac{1}{3}-4\frac{1}{2}$.

$$\begin{array}{rcr} 6\frac{1}{3}\cdot\frac{2}{2} & = & 6\frac{2}{6} \\ -\,4\frac{1}{2}\cdot\frac{3}{3} & = & -\,4\frac{3}{6} \\ \hline \end{array}$$

To subtract $\frac{3}{6}$ from $\frac{2}{6}$, we borrow 1, or $\frac{6}{6}$ from 6:

$$6\frac{2}{6}=5+1+\frac{2}{6}=5+\frac{6}{6}+\frac{2}{6}=5\frac{8}{6}$$

We can write this as

$$\begin{array}{rcr} 6\frac{2}{6} & = & 5\frac{8}{6} \\ -\,4\frac{3}{6} & = & -\,4\frac{3}{6} \\ \hline & & 1\frac{5}{6} \end{array}$$

Practice this exercise:

4. Subtract: $9\frac{3}{8}-3\frac{3}{4}$.

Objective D Multiply using mixed numerals.

Review this example for Objective D:

5. Multiply: $1\frac{7}{8}\cdot5\frac{2}{3}$.

$$1\frac{7}{8}\cdot5\frac{2}{3}=\frac{15}{8}\cdot\frac{17}{3}=\frac{15\cdot17}{8\cdot3}=\frac{\cancel{3}\cdot5\cdot17}{2\cdot2\cdot2\cdot\cancel{3}}=\frac{85}{8}=10\frac{5}{8}$$

Practice this exercise:

5. Multiply: $6\frac{2}{5}\cdot2\frac{3}{4}$.

Name: Date:
Instructor: Section:

Objective E Divide using mixed numerals.

Review this example for Objective E:

6. Divide: $4\frac{2}{3} \div 2\frac{2}{5}$.

$$4\frac{2}{3} \div 2\frac{2}{5} = \frac{14}{3} \div \frac{12}{5} = \frac{14}{3} \cdot \frac{5}{12}$$
$$= \frac{14 \cdot 5}{3 \cdot 12} = \frac{\cancel{2} \cdot 7 \cdot 5}{3 \cdot 2 \cdot \cancel{2} \cdot 3}$$
$$= \frac{35}{18} = 1\frac{17}{18}$$

Practice this exercise:

6. Divide: $9\frac{1}{8} \div 3\frac{1}{4}$.

ADDITIONAL EXERCISES

Objective A Convert between mixed numerals and fraction notation.

For extra help, see Examples 1–8 on pages 133–135 of your text and the Section 2.4 lecture video.

Convert to fraction notation.

1. $4\frac{3}{5}$

2. $3\frac{5}{9}$

Convert to a mixed numeral.

3. $\frac{37}{15}$

4. $\frac{47}{6}$

Objective B Add using mixed numerals.

For extra help, see Examples 9–10 on page 135–136 of your text and the Section 2.4 lecture video.

Add. Write a mixed numeral for each answer.

5. $8\frac{5}{7} + 2\frac{3}{7}$

6. $12\frac{3}{8} + 3\frac{7}{12}$

7. $5\frac{5}{6} + 11\frac{5}{8}$

8. $3\frac{4}{5} + 2\frac{3}{10}$

Objective C Subtract using mixed numerals.
For extra help, see Examples 11–14 on pages 136–137 of your text and the Section 2.4 lecture video.
Subtract. Write a mixed numeral for each answer.

9. $\begin{array}{r} 7\frac{2}{5} \\ -\ 3\frac{4}{5} \\ \hline \end{array}$

10. $\begin{array}{r} 45\frac{5}{12} \\ -\ 13\frac{7}{8} \\ \hline \end{array}$

11. $\begin{array}{r} 22\frac{1}{6} \\ -\ 18\frac{5}{9} \\ \hline \end{array}$

12. $32\frac{7}{12} - 18\frac{5}{6}$

Objective D Multiply using mixed numerals.
For extra help, see Examples 15–18 on pages 137–138 of your text and the Section 2.4 lecture video.
Multiply. Write a mixed numeral for each answer whenever possible.

13. $6 \cdot 3\frac{1}{4}$

14. $4\frac{3}{5} \cdot 2\frac{2}{3}$

15. $8\frac{2}{9} \cdot 3\frac{2}{5}$

16. $4\frac{7}{8} \cdot 5\frac{1}{3}$

Objective E Divide using mixed numerals.
For extra help, see Examples 19–22 on page 138 of your text and the Section 2.4 lecture video.
Divide. Write a mixed numeral for each answer whenever possible.

17. $12 \div 7\frac{1}{2}$

18. $6\frac{2}{3} \div 5$

19. $4\frac{1}{5} \div 5\frac{1}{3}$

20. $10\frac{1}{2} \div 7$

Name: Date:
Instructor: Section:

Chapter 2 FRACTION NOTATION

2.5 Applications and Problem Solving

Learning Objectives

A Solve applied problems involving addition, subtraction, multiplication, and division using fraction notation and mixed numerals.

Key Terms

Use the vocabulary terms listed below to complete each statement in Exercises 1–3.

area **solve** **check**

1. To ________________ an equation means to find all values of a variable that make the equation true.

2. The __________________ of a rectangle is length times width.

3. We can sometimes __________________ a solution by repeating calculations, finding the solution in a different manner, or seeing if the solution is reasonable.

GUIDED EXAMPLES AND PRACTICE

Objective A Solve applied problems involving addition, subtraction, multiplication, and division using fraction notation and mixed numerals.

Review these examples for Objective A:

1. A cookie recipe calls for $1\frac{3}{4}$ cups of sugar. How much sugar is needed to double the recipe?

1. *Familiarize*. Let s = the number of cups of sugar in two recipes.
2. *Translate*. The situation translates to the multiplication sentence

$$c = 2 \cdot 1\frac{3}{4}.$$

3. *Solve*. We carry out the multiplication.

$$c = 2 \cdot 1\frac{3}{4} = 2 \cdot \frac{7}{4}$$
$$= \frac{2 \cdot 7}{4} = \frac{\not{2} \cdot 7}{\not{2} \cdot 2}$$
$$= \frac{7}{2} = 3\frac{1}{2}$$

Practice these exercises:

1. A car travels 276 mi on $11\frac{5}{10}$ gal of gas. How many miles per gallon did it get?

4. *Check*. We can do a partial check by estimating: $2\cdot 1\frac{3}{4} \approx 2\cdot 2 = 4 \approx 3\frac{1}{2}$. We can also repeat the calculation. The answer checks.

5. *State*. To double the recipe, $3\frac{1}{2}$ cups of sugar are needed.

2. Melanie bought $1\frac{1}{2}$ lb of apples and $2\frac{3}{4}$ lb of pears. What was the total weight of the fruit?

1. *Familiarize*. Let w = the total weight of the fruit, in pounds.
2. *Translate*.

$$\underbrace{\text{Weight of apples}}_{\downarrow\; 1\frac{1}{2}} \;\underset{\downarrow\; +}{\text{plus}}\; \underbrace{\text{Weight of pears}}_{\downarrow\; 2\frac{3}{4}} \;\underset{\downarrow\; =}{\text{is}}\; \underbrace{\text{Total weight}}_{\downarrow\; w}$$

3. *Solve*. The translation tells us what to do. We add. The LCD is 4.

$$\begin{array}{rcr} 1\frac{1}{2}\cdot\frac{2}{2} & = & 1\frac{2}{4} \\ +\,2\frac{3}{4} & = & +\,2\frac{3}{4} \\ \hline & & 3\frac{5}{4} = 4\frac{1}{4} \end{array}$$

4. *Check*. We repeat the calculation. The answer checks.

5. *State*. The total weight of the fruit was $4\frac{1}{4}$ lb.

2. Sam is $73\frac{1}{4}$ in. tall, and Ray is $70\frac{1}{2}$ in. tall. How much taller is Sam?

3. Bert spent $\frac{7}{4}$ hr doing his chemistry and English assignments. He spent $\frac{5}{6}$ hr on the chemistry assignment. How long did he spend on the English assignment?

1. *Familiarize*. Let t = the number hours Bert spent on his English assignment.

3. Mary has walked $\frac{3}{4}$ mi and will stop walking when she has walked $\frac{9}{8}$ mi. How much farther does she have to walk?

Name: Date:
Instructor: Section:

2. *Translate*. This is a "how much more" situation.

$$\underbrace{\text{Chemistry time}} \quad \text{plus} \quad \underbrace{\text{English time}} \quad \text{is} \quad \underbrace{\text{Total time}}$$

$$\downarrow \quad \downarrow \quad \downarrow \quad \downarrow \quad \downarrow$$

$$\frac{5}{6} \quad + \quad t \quad = \quad \frac{7}{4}$$

3. *Solve*. We subtract $\frac{5}{6}$ on both sides of the equation.

$$\frac{5}{6}+t=\frac{7}{4}$$

$$\frac{5}{6}+t-\frac{5}{6}=\frac{7}{4}-\frac{5}{6}$$

$$t+0=\frac{7}{4}\cdot\frac{3}{3}-\frac{5}{6}\cdot\frac{2}{2}$$ Multiplying by 1 to obtain the LCD, 12

$$t=\frac{21}{12}-\frac{10}{12}=\frac{11}{12}$$

4. *Check*. We return to the original problem and add:

$$\frac{5}{6}+\frac{11}{12}=\frac{5}{6}\cdot\frac{2}{2}+\frac{11}{12}$$

$$=\frac{10}{12}+\frac{11}{12}$$

$$=\frac{21}{12}=\frac{3}{3}\cdot\frac{7}{4}$$

$$=\frac{7}{4}.$$

The answer checks.

5. *State*. Bert spent $\frac{11}{12}$ hr on his English assignment.

ADDITIONAL EXERCISES

Objective A Solve applied problems involving addition, subtraction, multiplication, and division using fraction notation and mixed numerals.

For extra help, see Examples 1–12 on pages 142–151 of your text and the Section 2.5 lecture video.

Solve.

1. Find the area of a rectangle with length $6\frac{1}{3}$ ft and width $3\frac{3}{4}$ ft.

2. A satellite orbits Earth every $1\frac{1}{2}$ hr. How many orbits are made every 24 hr?

3. On Tuesday, Clark walked $\frac{7}{8}$ mi, ran $\frac{5}{6}$ mi, and swam $\frac{9}{10}$ mi. How many miles did he cover altogether?

4. A $\frac{2}{3}$-cup mixture of cinnamon and sugar contains $\frac{1}{4}$ cup of cinnamon. How much sugar is in the mixture?

5. A plumber uses pipes of lengths $5\frac{5}{8}$ ft and $8\frac{3}{4}$ ft in the installation of a spa. How much pipe was used?

Name: Date:
Instructor: Section:

Chapter 2 FRACTION NOTATION

2.6 Order of Operations; Estimation

Learning Objectives
A Simplify expressions using the rules for order of operations.
B Estimate with fraction notation and mixed numerals.

Key Terms
Use the vocabulary terms listed below to complete each statement in Exercises 1–2.

simplify **order of operations** **average**

1. The rules for ________________ allow us to ________________ an expression by indicating which calculation to perform first, second, and so on to find an equivalent expression.

2. To compute a(n) ________________ , we find the sum of the numbers, then divide by the number of addends.

GUIDED EXAMPLES AND PRACTICE

Objective A Simplify expressions using the rules for order of operations.

Review these examples for Objective A:

1. Simplify: $\frac{5}{8} \div \frac{6}{7} - \left(\frac{2}{3} - \frac{4}{9}\right)$.

$\frac{5}{9} \div \frac{5}{7} - \left(\frac{2}{3} - \frac{4}{9}\right)$

$= \frac{5}{9} \div \frac{5}{7} - \left(\frac{6}{9} - \frac{4}{9}\right)$ Writing with the LCD, 9, in order to subtract

$= \frac{5}{9} \div \frac{5}{7} - \frac{2}{9}$ Subtracting within the parentheses

$= \frac{5}{9} \cdot \frac{7}{5} - \frac{2}{9}$ Doing the division by multiplying by the reciprocal of $\frac{5}{7}$

$= \frac{\not{5} \cdot 7}{9 \cdot \not{5}} - \frac{2}{9}$ Multiplying and factoring

$= \frac{7}{9} - \frac{2}{9}$ Simplifying

$= \frac{5}{9}$ Subtracting

Practice these exercises:

1. Simplify: $\frac{7}{12} \div \frac{1}{6} - \frac{3}{4} \cdot \frac{12}{5}$.

2. Find the average of $\frac{5}{12}$ and $\frac{3}{16}$.

To compute an average, we add the numbers and divide by the number of addends.

$$\frac{\frac{5}{12}+\frac{3}{16}}{2}=\frac{\frac{20}{48}+\frac{9}{48}}{2}=\frac{\frac{29}{48}}{2}$$ Adding, simplifying and converting to fraction notation

$$=\frac{29}{48}\cdot\frac{1}{2}$$ Multiplying by the reciprocal

$$=\frac{29}{96}$$

2. Find the average of $\frac{3}{10}$ and $\frac{9}{14}$.

Objective B Estimate with fraction notation and mixed numerals.

Review this example for Objective B:

3. Estimate as 0, $\frac{1}{2}$, or 1: $\frac{15}{17}$.

A fraction is close to 1 when the numerator is nearly equal to the denominator. Thus, 1 is an estimate for $\frac{15}{17}$ because 15 is nearly equal to 17.

Thus, $\frac{15}{17}\approx 1$.

Practice this exercise:

3. Estimate as 0, $\frac{1}{2}$, or 1: $\frac{3}{50}$.

ADDITIONAL EXERCISES

Objective A Simplify expressions using the rules for order of operations.

For extra help, see Examples 1–4 on pages 161–162 of your text and the Section 2.6 lecture video.

Simplify.

1. $\frac{5}{9}-\frac{1}{3}\left(\frac{1}{6}+\frac{5}{9}\right)$

2. $\left(\frac{2}{3}\right)^2-\frac{1}{21}\cdot 2\frac{5}{8}$

3. $\left(\frac{3}{8}+\frac{7}{6}\right)\div\left(\frac{1}{4}-\frac{2}{11}\right)$

4. Find the average of $4\frac{1}{2}$, $3\frac{3}{4}$ and $5\frac{1}{8}$.

Objective B Estimate with fraction notation and mixed numerals.

For extra help, see Examples 5–9 on page 163 of your text and the Section 2.6 lecture video.

Estimate each of the following as 0, 1/2, *or* 1.

5. $\frac{11}{25}$

6. $\frac{37}{35}$

Estimate each part of the following as a whole number, as 1/2, *or as a mixed number where the fractional part is* 1/2.

7. $\frac{3}{5}\cdot\frac{9}{10}$

8. $22\frac{7}{8}\cdot 4\frac{1}{8}-5\frac{4}{9}$

Name: Date:
Instructor: Section:

Chapter 3 DECIMAL NOTATION

3.1 Decimal Notation, Order, and Rounding

Learning Objectives
A Given decimal notation, write a word name.
B Convert between decimal notation and fraction notation.
C Given a pair of numbers in decimal notation, tell which is larger.
D Round decimal notation to the nearest thousandth, hundredth, tenth, one, ten, hundred, or thousand.

Key Terms
Use the vocabulary terms listed below to complete each statement in Exercises 1–3.

arithmetic number(s) **decimal point** **decimal notation**
whole numbers **fraction(s)**

1. An arithmetic number may be written in ________________, such as 3.25, or as a mixed numeral, such as $3\frac{1}{4}$, or a fraction, such as $\frac{13}{4}$.

2. When writing a decimal numbers as a word, we write the word "and" for the ________________.

3. ______________ consist of ______________ and nonnegative ______________.

GUIDED EXAMPLES AND PRACTICE

Objective A Given decimal notation, write a word name.

Review this example for Objective A:

1. Write a word name for 306.845.

a) Write a word name for the whole number.

Three hundred six

b) Write "and" for the decimal point.

Three hundred six and

c) Write a word name for the number named to the right of the decimal point, followed by the place value of the last digit.

Three hundred six and

eight hundred forty-five thousandths

A word name for 306.845 is three hundred six and eight hundred forty-five thousandths.

Practice this exercise:

1. Write a word name for 59.07.

Objective B Convert between decimal notation and fraction notation.

Review these examples for Objective B:

2. Write fraction notation for 3.471.

3.471 — 3 places; 3.471. — move 3 places; $\frac{3471}{1000}$ — 3 zeros

$3.471 = \frac{3471}{1000}$

3. Write decimal notation for $\frac{61}{1000}$.

$\frac{61}{1000}$ — 3 zeros; 0.061. — Move 3 places; $\frac{61}{1000} = 0.061$

4. Write decimal notation for $5\frac{7}{10}$.

$5\frac{7}{10} = 5 + \frac{7}{10} = 5$ and $\frac{7}{10} = 5.7$

Practice these exercises:

2. Write fraction notation for 16.09.

3. Write decimal notation for $\frac{259}{100}$.

4. Write decimal notation for $91\frac{23}{100}$.

Objective C Given a pair of numbers in decimal notation, tell which is larger.

Review this example for Objective C:

5. Which is larger: 0.01 or 0.009?

0.01 ↕ 0.009 — Starting at the left, these digits are the first to differ; 1 is larger than 0.

Thus, 0.01 is larger. In symbols, $0.01 > 0.009$.

Practice this exercise:

5. Which is larger: 2.08 or 2.11?

Objective D Round decimal notation to the nearest thousandth, hundredth, tenth, one, ten, hundred, or thousand.

Review this example for Objective D:

6. Round 46.1938 to the nearest hundredth.

46.19[3]8 — Thousandths digit is 4 or lower. Round down.

↓

46.19

Practice this exercise:

6. Round 327.249 to the nearest tenth.

Name: Date:
Instructor: Section:

ADDITIONAL EXERCISES

Objective A Given decimal notation, write a word name.

For extra help, see Examples 1–4 on pages 179–180 of your text and the Section 3.1 lecture video.

Write a word name for the number in each sentence.

1. The Suez Canal is 119.9 miles long.

2. One meter is equivalent to about 1.094 yards.

3. The gasoline costs $2.789 per gallon.

4. Write a word name for 5.0801.

Objective B Convert between decimal notation and fraction notation.

For extra help, see Examples 5–13 on pages 181–182 of your text and the Section 3.1 lecture video.

Write fraction notation. Do not simplify.

5. 4.1376

6. 5.039

Write decimal notation for each number.

7. $\frac{213}{100}$

8. $417\frac{496}{10,000}$

Objective C Given a pair of numbers in decimal notation, tell which is larger.
For extra help, see Examples 14–15 on page 183 of your text and the Section 3.1 lecture video.
Which number is larger?

9. 0.9, 0.85

10. 0.143, 0.134

11. 146.19, 147.18

12. 0.008, $\frac{8}{100}$

Objective D Round decimal notation to the nearest thousandth, hundredth, tenth, one, ten, hundred, or thousand.
For extra help, see Examples 16–20 on pages 183–184 of your text and the Section 3.1 lecture video.
Round 234.0645 to the nearest:

13. Tenth

14. Hundredth

15. Thousandth

16. Round 36.4519 to the nearest one.

Name: Date:
Instructor: Section:

Chapter 3 DECIMAL NOTATION

3.2 Addition and Subtraction

Learning Objectives
A Add using decimal notation.
B Subtract using decimal notation.
C Solve equations of the type $x + a = b$ and $a + x = b$, where a and b may be in decimal notation.
D Balance a checkbook.

Key Terms
Use the vocabulary terms listed below to complete each statement in Exercises 1–4. Terms may be used more than once.

balance forward **debit(s)** **credit(s)** **place-value digits**

1. A ________________ increases the balance in a checking account.
2. A ________________ decreases the balance in a checking account.
3. The ________________ column in a check book indicates how much money is in the account after all the ________________ and ________________ up to that point have been calculated.
4. When we add numbers in decimal notation, we must add corresponding ________________, so we line up the decimal points.

GUIDED EXAMPLES AND PRACTICE

Objective A Add using decimal notation.

Review this example for Objective A:
1. Add: 14.26 + 63.589.

$$\begin{array}{r} \overset{1}{1}4.\overset{1}{2}60 \\ +\ 63.589 \\ \hline 77.849 \end{array}$$

Lining up the decimals and writing and extra zero
Adding

Practice this exercise:
1. Add: 3.08 + 25.962.

Objective B Subtract using decimal notation.

Review this example for Objective B:
2. Subtract: 67.345 – 24.28.

$$\begin{array}{r} 67.\overset{2}{\not3}\overset{14}{\not4}5 \\ -\ 24.280 \\ \hline 43.065 \end{array}$$

Writing an extra zero
Subtracting

Practice this exercise:
2. Subtract: 221.04 – 13.192.

Objective C Solve equations of the type $x + a = b$ and $a + x = b$, where a and b may be in decimal notation.

Review this example for Objective C:

3. Solve: $x + 15.4 = 38.32$.

$$x + 15.4 - 15.4 = 38.32 - 15.4$$
$$x = 22.92$$

$$\begin{array}{r} \overset{7}{\cancel{8}}\;\overset{13}{\cancel{3}} \\ 38.32 \\ -\,15.40 \\ \hline 22.92 \end{array}$$

Practice this exercise:

3. Solve: $y + 0.88 = 115.8$.

Objective D Balance a checkbook.

Review this example for Objective D:

4. Find the errors, if any, in the balances of the checkbook.

Transaction Description	Payment/ Debit	Deposit/ Credit	Balance Forward 3657.42
Rent	780.00		2787.42
Deposit		382.52	3169.94
Phone Bill	59.95		3109.95
Deposit		50.00	3159.95

We successively add or subtract deposit/credits and debits, and check the result in the "Balance forward" column.

3657.42 – 780.00 = 2877.42

We found an error. The subtraction was incorrect. We have to correct all following lines.

2877.42 + 382.52 = 3259.94

3259.94 – 59.95 = 3199.99

3199.99 + 50.00 = 3249.99

The corrected checkbook is below.

Transaction Description	Payment/ Debit	Deposit/ Credit	Balance Forward 3657.42
Rent	780.00		2877.42
Deposit		382.52	3259.94
Phone Bill	59.95		3199.99
Deposit		50.00	3249.99

Practice this exercise:

4. Tori had $580.66 in her checking account before she wrote checks for payments of $36.43, $280.75, and $128.64, and made a deposit of $289.08. Find her balance after making these transactions.

Name: Date:
Instructor: Section:

ADDITIONAL EXERCISES

Objective A Add using decimal notation.

For extra help, see Examples 1–3 on pages 188–189 of your text and the Section 3.2 lecture video.

Add.

1. $\begin{array}{r} 19.74 \\ +\ 21.43 \\ \hline \end{array}$

2. $2.306 + 5.829$

3. $0.347 + 0.16$

4. $157 + 0.45 + 2.9$

Objective B Subtract using decimal notation.

For extra help, see Examples 4–7 on pages 189–190 of your text and the Section 3.2 lecture video.

Subtract.

5. $\begin{array}{r} 34.192 \\ -\ 8.341 \\ \hline \end{array}$

6. $60.55 - 0.493$

7. $5 - 3.1405$

8. $125.3 - 16.71$

Objective C Solve equations of the type $x + a = b$ and $a + x = b$, where a and b may be in decimal notation.

For extra help, see Examples 8–10 on page 191 of your text and the Section 3.2 lecture video.

Solve.

9. $w + 0.0054 = 20$

10. $403.962 + a = 811.243$

11. $23{,}009.2 = x + 24.51$

Objective D Balance a checkbook.

For extra help, see Example 11 on pages 191–192 of your text and the Section 3.2 lecture video.

Find the errors, if any, in the checkbook.

12.

Date	Check Number	Transaction Description	Payment/ Debit	Deposit/ Credit	Balance Forward
					4652.38
9/6	426	Safeway	46.92		4605.46
9/7		Deposit		200.00	4805.46
9/13	427	Leppert's Video	16.95		4822.41
9/14		Deposit		55.81	4766.60
9/15	428	Utility Company	42.95		4723.65
9/15	429	Water District	54.23		4669.42

Name: Date:
Instructor: Section:

Chapter 3 DECIMAL NOTATION

3.3 Multiplication

Learning Objectives

A Multiply using decimal notation.

B Convert from notation like 45.7 million to standard notation, and convert between dollars and cents.

Key Terms

Use the terms listed below to complete each statement in Exercises 1–3. Terms may be used more than once.

cent(s) **dollar(s)** **$** **¢**

1. There are 100 ________________ in one ________________ .

2. The symbol ________________ follows a number to the right and indicates how many ________________ there are.

3. The ________________ symbol precedes a number to the left and indicates how many ________________ there are.

GUIDED EXAMPLES AND PRACTICE

Objective A Multiply using decimal notation.

Review these examples for Objective A:

1. Multiply: 2.8×0.03.

$$\begin{array}{rl} 2.8 & \text{(1 decimal place)} \\ \times\ 0.0\,3 & \text{(2 decimal places)} \\ \hline 0.0\,8\,4 & \text{(3 decimal places)} \end{array}$$

2. Multiply: 0.001×72.4.

$0.001 \times 72.4 = 0.\underbrace{072}_{\leftarrow}.4$ Moving the decimal point 3 places to the left. This requires adding one extra zero.

$= 0.0724$

3. Multiply: 34.6×100.

$34.6 \times 100 = 3460$ Moving the decimal point 2 places to the right and using one zero as a placeholder.

Practice these exercises:

1. Multiply: 4.63×2.5.

2. Multiply: 14.3×0.01.

3. Multiply: 1000×85.043.

Objective B Convert from notation like 45.7 million to standard notation, and convert between dollars and cents.

Review these examples for Objective B:

4. Convert 45.7 million to standard notation.

$$\begin{aligned} 45.7 \text{ million} &= 45.7 \times 1 \text{ million} \\ &= 45.7 \times 1,\underline{000,000} \quad \text{6 zeros} \\ &= 45,700,000 \quad \text{Moving the decimal 6 places to the right} \end{aligned}$$

5. Convert \$63.42 to cents.

$$\begin{aligned} \$63.42 &= 63.42 \times \$1 \\ &= 63.42 \times 100¢ \quad \text{Substituting } 100¢ \text{ for } \$1 \\ &= 6342¢ \quad \text{Multiplying} \end{aligned}$$

6. Convert 9168¢ to dollars.

$$\begin{aligned} 9168¢ &= 9168 \times 1¢ \\ &= 9168 \times \$0.01 \quad \text{Substituting } \$0.01 \text{ for } 1¢ \\ &= \$91.68 \quad \text{Multiplying} \end{aligned}$$

Practice these exercises:

4. Convert 6.2 billion to standard notation.

5. Convert \$125.49 to cents.

6. Convert 245¢ to dollars.

ADDITIONAL EXERCISES

Objective A Multiply using decimal notation.

For extra help, see Examples 1–11 on pages 196–198 of your text and the Section 3.3 lecture video.

Multiply.

1. $\begin{array}{r} 4.6 \\ \times \ \ 8 \\ \hline \end{array}$

2. 0.56×1000

3. 0.0675×0.01

4. 15.3×45.1

Objective B Convert from notation like 45.7 million to standard notation, and convert between dollars and cents.

For extra help, see Examples 12–16 on page 200 of your text and the Section 3.3 lecture video.

Convert from dollars to cents.

5. \$47.13

6. \$0.79

Convert from cents to dollars.

7. 4351¢

8. 26¢

9. Convert the number in this sentence to standard notation:
The number of real Christmas tress purchased in a recent year was 32.8 million.

Name: Date:
Instructor: Section:

Chapter 3 DECIMAL NOTATION

3.4 Division

Learning Objectives
A Divide using decimal notation.
B Solve equations of the type $a \cdot x = b$, where a and b may be in decimal notation.
C Simplify expressions using the rules for order of operations.

Key Terms
Write the letters A, B, C, and D in the blanks to indicate the order in which operations should be performed when simplifying expressions.

1. ______________

2. ______________

3. ______________

4. ______________

A. Evaluate all exponential expressions.
B. Do all additions and subtractions in order from left to right.
C. Do all calculations within grouping symbols.
D. Do all multiplications and divisions in order from left to right.

GUIDED EXAMPLES AND PRACTICE

Objective A Divide using decimal notation.

Review these examples for Objective A:

1. Divide: $36.8 \div 8$.

$$\begin{array}{r} 4.6 \\ 8 \overline{)3\,6.8} \\ \underline{3\,2} \\ 4\,8 \\ \underline{4\,8} \\ 0 \end{array}$$

2. Divide: $21.35 \div 6.1$.

First we find $21.35 \div 6.1$.

$$\begin{array}{r} 3.5 \\ 6.1_{\wedge} \overline{)2\,1.3_{\wedge}5} \\ \underline{1\,8\,3} \\ 3\,0\,5 \\ \underline{3\,0\,5} \\ 0 \end{array}$$

Practice these exercises:

1. Divide: $615.6 \div 12$.

2. Divide: $24.07 \div 2.9$.

3. Divide: $\frac{16.7}{1000}$.

$$\frac{16.7}{1000} = \frac{0.016.7}{1000} = \frac{0.0167}{1.000} = 0.0167$$

3 zeros 3 places to the left to change 1000 to 1

The answer is 0.0167.

4. Divide: $\frac{42.93}{0.001}$.

$$\frac{42.93}{0.001} = \frac{42.930}{0.001} = \frac{42{,}930}{1.} = 42{,}930$$

3 zeros 3 places to the right to change 0.001 to 1

The answer is 42,930.

3. Divide: $\frac{3.9}{100}$.

4. Divide: $\frac{123.4}{0.01}$.

Objective B Solve equations of the type *a* · *x* = *b*, where *a* and *b* may be in decimal notation.

Review this example for Objective B:

5. Solve: $5.4 \cdot x = 38.88$

We have

$\frac{5.4 \cdot x}{5.4} = \frac{38.88}{5.4}$ Dividing by 5.4 on both sides

$x = 7.2$

$$\begin{array}{r} 7.2 \\ 5.4_{\wedge}\overline{)38.8_{\wedge}8} \\ \underline{378} \\ 108 \\ \underline{108} \\ 0 \end{array}$$

Practice this exercise:

5. Solve: $1.4 \cdot y = 13.02$.

Objective C Simplify expressions using the rules for order of operations.

Review these examples for Objective B:

6. Simplify: $(8-2.4) \div 2^2 + 4.9 \times 10$.

$(8-2.4) \div 2^2 + 4.9 \times 10$

$= 5.6 \div 2^2 + 4.9 \times 10$ Working inside the parentheses

$= 5.6 \div 4 + 4.9 \times 10$ Evaluating the exponential term

$= 1.4 + 49$ Multiplying and dividing in order from left to right

$= 50.4$ Adding

Practice these exercises:

6. Simplify:

$5 \div 0.5 + 2.1 \times 8 - (1-0.9)^3$.

Name: Date:
Instructor: Section:

7. While on a weekend trip, the Hernandez family drove 250.6 mi on the first day, 164.5 mi on the second day, and 203.8 mi the third day. Find the average distance driven per day.

To find the average distance, find the sum of the distances and divide by the number of addends, 3.

$(250.6+164.5+203.8)\div 3=618.9\div 3=206.3$

The average distance is 206.3 mi.

7. Keisha walked 3.5 mi on Tuesday, 1.2 mi on Wednesday, 2.9 mi on Thursday, and 4 mi on Friday. Find the average distance walked per day.

ADDITIONAL EXERCISES

Objective A Divide using decimal notation.

For extra help, see Examples 1–8 on pages 204–208 of your text and the Section 3.4 lecture video.

Divide.

1. $21\overline{)70.35}$

2. $0.48\overline{)0.1728}$

3. $\dfrac{345}{1000}$

4. $\dfrac{35.17}{0.01}$

Objective B Solve equations of the type $a \cdot x = b$, where a and b may be in decimal notation.

For extra help, see Examples 9–10 on pages 208–209 of your text and the Section 3.4 lecture video.

Solve.

5. $100\cdot w=5.702$

6. $1809.75=0.38\cdot p$

7. $508=20.32\cdot h$

Objective C Simplify expressions using the rules for order of operations.
For extra help, see Examples 11–14 on pages 209–210 of your text and the Section 3.4 lecture video.
Simplify.

8. $15 \times (6.4 - 3.8)$

9. $10 \div 0.01 - 0.3 \times 7 + 0.2^2$

10. $13^2 \div (11 + 5.9) - [(3.6 - 3) \div 0.3]$

11. Marco earned $25.50 in tips on Monday, $21.40 on Tuesday, $24.75 on Wednesday, $35 on Thursday, and $42.90 on Friday. Find the average amount earned in tips per day during this period.

Name: Date:
Instructor: Section:

Chapter 3 DECIMAL NOTATION

3.5 Converting from Fraction Notation to Decimal Notation

Learning Objectives
A Convert from fraction notation to decimal notation.
B Round numbers named by repeating decimals in problem solving.
C Calculate using fraction notation and decimal notation together.

Key Terms
Use the vocabulary terms listed below to complete each statement in Exercises 1–2.

terminating decimal **repeating decimal**

1. A ________________ occurs if division leads to a repeating pattern of nonzero remainders.

2. A ________________ occurs if division leads to a remainder of zero.

GUIDED EXAMPLES AND PRACTICE

Objective A Convert from fraction notation to decimal notation.

Review these examples for Objective A:

1. Find decimal notation for $\frac{13}{20}$.

   Since $20 \cdot 5 = 100$, we use $\frac{5}{5}$ for 1 to get a denominator of 100.

   $$\frac{13}{20} = \frac{13}{20} \cdot \frac{5}{5} = \frac{65}{100} = 0.65$$

2. Find decimal notation for $\frac{1}{6}$.

$$\begin{array}{r} 0.166 \\ 6\overline{)1.000} \\ \underline{6} \\ 40 \\ \underline{36} \\ 40 \\ \underline{36} \\ 4 \end{array}$$

   Since 4 keeps reappearing as a remainder, the digit repeats and will continue to do so. Thus, $\frac{1}{6} = 0.1\overline{6}$.

Practice these exercises:

1. Find decimal notation for $\frac{17}{25}$.

2. Find decimal notation for $\frac{5}{9}$.

Objective B **Round numbers named by repeating decimals in problem solving.**

Review this example for Objective B:

3. Round $0.\overline{37}$ to the nearest thousandth.

$0.\overline{37} = 0.3737...$
The digit in the ten-thousandths place, 7, is 5 or higher, so we round up and get 0.374.

Practice this exercise:

3. Round $0.\overline{15}$ to the nearest hundredth.

Objective C **Calculate using fraction notation and decimal notation together.**

Review this example for Objective C:

4. Calculate: $1.512 \times \frac{4}{3}$.

We will use the first method.

$$\begin{aligned}1.512 \times \frac{4}{3} &= \frac{1.512}{1} \times \frac{4}{3} \\ &= \frac{1.512 \times 4}{3} \\ &= \frac{6.048}{3} \\ &= 2.016\end{aligned}$$

Practice this exercise:

4. Calculate: $5.32 \div \frac{4}{5}$.

ADDITIONAL EXERCISES

Objective A **Convert from fraction notation to decimal notation.**

For extra help, see Examples 1–9 on pages 217–219 of your text and the Section 3.5 lecture video.

Find decimal notation for each number.

1. $\frac{3}{4}$

2. $\frac{21}{25}$

3. $\frac{1}{8}$

4. $\frac{21}{16}$

5. $\frac{3}{11}$

Name: Date:
Instructor: Section:

Objective B Round numbers named by repeating decimals in problem solving.
For extra help, see Examples 10–15 pages 219–220 of your text and the Section 3.5 lecture video.
Round the decimal notation for each number to the nearest tenth, hundredth, and thousandth.

6. $\frac{7}{11}$

7. $\frac{4}{7}$

Find the gas mileage rounded to the nearest tenth.

8. 326 mi; 14 gal

9. 502 mi; 15.3 gal

Objective C Calculate using fraction notation and decimal notation together.
For extra help, see Examples 16–17 on page 221 of your text and the Section 3.5 lecture video.
Calculate and write the result as a decimal.

10. $\frac{3}{5}(439.6)$

11. $\frac{8}{3}(-24.12)$

12. $\frac{7}{8}\times 0.1088+\frac{5}{6}\times 0.4332$

13. $\left(\frac{3}{4}\right)123.62-\left(\frac{7}{10}\right)435.7$

Name: Date:
Instructor: Section:

Chapter 3 DECIMAL NOTATION

3.6 Estimating

Learning Objectives
A Estimate sums, differences, products, and quotients.

GUIDED EXAMPLES AND PRACTICE

Objective A Estimate sums, differences, products, and quotients.

Review this example for Objective A:

1. Estimate 4.25 + 6.91 + 1.046 by rounding to the nearest tenth.

$$4.25 + 6.91 + 1.046 \approx 4.3 + 6.9 + 1.0 = 12.2$$

Practice this exercise:

1. Estimate 68×4.2. Which of the following is an appropriate estimate?
a) 17 b) 280
c) 650 d) 700

ADDITIONAL EXERCISES

Objective A Estimate sums, differences, products, and quotients.

For extra help, see Examples 1–9 on pages 226–228 of your text and the Section 3.6 lecture video.

Estimate by rounding, as directed.

1. $16.24 + 84.53 + 7.67$; nearest one

2. $21.319 - 17.481$; nearest tenth

Estimate. Choose a rounding digit that gives one or two nonzero digits. Indicate which choice is an appropriate estimate.

3. 563.40 – 256.49
a) 820 b) 310 c) 300 d) 270

4. $75.21 \div 24.6$
a) 0.3 b) 0.4 c) 3 d) 4

An iPod sells for $197.99 and a GPS navigation system sells for $349.99.

5. About how much more does the GPS system cost than the iPod?

Name: Date:
Instructor: Section:

Chapter 3 DECIMAL NOTATION

3.7 Applications and Problem Solving

Learning Objectives
A Solve applied problems involving decimals.

Key Terms
Use the vocabulary terms listed below to complete each statement in Exercises 1–2.

variable **estimate**

1. A(n) ________________ is a letter used to represent an unknown quantity.

2. A partial check of a solution can often be done by finding a(n) ________________ .

GUIDED EXAMPLES AND PRACTICE

Objective A Solve applied problems involving decimals.

Review this example for Objective B:

2. Erik's odometer read 45,918.7 mi at the beginning of a trip. It read 47,304.2 mi at the end of the trip. How far did Erik drive?

1. *Familiarize*. We make a drawing. Let m = the number of miles Erik drove.

45,918.7 mi	m
47,304.2 mi	

2. *Translate*. This is a "how much more" situation.

$$\underbrace{\text{First reading}} + \underbrace{\text{Additional number of miles}} = \underbrace{\text{Final reading}}$$

$$45{,}918.7 + m = 47{,}304.2$$

3. *Solve*. We subtract 45,918.7 on both sides of the equation.

$$45{,}918.7 + m = 47{,}304.2$$
$$45{,}918.7 + m - 45{,}918.7 = 47{,}304.2 - 45{,}918.7$$
$$m = 1385.5$$

4. *Check*. We can check by adding: 45,918.7 + 1385.5 = 47,304.2. The result checks.

5. *State*. Erik drove 1385.5 mi.

Practice this exercise:

2. A car loan of $7791.60 is to be paid off in 24 equal monthly payments. How much is each payment?

ADDITIONAL EXERCISES

Objective A Solve applied problems involving decimals.

For extra help, see Examples 1–8 on pages 232–238 of your text and the Section 3.7 lecture video.

Solve using the five-step problem-solving procedure.

1. Normal body temperature is 98.6°F. At one point during an illness, Jessica's temperature was 101.4°F. How much did her temperature rise?

2. The four Maloney children buy their parents an anniversary gift that costs $139.80 and split the cost equally. How much does each child pay?

3. Find the perimeter and the area of a rectangle with length 15.3 cm and width 7.6 cm.

4. Computer Wizard charges $40 for a house call plus $45.50 for each hour a job takes. How long did an employee of Computer Wizard work on a house call if the bill totals $108.25?

Name: Date:
Instructor: Section:

Chapter 4 PERCENT NOTATION

4.1 Ratio and Proportion

Learning Objectives
A Find fraction notation for ratios.
B Give the ratio of two different measures as a rate.
C Determine whether two pairs of numbers are proportional.
D Solve proportions.
E Solve applied problems involving proportions.

Key Terms
Use the vocabulary terms listed below to complete each statement in Exercises 1–3.

rate **cross products** **proportions**

1. Equations that state that two ratios are equal are called ________________ .

2. When a ratio is used to compare two different kinds of measure, it is called a ________________ .

3. To solve $\frac{a}{b}=\frac{c}{d}$ for a specific variable, equate ________________ and then divide on both sides to get that variable alone. (Assume $b, d \neq 0$.)

GUIDED EXAMPLES AND PRACTICE

Objective A Find fraction notation for ratios.

Review these examples for Objective A:

1. Find fraction notation for the ratio of 5 to 12.

Write a fraction with a numerator of 5 and a denominator of 12: $\frac{5}{12}$.

2. A sweater that originally sold for $35.95 was marked down to a sale price of $29.99. What is the ratio of the original price to the sale price? of the sale price to the original price?

The ratio of the original price to the sale price is

$$\frac{\$35.95 \leftarrow \text{original price}}{\$29.99 \leftarrow \text{sale price}}$$

The ratio of the sale price to the original price is

$$\frac{\$29.99 \leftarrow \text{sale price}}{\$35.95 \leftarrow \text{original price}}$$

Practice these exercises:

1. Find fraction notation for the ratio of 8 to 3.

2. A rectangular carpet measures 12 ft by 10 ft. What is the ratio of the length to width? of the width to length?

3. Simplify the ratio of 4 to 2.5.

We write the ratio in fraction notation. Next, we multiply by 1 to clear the decimal from the numerator. Then we simplify.

$$\frac{4}{2.5}=\frac{4}{2.5}\cdot\frac{10}{10}=\frac{40}{25}=\frac{5\cdot 8}{5\cdot 5}=\frac{5}{5}\cdot\frac{8}{5}=\frac{8}{5}.$$

3. Simplify the ratio of 1.6 to 3.6.

Objective B Give the ratio of two different measures as a rate.

Review this example for Objective B:

4. A driver travels 132 mi on 5.5 gal of gas. What is the rate in miles per gallon?

$\frac{132 \text{ mi}}{5.5 \text{ gal}}$, or $\frac{132 \text{ mi}}{5.5 \text{ gal}}=\frac{24 \text{ mi}}{1 \text{ gal}}$, or 24 mpg.

Practice this exercise:

4. A student earned \$91 for working 14 hr. What was the rate of pay per hour?

Objective C Determine whether two pairs of numbers are proportional.

Review this example for Objective C:

5. Determine whether 3, 4, and 7, 9 are proportional.

$3\cdot 9=27 \quad \frac{3}{4} \overset{?}{=} \frac{7}{9} \quad 4\cdot 7=28$

Since the cross products are not the same ($27 \neq 28$), then $\frac{3}{4}\neq\frac{7}{9}$ and the numbers are not proportional.

Practice this exercise:

5. Determine whether 5, 9 and 20, 36 are proportional.

Objective D Solve proportions.

Review these examples for Objective D:

6. Solve: $\frac{5}{4}=\frac{y}{11}$.

$\frac{5}{4}=\frac{y}{11}$

$5\cdot 11=4\cdot y$ Equating cross products

$\frac{5\cdot 11}{4}=\frac{4\cdot y}{4}$ Dividing both sides by 4

$\frac{55}{4}=y$ Simplifying

Practice these exercises:

6. Solve: $\frac{12}{10}=\frac{x}{15}$.

Name: Date:
Instructor: Section:

7. Solve: $\frac{7}{3}=\frac{98}{x}$.

$$\frac{7}{3}=\frac{98}{x}$$

$7\cdot x=3\cdot 98$ Equating cross products

$\frac{7\cdot x}{7}=\frac{3\cdot 98}{7}$ Dividing both sides by 7

$x=\frac{294}{7}$ Simplifying

$x=42$ Dividing

7. Solve: $\frac{6}{x}=\frac{5}{3}$.

Objective E Solve applied problems involving proportions.

Review this example for Objective E:

8. Louis bought 3 tickets to a campus theater production for \$16.50. How much would 8 tickets cost?

1. *Familiarize*. Let c = the cost of 8 tickets.
2. *Translate*. We translate to a proportion, keeping the number of tickets in the numerator.

$$\begin{array}{r} \text{Tickets} \rightarrow \\ \text{Cost} \rightarrow \end{array} \frac{3}{16.50}=\frac{8}{c} \begin{array}{l} \leftarrow \text{Tickets} \\ \leftarrow \text{Cost} \end{array}$$

3. *Solve*. We solve the proportion.

$$\frac{3}{16.50}=\frac{8}{c}$$

$3\cdot c=16.50\cdot 8$ Equating cross products

$c=\frac{16.50\cdot 8}{3}$ Dividing both sides by 3

$c=44$ Multiplying and dividing

4. *Check*. We use a different approach as a check. Find the cost per ticket and then multiply it by 8:
$\$16.50\div 3=\$5.50$ and
$\$5.50\cdot 8=\$44$.
The answer checks.
5. *State*. Eight tickets would cost \$44.

Practice this exercise:

8. On a map, $\frac{1}{2}$ in. represents 40 mi. If two cities are $2\frac{1}{4}$ in. apart on the map, how far apart are they in reality?

ADDITIONAL EXERCISES

Objective A Find fraction notation for ratios.

For extra help, see Examples 1–6 on pages 256–258 of your text and the Section 4.1 lecture video.

Find the fraction notation for each ratio. You need not simplify.

1. 7 to 9

2. $6\frac{2}{3}$ to $34\frac{5}{6}$

3. Monica has 6 grandsons and 5 granddaughters. What is the ratio of granddaughters to grandsons? Of grandsons to granddaughters?

Find the ratio of the first number to the second number and simplify.

4. 5.6 to 6.4

5. 270 to 162

Objective B Give the ratio of two different measures as a rate.

For extra help, see Examples 7–9 on page 259 of your text and the Section 4.1 lecture video.

6. Find the rate or speed as a ratio of distance to time of 130 mi, 2 hr.

7. A leaky faucet can lose 21 gal of water in a week. What is the rate in gal per day?

8. It takes 60 oz of grass seed to seed a 5000 ft^2 lawn. What is the rate in ounce per square foot?

9. Dale burn 159 calories walking for 30 min. What is the rate in calories burned per minute?

Objective C Determine whether two pairs of numbers are proportional.

For extra help, see Examples 10–11 on pages 260–261 of your text and the Section 4.1 lecture video.

Determine whether the two pairs of numbers are proportional.

10. 9, 5 and 6, 4

11. 2, 3 and 5, 8

12. 4.8, 1.4 and 2.4, 0.7

13. $4\frac{2}{3}$, $8\frac{7}{10}$ and 7, $12\frac{3}{10}$

Name: Date:
Instructor: Section:

Objective D Solve proportions.

For extra help, see Examples 12–15 on pages 262–263 of your text and the Section 4.1 lecture video.

Solve.

14. $\frac{n}{9}=\frac{8}{3}$

15. $\frac{3}{51}=\frac{n}{68}$

16. $\frac{18}{y}=\frac{30}{25}$

17. $\frac{2.36}{4.18}=\frac{9.44}{y}$

18. $\frac{1\frac{3}{4}}{3\frac{1}{2}}=\frac{x}{6\frac{5}{8}}$

Objective E Solve applied problems involving proportions.

For extra help, see Examples 16–20 on pages 263–266 of your text and the Section 4.1 lecture video.

Solve.

19. A train traveled 312 mi in 4 hr. At this rate, how far will it travel in 10 hr?

20. In a metal alloy, the ratio of copper to zinc is 15 to 8. If there are 360 lb of zinc, how many pounds of copper are there?

21. To determine the number of deer in a game preserve, a forest ranger catches 120 deer, tags them, and then releases them. Later 75 deer are caught and it is found that 45 of them are tagged. Estimate the number of deer in the game preserve.

22. A 5-lb ham contains 20 servings of meat. How many pounds of ham would be needed for 12 servings?

Name: Date:
Instructor: Section:

Chapter 4 PERCENT NOTATION

4.2 Percent Notation

Learning Objectives
A Write three kinds of notation for a percent.
B Convert between percent notation and decimal notation.

Key Terms
Use the vocabulary terms listed below to complete each statement in Exercises 1–2.

left **right**

1. When we convert from percent notation to decimal notation, we move the decimal point two places to the ________________ .

2. When we convert from decimal notation to percent notation, we move the decimal point two places to the ________________ .

GUIDED EXAMPLES AND PRACTICE

Objective A Write three kinds of notation for a percent.

Review this example for Objective A:

1. Write three kinds of notation for 67%.

Ratio:

$67\% = \frac{67}{100}$ A ratio of 67 to 100

Fraction notation:

$67\% = 67 \times \frac{1}{100}$ Replacing % with $\times \frac{1}{100}$

Decimal notation:

$67\% = 67 \times 0.01$ Replacing % with $\times 0.01$

Practice this exercise:

1. Write three kinds of notation for 13%.

Objective B Convert between percent notation and decimal notation.

Review these examples for Objective B:

2. Find decimal notation for 4.3%.

a) Replace the decimal symbol with $\times 0.01$

4.3×0.01

b) Move the decimal point two places to the left.

0.04.3

Thus, 4.3% = 0.043.

Practice these exercises:

2. Find decimal notation for 54.8%

3. Find percent notation for 0.09.

a) Move the decimal point two places to the right.

0.09.

b) Write a percent symbol: 9%

Thus, 0.09 = 9%.

3. Find percent notation for 1.5.

ADDITIONAL EXERCISES

Objective A Write three kinds of notation for a percent.

For extra help, see Examples 1–2 on page 273 of your text and the Section 4.2 lecture video.

Write three kinds of notation, as in Examples 1 and 2 on p. 273 of the text.

1. 85%

2. 126%

3. 4.8%

4. 27.3%

Objective B Convert between percent notation and decimal notation.

For extra help, see Examples 3–8 on page 274–275 of your text and the Section 4.2 lecture video.

Write each percent as an equivalent decimal.

5. 150%

6. $93\frac{1}{2}\%$

Write each decimal as an equivalent percent.

7. 0.83

8. 0.001

Name: Date:
Instructor: Section:

Chapter 4 PERCENT NOTATION

4.3 Percent Notation and Fraction Notation

Learning Objectives
A Convert from fraction notation to percent notation.
B Convert from percent notation to fraction notation.

Key Terms
Use the vocabulary terms listed below to complete each statement in Exercises 1–2.

decimal equivalent **fraction equivalent**

1. The ___________________ to 0.4 is $\frac{2}{5}$.

2. The ___________________ $\frac{1}{6}$ is $0.1\overline{6}$.

GUIDED EXAMPLES AND PRACTICE

Objective A Convert from fraction notation to percent notation.

Review this example for Objective A:

1. Find percent notation for $\frac{7}{20}$.

Find decimal notation by division.

$$\begin{array}{r} 0.35 \\ 20\overline{)7.00} \\ \underline{60} \\ 100 \\ \underline{100} \\ 0 \end{array}$$

$\frac{7}{20} = 0.35$

Convert to percent notation by moving the decimal point two places to the right and write a % symbol.

$\frac{7}{20} = 35\%$

When the denominator of the fraction is a factor of 100, we can also find percent notation by first multiplying by 1 to get 100 in the denominator:

$\frac{7}{20} = \frac{7}{20} \cdot \frac{5}{5} = \frac{35}{100} = 35\%$.

Practice this exercise:

1. Find percent notation for $\frac{7}{8}$.

Objective B Convert from percent notation to fraction notation.

Review this example for Objective B:

2. Find fraction notation for 40%.

$$40\% = \frac{40}{100} \quad \text{Using the definition of percent}$$
$$= \frac{20\cdot 2}{20\cdot 5}$$
$$= \frac{20}{20}\cdot\frac{2}{5} \quad \text{Simplifying by removing a factor equal to 1: } \frac{20}{20}=1$$
$$= \frac{2}{5}$$

Practice this exercise:

2. Find fraction notation for 65%.

ADDITIONAL EXERCISES

Objective A Convert from fraction notation to percent notation.

For extra help, see Examples 1–5 on pages 279–281 of your text and the Section 4.3 lecture video.

Write each fraction as a equivalent percent.

1. $\frac{14}{25}$

2. $\frac{7}{12}$

3. $\frac{5}{8}$

4. $\frac{15}{16}$

Objective B Convert from percent notation to fraction notation.

For extra help, see Examples 6–9 on page 281–282 of your text and the Section 4.3 lecture video.

For each percent, write an equivalent fraction. When possible, simplify.

5. 37.5%

6. $16\frac{2}{3}\%$

7. $83.\overline{3}\%$

8. 350%

Name: Date:
Instructor: Section:

Chapter 4 PERCENT NOTATION

4.4 Solving Percent Problems Using Percent Equations

Learning Objectives
A Translate percent problems to percent equations.
B Solve basic percent problems.

Key Terms
Use the terms listed below to complete each statement in Exercises 1–4.

% **is** **of** **what**

1. ________________ translates to "=."
2. ________________ translates to any letter.
3. ________________ translates to "$\times\frac{1}{100}$" or "$\times 0.01$."
4. ________________ translates to "$\cdot$" or "$\times$."

GUIDED EXAMPLES AND PRACTICE

Objective A Translate percent problems to percent equations.

Review this example for Objective A:

1. Translate to an equation: What is 12% of 84?

What is 12% of 84?
↓ ↓ ↓ ↓ ↓
$a = 12\% \cdot 84$

Practice this exercise:

1. Translate to an equation: 16% of what is 224?

Objective B Solve basic percent problems.

Review these examples for Objective B:

2. What percent of 60 is 21?

Translate: What percent of 60 is 21?
↓ ↓ ↓ ↓ ↓
$p \times 60 = 21$

Divide by 60 on both sides and convert to percent.

$$\frac{p \times 60}{60} = \frac{21}{60}$$
$$p = 0.35 = 35\%$$

Practice these exercises:

2. What percent of 125 is 50?

3. 84% of 60 is what?

Translate: 84% of 60 is what?

$$0.84 \times 60 = a$$

We carry out the calculation.

$$a = 0.84 \times 60 = 50.4$$

4. 51 is 30% of what?

Translate: 51 is 30% of what?

$$51 = 0.30 \times b$$

Divide by 0.30 on both sides.

$$\frac{51}{0.30} = \frac{0.30 \times b}{0.30}$$

$$b = 170$$

3. 120% of $80 is what?

4. 9 is 15% of what?

ADDITIONAL EXERCISES

Objective A Translate percent problems to percent equations.

For extra help, see Examples 1–6 on page 287 of your text and the Section 4.4 lecture video.

Translate to an equation. Do not solve.

1. What is 18% of 93?

2. 5% of what is 36?

3. 65 is what percent of 150?

Objective B Solve basic percent problems.

For extra help, see Examples 7–12 on pages 288–290 of your text and the Section 4.4 lecture video.

Translate to an equation and solve.

4. $9.90 is what percent of $45?

5. 16% of 130 is what?

6. 32% of what is 64?

7. 44 is 55% of what?

8. What percent of 140 is 84?

Name: Date:
Instructor: Section:

Chapter 4 PERCENT NOTATION

4.5 Solving Percent Problems Using Proportions

Learning Objectives
A Translate percent problems to proportions.
B Solve basic percent problems.

Key Terms
Use the vocabulary terms listed below to complete the statement in Exercises 1–2.

part **whole**

1. When translating a percent problem to a proportion, the ________________ corresponds to the base.

2. When translating a percent problem to a proportion, the ________________ corresponds to the amount.

GUIDED EXAMPLES AND PRACTICE

Objective A Translate percent problems to proportions.

Review these examples for Objective A:

1. Translate to a proportion: 7 is 15% of what?

$$\frac{15}{100}=\frac{7}{b}$$

2. Translate to a proportion:
14 is what percent of 25?

$$\frac{N}{100}=\frac{14}{25}$$

3. Translate to a proportion:
What is 24% of 96?

$$\frac{24}{100}=\frac{a}{96}$$

Practice these exercises:

1. Translate to a proportion:
34 is 85% of what?

2. Translate to a proportion:
30 is what percent of 45?

3. Translate to a proportion:
What is 61% of 320?

Objective B Solve basic percent problems.

Review this example for Objective B:

4. 75% of 150 is what?

Translate: $\frac{75}{100} = \frac{a}{150}$

$75 \cdot 150 = 100 \cdot a$ Equating cross products

$\frac{75 \cdot 150}{100} = \frac{100 \cdot a}{100}$ Dividing by 100

$\frac{11,250}{100} = a$

$112.5 = a$ Simplifying

Thus 75% of 150 is 112.5. The answer is 112.5.

Practice this exercise:

4. 4% of what is 3.6?

ADDITIONAL EXERCISES

Objective A Translate percent problems to proportions.

For extra help, see Examples 1–6 on page 294–295 of your text and the Section 4.5 lecture video.

Translate to an equation. Do not solve.

1. What percent of 80 is 16?

2. 30% of 63 is what?

3. 20 is 45% of what?

Objective B Solve basic percent problems.

For extra help, see Examples 7–11 page 295–296 of your text and the Section 4.5 lecture video.

Translate to an equation and solve.

4. What percent of 75 is 15?

5. 120% of 95 is what?

6. 8% of what is 64?

7. 6.65% of 200 is what?

8. 21 is what percent of 84?

Name: Date:
Instructor: Section:

Chapter 4 PERCENT NOTATION

4.6 Applications of Percent

Learning Objectives
A Solve applied problems involving percent.
B Solve applied problems involving percent of increase or percent of decrease.

Key Terms
Use the vocabulary terms listed below to complete Exercise 1. One of the terms will not be used.

original amount **new amount** **amount of increase/decrease**

1. We find a percent of increase/decrease by dividing the ________________ by the ________________.

GUIDED EXAMPLES AND PRACTICE
Objective A Solve applied problems involving percent.

Review this example for Objective A:

1. On a test of 40 items, James had 34 correct. What percent were correct?

1. *Familiarize*. The problem asks for a percent. Let n = the percent of test items that were correct.
2. *Translate*. Rephrase the question and translate.

$$\underbrace{34}_{\downarrow\ 34} \text{ is } \underbrace{\text{what percent}}_{\downarrow\ p} \text{ of } \underbrace{40?}_{\downarrow\ 40}$$

$$34 = p \times 40$$

3. *Solve*. We divide by 40 on both sides and convert the answer to percent notation.

$$34 = p \times 40$$
$$\frac{34}{40} = \frac{p \times 40}{40}$$
$$0.85 = p$$
$$85\% = p$$

4. *Check*. We can repeat the calculation. The answer checks.
5. *State*. 85% of the test items were correct.

Practice this exercise:

1. The Collins spend 5% of their income on clothing. If their annual income is $43,000, how much is spent on clothing in a year?

Objective B Solve applied problems involving percent of increase or percent of decrease.

Review this example for Objective B:

2. Jo's supervisor tells her that her weekly salary of \$450 will be increased to \$477. What is the percent of increase?

1. *Familiarize*. We find the amount of increase and then make a drawing.

$$\begin{array}{rl} 477 & \text{New salary} \\ -\ 450 & \text{Original salary} \\ \hline 27 & \text{Increase} \end{array}$$

\$450	
\$450	\$27
100%	
100%	?%

2. *Translate*. Rephrase the question and translate.

$$\underbrace{\$27}\ \text{is}\ \underbrace{\text{what percent}}\ \text{of}\ \underbrace{450?}$$

$$27 = p \times 450$$

3. *Solve*. We divide by 450 on both sides and convert the answer to percent notation.

$$27 = p \times 450$$
$$\frac{27}{450} = \frac{p \times 450}{450}$$
$$0.06 = p$$
$$6\% = p$$

4. *Check*. Note that with a 6% increase, the new salary would be 106% of the original salary. Since 106% of $\$450 = 1.06 \times \$450 = \$477$, the answer checks.

5. *State*. The percent increase is 6%.

Practice this exercise:

2. Sales of gift cards for Christmas gifts increased from \$18.5 billion in 2005 to \$24.8 billion in 2006. What was the percent of increase? Round to the nearest percent.

ADDITIONAL EXERCISES

Objective A Solve applied problems involving percent.

For extra help, see Examples 1–2 on pages 301–303 of your text and the Section 4.6 lecture video.

Solve.

1. On a test of 60 items, Aaron got 80% correct. How many items did he get correct?

2. On a test, Elena got 90%, or 72, of the items correct. How many items were on the test?

Name: Date:
Instructor: Section:

3. American pet owners have about 164 million dogs and cats. Of this total, 90.2 million are cats. What percent are cats?

4. Amelia estimates that about 35% of the e-mail she receives is spam. If she receives 200 e-mails one week, how many are spam?

Objective B Solve applied problems involving percent of increase or percent of decrease.

For extra help, see Examples 3–4 on pages 304–306 of your text and the Section 4.6 lecture video.

Solve.

5. About 198.2 million pounds of honey were produced in the U.S. in 1997. Production fell to 154.8 million pounds in 2006. Find the percent of decrease. Round to the nearest percent.

6. Total sales of fax machines decreased from $647 million in 1998 to $128 million in 2005. What was the percent of decrease? Round to the nearest percent.

7. Individuals and companies spent about $80 billion on tax preparation costs in 1990. This amount increased to about $300 billion in 2007. What is the percent of increase?

8. It is estimated that the population age 65 and older in the U.S. will increase from 35 million in 2000 to 71.5 million in 2030. What is the percent of increase? Round to the nearest percent.

Name: Date:
Instructor: Section:

Chapter 4 PERCENT NOTATION

4.7 Sales Tax, Commission, and Discount

Learning Objectives
A Solve applied problems involving sales tax and percent.
B Solve applied problems involving commission and percent.
C Solve applied problems involving discount and percent.

Key Terms
Use the vocabulary terms listed below to complete each statement in Exercises 1–2.

Sales **Sale price** **Sales tax**

1. Commission = Commission rate · ________________

2. Total price = Purchase price + ________________

3. ________________ = Original price - Discount

GUIDED EXAMPLES AND PRACTICE

Objective A Solve applied problems involving sales tax and percent.

Review this example for Objective A:

1. The sales tax in Indiana is 5%. How much tax is charged on a bread machine costing $150? What is the total price?

The sales tax is

$$\underbrace{\text{Sales tax rate}}_{\downarrow\atop 5\%} \times \underbrace{\text{Purchase price}}_{\downarrow\atop \$150}$$

or 0.05×150, or 7.5. Thus the sales tax is $7.50.

The total price is the purchase price plus the sales tax:

$150 + $7.50, or $157.50.

Practice this exercise:

1. The sales tax rate in Connecticut is 8%. What is the total price of a digital camera that sells for $350?

Objective B Solve applied problems involving commission and percent.

Review this example for Objective B:

2. Celina earns a commission of $1750 selling $25,000 worth of office equipment. What is the commission rate?

$$\begin{array}{ccccc} \textit{Commission} & = & \textit{Commission rate} & \times & \textit{Sales} \\ 1750 & = & r & \cdot & 25{,}000 \end{array}$$

To solve this equation we divide by 25,000 on both sides.

$$\frac{1750}{25{,}000} = \frac{r \cdot 25{,}000}{25{,}000}$$

$$0.07 = r$$

$$7\% = r$$

The commission rate is 7%.

Practice this exercise:

2. Frank's commission rate is 12%. He receives a commission of $180 on the sale of sporting goods. How much did the sporting goods cost?

Objective C Solve applied problems involving discount and percent.

Review this example for Objective C:

3. A shirt marked $45 is on sale at 20% off. What is the discount? the sale price?

$$\begin{array}{ccccc} \textit{Discount} & = & \textit{Rate of discount} & \times & \textit{Original price} \\ D & = & 20\% & \times & 45 \end{array}$$

Convert 20% to decimal notation and multiply.

$D = 0.20 \times 45 = 9.00$

The discount is $9.00.

$$\begin{array}{ccccc} \textit{Sale price} & = & \textit{Original price} & - & \textit{Discount} \\ S & = & 45 & - & 9 \end{array}$$

$$S = 36$$

The sale price is $36.

Practice this exercise:

3. A sofa marked $650 is on sale at 30% off. What is the sale price?

ADDITIONAL EXERCISES

Objective A Solve applied problems involving sales tax and percent.

For extra help, see Examples 1–3 on pages 315–3164 of your text and the Section 4.7 lecture video.

Solve.

1. The sales tax is $22.50 on the purchase of a laptop computer that sells for $450. What is the sales tax rate?

2. The sales tax rate in Minnesota is 6.5%. How much sales tax would be charged on a dishwasher that sells for $499?

Name: Date:
Instructor: Section:

3. The sales tax on the purchase of a flat screen TV is $87.89 and the sales tax rate is 5.5%. Find the purchase price.

4. The sales tax rate in Ohio is 5.5%. How much tax is charged on a purchase of two gas grills at $179 apiece? What is the total price?

Objective B Solve applied problems involving commission and percent.

For extra help, see Example 4–6 on pages 317–318 of your text and the Section 4.7 lecture video.

Solve.

5. Audra's commission rate is 12%. What is the commission from the sale of $2500 worth of luggage?

6. Cornelia earns $900 selling $6000 worth of appliances. What is the commission rate?

7. An auction house's commission rate is 20%. A commission of $1300 is received. How much merchandise was sold?

8. Antonio earns a monthly salary of $1200, plus a 4% commission on sales. One month he sold $5500 worth of sound equipment. What were his wages that month?

Objective C Solve applied problems involving discount and percent.
For extra help, see Examples 7–8 on page 319 of your text and the Section 4.7 lecture video.
Find what is missing.

	Marked price	Rate of discount	Discount	Sale price
9.	\$500	25%		
10.		8%	\$9.60	
11.	\$1500		\$180	
12.	\$50	30%		
13.	\$250		\$15	

Name: Date:
Instructor: Section:

Chapter 4 PERCENT NOTATION

4.8 Simple Interest and Compound Interest; Credit Cards

Learning Objectives

A Solve applied problems involving simple interest.
B Solve applied problems involving compound interest.
C Solve applied problems involving interest rates on credit cards.

Key Terms

Use the vocabulary terms listed below to complete each statement in Exercises 1–2.

compound interest **simple interest**

1. The formula $I = P \cdot r \cdot t$ is used to find the ________________ I on principal P, invested at interest rate r for t years.

2. When interest is paid on interest we call it ________________ .

GUIDED EXAMPLES AND PRACTICE

Objective A Solve applied problems involving simple interest.

Review this example for Objective A:

1. What is the interest on $4400 invested at an interest rate of 8% for $\frac{1}{2}$ year?

Substitute $4400 for P, 8% for r, and $\frac{1}{2}$ for t in the simple interest formula.

$$\begin{aligned} I &= P \cdot r \cdot t \\ &= \$4400 \cdot 0.08 \cdot \frac{1}{2} \\ &= \frac{\$4400 \cdot 0.08}{2} \\ &= \$176 \end{aligned}$$

The interest for $\frac{1}{2}$ year is $176.

Practice this exercise:

1. What is the interest on $1200 invested at 6% for $\frac{1}{4}$ year?

Objective B Solve applied problems involving compound interest.

Review this example for Objective B:

2. The Jensens invest \$2000 in an account paying 12% interest, compounded semiannually. Find the amount in the account after $1\frac{1}{2}$ years.

Substitute \$2000 for P, 12% for r, 2 for n, and $1\frac{1}{2}$ for t in the compound interest formula.

$$\begin{aligned} A &= P\cdot\left(1+\frac{r}{n}\right)^{n\cdot t} \\ &= \$2000\cdot\left(1+\frac{0.12}{2}\right)^{2\cdot\frac{3}{2}} \\ &= \$2000\cdot(1.06)^{3} \\ &\approx \$2382.03 \end{aligned}$$

The amount in the account after $1\frac{1}{2}$ years is \$2382.03.

Practice this exercise:

2. The Shaws invest \$3600 in an account paying 8%, compounded quarterly. Find the amount in the account after 1 year.

Objective C Solve applied problems involving interest rates on credit cards.

Review this examples for Objective C:

3. Tom has a balance of \$2506.94 on a credit card with an annual percentage rate of 18.2%. His minimum payment is \$50. Find the amount of interest and the amount applied to reduce the principal when the minimum payment is made.

The amount of interest on \$2506.94 at 18.2% for 1 month is

$$\begin{aligned} I &= P\cdot r\cdot t \\ &= \$2506.94\cdot 0.182\cdot\frac{1}{12} \\ &\approx \$38.02 \end{aligned}$$

We subtract the interest from the minimum payment to find the amount applied to reduce the principal.

$$\$50 - \$38.02 = \$11.98$$

Practice this exercise:

3. Jody has a balance of \$1432.50 on a credit card with an annual percentage rate of 16.8%. Her minimum payment is \$50. Find the amount of interest and the amount applied to reduce the principal when the minimum payment is made.

Name: Date:
Instructor: Section:

ADDITIONAL EXERCISES

Objective A Solve applied problems involving simple interest.

For extra help, see Examples 1–3 on pages 323–324 of your text and the Section 4.8 lecture video.

Find the simple interest.

	Principal	Rate of interest	Time	Simple interest
1.	\$1200	3%	1 year	
2.	\$500	4.8%	$\frac{1}{2}$ year	
3.	\$25,000	$5\frac{3}{8}\%$	$\frac{1}{4}$ year	

Sheila borrows \$2400 at 6% for 90 days.

4. Find the amount of interest due.

5. Find the total amount that must be paid after 90 days.

Objective B Solve applied problems involving compound interest.

For extra help, see Examples 4–5 on pages 324–326 of your text and the Section 4.8 lecture video.

Solve.

6. The Hansons deposit \$1000 in an account paying 5% compounded monthly. How much is in the account after 6 months?

7. Corey invests \$3000 in an account paying 8%, compounded quarterly. How much is in the account after 2 years?

8. Lucas invests \$10,000 in an account paying 6%, compounded quarterly. How much is in the account after 18 months?

9. Skylar deposits \$5000 in an account paying 4% compounded daily. How much is in the account after 30 days?

Objective C Solve applied problems involving interest rates on credit cards.
For extra help, see Example 6 on pages 327–328 of your text and the Section 4.8 lecture video.

Solve.

Tina has a balance of $3976.84 on a credit card with an annual percentage rate (APR) of 18%. She decides not to make additional purchases with this card until she has paid off the balance.

10. Tina's credit card company requires a minimum monthly payment of 4% of the balance. What is Tina's minimum payment of her balance? Round to the nearest dollar.

11. Find the amount of interest and the amount applied to reduce the principal in the minimum payment found in Exercise #10.

12. If Tina had transferred her balance to a card with an APR of 13.2% and made the same payment as in Exercise #10, how much of her first payment would be interest and how much would be applied to the principal?

13. By how much does the amount for 13.2% from Exercise #12 differ from the amount for 18% from Exercise #11?

Name: Date:
Instructor: Section:

Chapter 5 DATA, GRAPHS, AND STATISTICS

5.1 Averages, Medians, and Modes

Learning Objectives
A Find the average of a set of numbers and solve applied problems involving averages.
B Find the median of a set of numbers and solve applied problems involving medians.
C Find the mode of a set of numbers and solve applied problems involving modes.
D Compare two sets of data using their means.

Key Terms
Use the vocabulary terms listed below to complete each statement in Exercises 1–4.

average **median** **mode** **statistic**

1. The ________________ is the middle number or average of the two middle numbers in a list that has been ordered from smallest to largest.
2. To find the ________________ of a set of numbers, add the numbers and then divide by the number of items of data.
3. The mean is an example of a ________________, a number describing a set of data.
4. The ________________ of a set of data is the number or numbers that occur most often.

GUIDED EXAMPLES AND PRACTICE

Objective A Find the average of a set of numbers and solve applied problems involving averages.

Review this example for Objective A:
1. A student's scores on four tests were 80, 64, 91, and 85. What was the average score?

$$\frac{80+64+91+85}{4}=\frac{320}{4}=80$$

The average score was 80.

Practice this exercise:
1. On 5 successive days, Morgan ran 4 mi, 2 mi, 10 mi, 3 mi, and 6 mi. What was the average number of miles per day?

Objective B Find the median of a set of numbers and solve applied problems involving medians.

Review these examples for Objective B:
2. Find the median of the set of hourly wages:
$6.50, $5.75, $7.25, $8.00, $7.40.

List the data in order from smallest to largest:
$5.75, $6.50, $7.25, $7.40, $8.00.
There is an odd number of data items. The middle number is $7.25, so the median wage is $7.25.

Practice these exercises:
2. Find the median of the following temperatures:
56°, 48°, 61°, 66°, 53°.

3. Find the median of the set of hourly wages:
$20, $15, $10, $12.

List the data in order from smallest to largest:
$10, $12, $15, $20.
There is an even number of items. The median is the average of the two middle numbers:

$$\text{Median} = \frac{\$12 + \$15}{2} = \frac{\$27}{2} = \$13.50$$

3. Find the median of the following temperatures:
86°, 79°, 90°, 84°, 91°, 78°.

Objective C Find the mode of a set of numbers and solve applied problems involving modes.

Review this example for Objective C:

4. Find the modes of each set of data.
a) 16, 23, 27, 27, 27
b) $34, $34, $51, $58, $58, $64
c) 7, 9, 15, 21, 45

a) The number that occurs most often is 27. Thus, the mode is 27.

b) The two numbers $34 and $58 occur most often. Thus, the modes are $34 and $58.

c) No number occurs more often than any other. Thus, there is no mode.

Practice this exercise:

4. Find the mode of these data:
$17, $28, $33, $41, $56, $56, $91.

Objective D Compare two sets of data using their means.

Review this example for Objective D:

5. Volunteers drank two brands of orange juice and rated their taste from 1 to 10, where 10 represents the best taste. The results are given below. On the basis of this test, which brand tastes better?

Brand A: 7, 8, 6, 4, 10, 5, 9, 8, 8, 7
Brand B: 6, 10, 9, 7, 8, 7, 4, 5, 6, 7

Brand A average:

$$\frac{7+8+6+4+10+5+9+8+8+7}{10} = \frac{72}{10} = 7.2$$

Brand B average:

$$\frac{6+10+9+7+8+7+4+5+6+7}{10} = \frac{69}{10} = 6.9$$

The average for Brand A is higher than for Brand B, so Brand A tastes better.

Practice this exercise:

5. Two brands of light bulbs were tested. The lives, in hours, of 8 bulbs of each brand are listed below. On basis of this test, which bulb is better?

Brand A:
950, 967, 835, 1214, 1130, 891, 1070, 998
Brand B:
1015, 898, 1147, 935, 946, 893, 1235, 842

Name: Date:
Instructor: Section:

ADDITIONAL EXERCISES

Objective A Find the average of a set of numbers and solve applied problems involving averages.

For extra help, see Examples 1–5 on pages 346–348 of your text and the Section 5.1 lecture video.

Objective B Find the median of a set of numbers and solve applied problems involving medians.

For extra help, see Examples 6–8 on pages 348–349 of your text and the Section 5.1 lecture video.

Objective C Find the mode of a set of numbers and solve applied problems involving modes.

For extra help, see Examples 9–10 on pages 349–350 of your text and the Section 5.1 lecture video.

For each set of numbers, find the average, the median, and any modes that exist.

1. 12, 18, 5, 12, 22, 12

2. 21, 7, 9, 33, 15, 20

3. 53, 49, 32, 53, 32

4. Martha drove 208 mi on 8 gal of gasoline. What is the average number of miles driven per gallon. That is, what is the gas mileage?

5. The following prices per pound of orange roughy were found at five fish markets.

$7.98, $7.49, $6.99, $7.49, $8.95

What is the average price per pound? the median price? the mode?

Objective D Compare two sets of data using their means.

For extra help, see Example 11 on pages 350–351 of your text and the Section 5.1 lecture video.

An experiment was performed to compare the lives of two brands of batteries. Several batteries of each type were tested and the results are listed in the following table. On the basis of this test, which battery is better?

6.

Dura Battery Times (in hours)			Vita Battery Times (in hours)		
125	102	132	104	141	129
140	113	105	127	111	135
136	121	109	139	123	131
122	108	130	100	107	123

An experiment was performed to determine which of two brands of peanut butter tasted better. Participants tasted each brand and rated it from 1 to 10, with 10 representing the best taste. The results are given in the following table. On the basis of this test, which peanut butter tastes better?

7.

Wynn's Peanut Butter				Penn's Peanut Butter			
9	8	7	6	8	7	10	5
6	10	8	7	9	8	8	6
5	6	9	5	10	9	7	7
10	8	7	8	6	5	8	7

Name: Date:
Instructor: Section:

Chapter 5 DATA, GRAPHS, AND STATISTICS

5.2 Tables and Pictographs

Learning Objectives
A Extract and interpret data from tables.
B Extract and interpret data from pictographs.

Key Terms
Use the vocabulary terms listed below to complete each statement in Exercises 1–2.

pictograph **table**

1. A ____________________ presents information using symbols to represent amounts.

2. A ____________________ presents data in rows and columns.

GUIDED EXAMPLES AND PRACTICE

Objective A Extract and interpret data from tables.

Review this example for Objective A:

1. The following table lists nutritional information for Fresh Stuffed Pitas at Wendy's.

Pita	Calories	Fat	Protein	Sodium
Garden Veggie	300	15 g	13 g	780 mg
Garden Ranch Chicken	480	17 g	32 g	1170 mg
Chicken Caesar	490	17 g	36 g	1300 mg
Classic Greek	430	19 g	17 g	1070 mg

Which pita contains the most sodium?

Look down the column headed "Sodium" until you find the largest number. That number is 1300 mg. Then look across that row to find the type of pita, Chicken Caesar.

Practice this exercise:

1. Use the table at left to determine which Fresh Stuffed Pita has the least fat.

Objective B Extract and interpret data from pictographs.

Review this example for Objective B:

2. The following pictograph lists the calories per tablespoon in various tablespreads.

Tablespread	
Jam	
Mayonaise	
Peanut butter	
Honey	
Syrup	

= 10 calories

a) Which tablespread contains the most calories per tablespoon?
b) How many calories per tablespoon does syrup contain?

a) Peanut butter has the largest number of symbols, so it contains the most calories per tablespoon.
b) Syrup is represented by 5 symbols, each of which represents 10 calories. Thus, syrup contains $5 \cdot 10$, or 50 calories per tablespoon.

Practice this exercise:

2. Use the pictograph at left to determine approximately how many more calories per tablespoon there are in peanut butter than in jam.

ADDITIONAL EXERCISES

Objective A Extract and interpret data from tables.

For extra help, see Examples 1–2 on pages 354–356 of your text and the Section 5.2 lecture video.

The following table lists the number of calories burned in 30 min of exercise for various types of activities and several weights.

	Calories burned in 30 min		
Activity	110 lb	132 lb	154 lb
Aerobic dance	201	237	282
Calisthenics	216	261	351
Racquetball	213	252	294
Tennis	165	192	222
Moderate bicycling	138	171	198
Moderate jogging	321	378	453
Moderate walking	111	132	159

1. How many calories are burned by a 154-lb person during 30 min of moderate bicycling?

2. Which activity burns the most calories for a 110-lb person?

3. Which activity burns the fewest calories for a 132-lb person?

4. Which activity burns calories at a rate of 252 every 30 min for a 132-lb person?

Name: Date:
Instructor: Section:

Objective B Extract and interpret data from pictographs.

For extra help, see Example 3 on page 356 of your text and the Section 5.2 lecture video.

The following pictograph shows sales of shampoo for the Natural Extracts Company for five consecutive years.

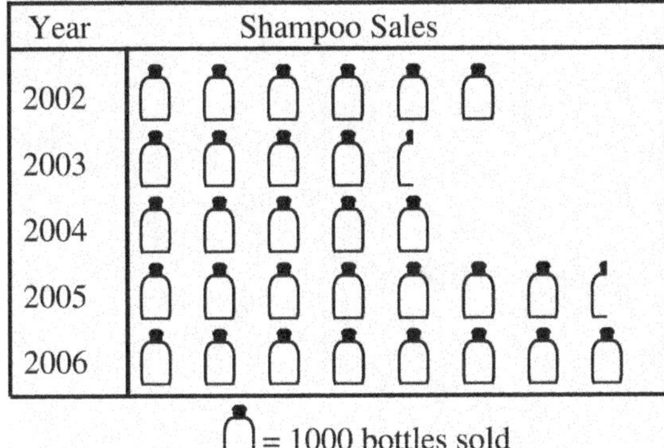

5. How many bottles of shampoo were sold in 2005?

6. In which year were the fewest bottles sold?

7. Between which two consecutive years was there a decline in sales?

8. In which year were 5000 bottles sold?

Name: Date:
Instructor: Section:

Chapter 5 DATA, GRAPHS, AND STATISTICS

5.3 Bar Graphs and Line Graphs

Learning Objectives

A Extract and Interpret data from bar graphs.
B Draw bar graphs.
C Extract and interpret data from line graphs.
D Draw line graphs.

Key Terms

Use the vocabulary terms listed below to complete each statement in Exercises 1–2.

bar graph **line graph**

1. A _________________ is convenient for showing comparisons because you can tell at a glance which amount represents the smallest or largest quantity.

2. A _________________ is often used to show change over time as well as to indicate patterns or trends.

GUIDED EXAMPLES AND PRACTICE

Objective A Extract and interpret data from bar graphs.

Review this example for Objective A:

1. The following bar graph shows the number of calories burned per hour by a 152 lb person during various activities.

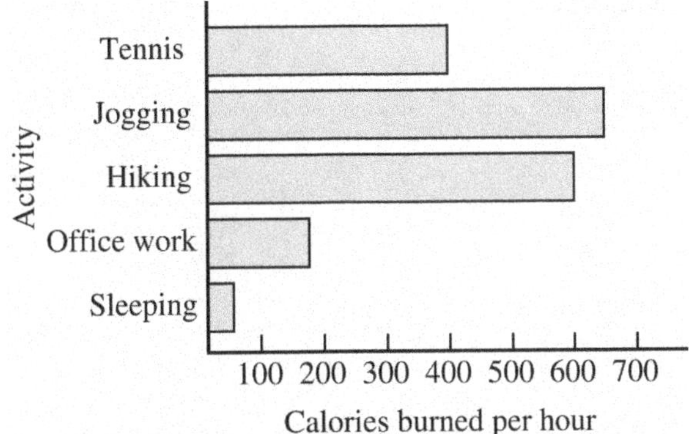

a) Which activity burns the fewest calories?
b) About how many calories are burned per hour by jogging?

a) The shortest bar is for sleeping. Thus, sleeping burns the fewest calories.
b) Move to the right end of the bar representing jogging and then go down to the horizontal scale. Jogging burns about 650 calories per hour.

Practice this exercise:

1. Use the bar graph at left to determine which activity burns about 200 calories per hour.

Objective B   Draw bar graphs.

Review this example for Objective B:

2. Listed below are the reasons adult workers give for not going into business for themselves. Make a horizontal bar graph of the data.

Lack of benefits: 34%
Lack of security: 29%
Reduced leisure time: 22%
Lower salary: 11%
Don't know: 4%

First, on the vertical scale label the reasons given in five equally spaced intervals, and the title scale "Reason." Then mark and label the horizontal scale by 5's, and title this scale "Percent." Draw a horizontal bar for each reason to show the corresponding percent. Finally, give the graph an appropriate title, such as "Barriers to Being Own Boss."

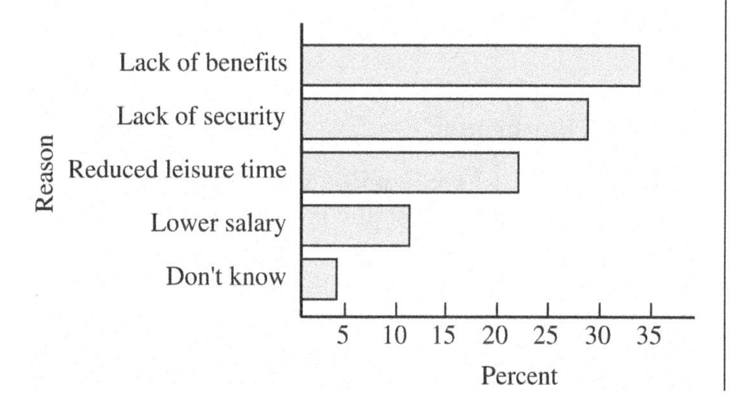

Practice this exercise:

2. The number of units of a popular software product sold in four recent years are listed below.

1995: 30 million
1996: 63 million
1997: 84 million
1998: 110 million

Make a horizontal bar graph of the data.

Name: Date:
Instructor: Section:

Objective C Extract and interpret data from line graphs.

Review this example for Objective C:

3. The following line graph shows the number of cars passing through an intersection during various hours of the day.

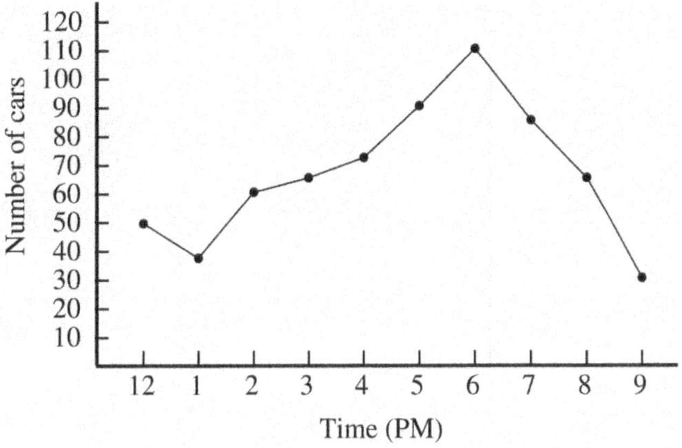

a) During which hour was traffic the heaviest?
b) During which hour did about 80 cars pass through the intersection?

a) Find the highest point on the graph and then go down to the horizontal scale to read the corresponding hour. We see that traffic was heaviest during the 6 PM hour.
b) We locate 80 on the vertical scale and go to the right until we come to a point (a dot) on the graph. Then go down to the horizontal scale to read the corresponding hour. We see that about 80 cars passed through the intersection during the 7 PM hour.

Practice this exercise:

3. Use the line graph at left to determine whether the traffic decreased, stayed the same, or increased between 3 PM and 4 PM.

Objective D Draw line graphs.

Review this example for Objective D:

4. Listed below is the total revenue for the Uptown Boutique for several years. Make a line graph of the data.

   2005: $95,000
   2006: $120,000
   2007: $100,000
   2008: $125,000

Indicate the years on the horizontal scale and label it "Year." We will scale the vertical axis in

Practice this exercise:

4. Listed below are temperatures during a fall afternoon. Make a line graph of the data.

   1 PM: 55°F
   3 PM: 59°F
   5 PM: 56°F
   7 PM: 50°F

thousands. Mark the vertical scale by 10's starting with $90. Use a jagged line to indicate the missing numbers. Label the vertical scale "Total revenue (in thousands)." Draw points representing the data and connect them with line segments. Finally, give the graph in appropriate title, such as "Uptown Boutique."

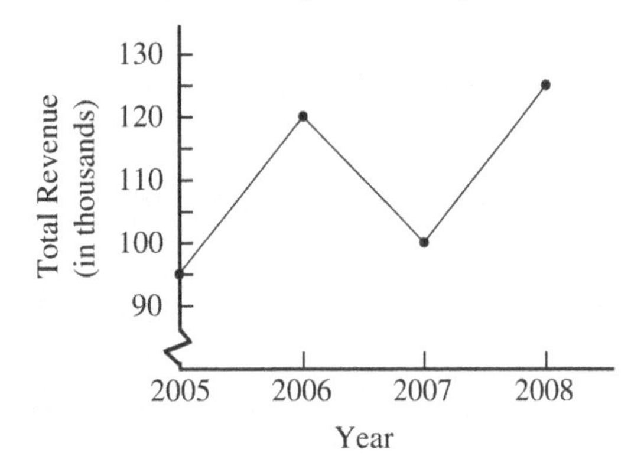

ADDITIONAL EXERCISES

Objective A Extract and interpret data from bar graphs.

For extra help, see Examples 1–2 on pages 363–365 of your text and the Section 5.3 lecture video.

The following horizontal bar graph shows the maximum consecutive hours of work allowed in various occupations.

Maximum Consecutive Hours Allowed

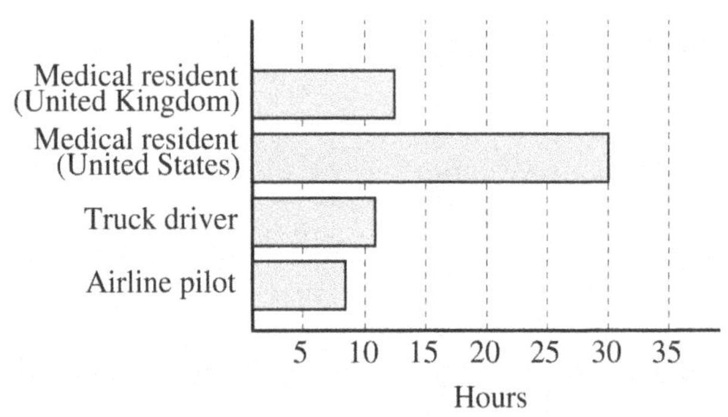

1. Which occupation allows the fewest consecutive hours of work?

2. Estimate the maximum number of consecutive hours a truck driver can work.

3. Which occupation allows only 8 consecutive hours of work?

4. How many more consecutive hours of work is a medical resident in the United States allowed than a medical resident in the United Kingdom?

Name: Date:
Instructor: Section:

Objective B Draw bar graphs.

For extra help, see Example 3 on page 365 of your text and the Section 5.3 lecture video.

The following table lists the average amounts of several fruits consumed per person in the United States in 2005, rounded to the nearest pound. Make a horizontal bar graph to illustrate the data.

Fruit	Amount Consumed (in pounds)
Apples	17
Bananas	25
Grapes	9
Oranges	11
Strawberries	6

5.

Use the data and the bar graph from Exercise #5.

6. Which was the most heavily consumed fruit?

7. How many more pounds of apples were consumed on average per person than oranges?

8. Which single fruit's consumption is the same as the total amount of oranges and strawberries consumed per person?

Objective C Extract and interpret data from line graphs.

For extra help, see Examples 4–5 on pages 366–367 of your text and the Section 5.3 lecture video.

Job satisfaction is declining in the United States. The line graph below shows the number of workers, per 100, who are satisfied with their jobs in recent years.

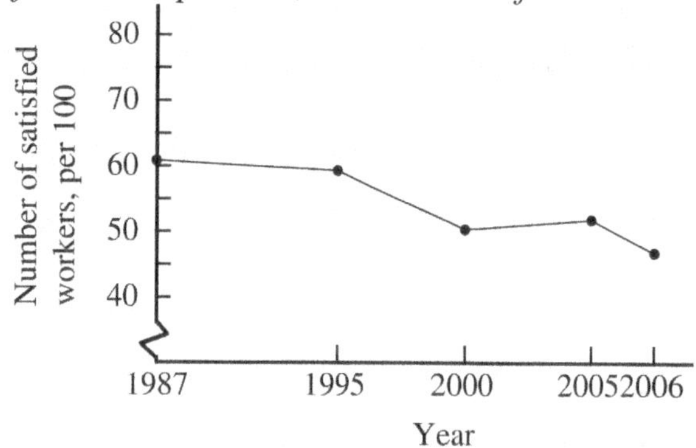

9. About how many fewer people, per hundred, were satisfied with their jobs in 2006 than in 1987?

10. About how many fewer people, per hundred, were satisfied with their jobs in 2000 than in 1995?

11. In which year were 59 people, per hundred, satisfied with their job?

12. In which year were 47 people, per hundred, satisfied with their job?

Objective D Draw line graphs.

For extra help, see Example 6 on page 368 of your text and the Section 5.3 lecture video.

The following table lists the average verbal scores on a Scholastic Aptitude Test (SAT) for several years. Make a line graph of the data, listing years on the horizontal axis.

Year	Score
1996	505
1998	505
2000	505
2002	504
2004	508
2005	508

13.

Use the data and the line graph from Exercise #13.

14. Between which years did score decrease?

15. Between which years did the scores remain unchanged?

16. In which year did the lowest score occur?

Name: Date:
Instructor: Section:

Chapter 5 DATA, GRAPHS, AND STATISTICS

5.4 Circle Graphs

Learning Objectives
A Extract and interpret data from circle graphs.
B Draw circle graphs.

Key Terms
Use the vocabulary terms listed below to complete each statement in Exercises 1–2.

pie chart **100%** **wedge**

1. Another name for a circle graph is a ________________ .

2. The total of all the sections, or ________________s, of a circle graph should be ________________ .

GUIDED EXAMPLES AND PRACTICE

Objective A Extract and interpret data from circle graphs.

Review this example for Objective A:

1. The following circle graph shows the distribution of types of material used for the exterior walls of newly built single-family homes.

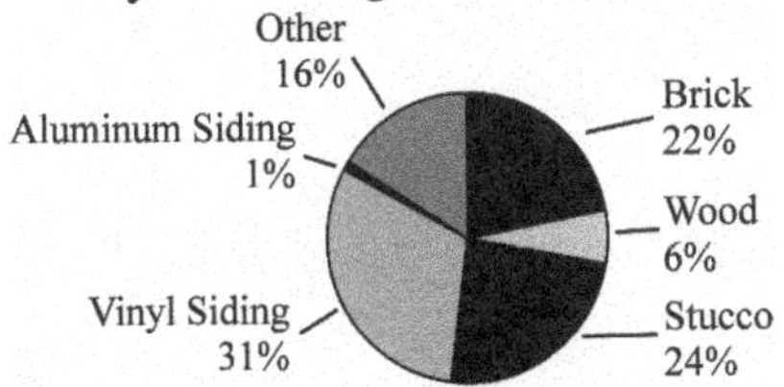

a) Together what percent of homes have exterior walls that are vinyl or aluminum siding?
b) If 20,000 homes are built in a month, how many would be expected to have wood exterior walls?

We look at the sections of the graph to find the answers.
a) We add the percents representing vinyl and aluminum siding:
$31\% + 1\% = 32\%$
b) The section representing wood is 6%;
6% of 20,000 is 1200. Thus 1200 homes would be expected to have wood exterior walls.

Practice this exercise:

1. Use the circle graph at left to determine
a) What percent of homes have exterior walls that are stucco?
b) If 46,000 homes are built in a year, how many are expected to have brick exterior walls?

Objective B Draw circle graphs.

Review this example for Objective B:

2. The table below lists how often people claim they exercise. Use the information to draw a circle graph.

Frequency	Percent
Daily	11.9%
More than 3 times per week, but less than daily	20.6%
2–3 times per week	29.3%
Once per week	19.2%
Never	16.2%
Other	2.8%

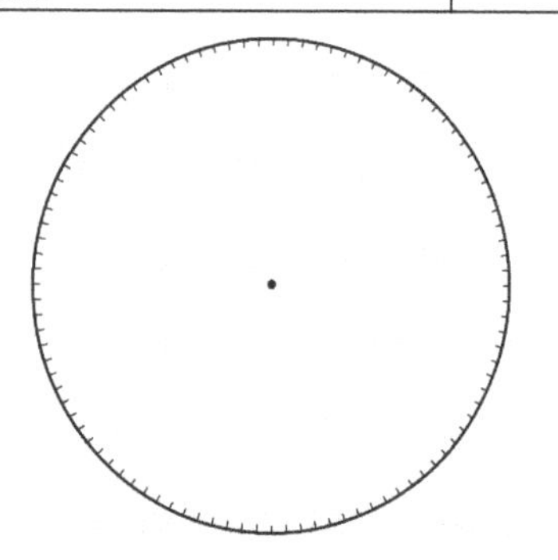

Using the circle with 100 equally spaced tick marks, we start with the 11.9% given for Daily. We draw a line from the center to the tick mark at "noon" or 12:00 position. Then we count off 11.9 ticks and draw another line, and label the wedge. To do the next wedge, at 20.6%, we start at one side of "Daily" wedge, count off 20.6%, and draw another line and label the wedge. Continue in this manner we obtain the final graph shown below.

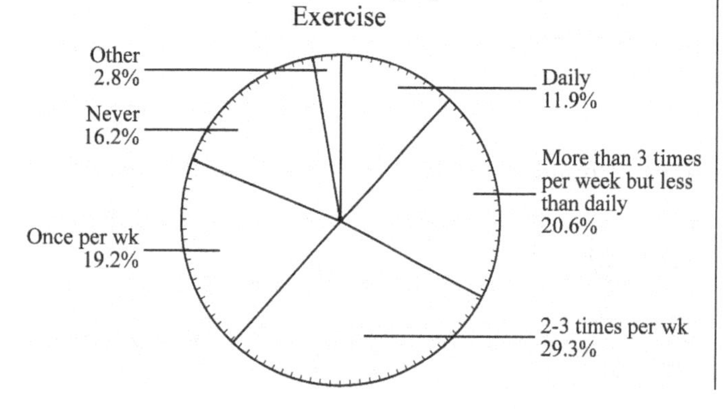

Practice this exercise:

2. A survey was conducted and the table below lists respondents' favorite colors. Use the information to draw a circle graph.

Color	Percent
Blue	42%
Purple	14%
Green	14%
Red	8%
Black	7%
Other	15%

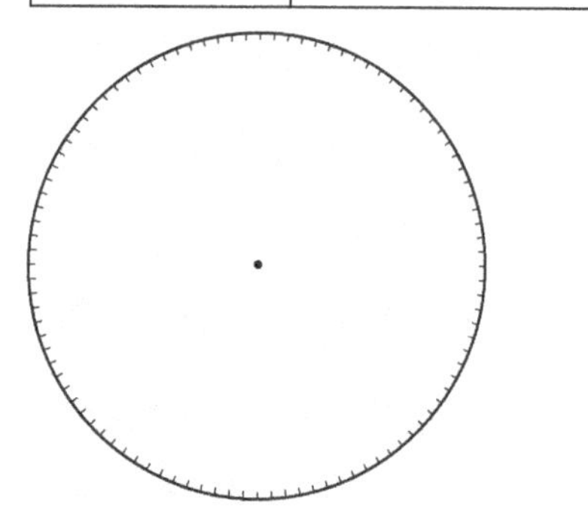

Name: Date:
Instructor: Section:

ADDITIONAL EXERCISES

Objective A Extract and interpret data from circle graphs.

For extra help, see Example 1 on page 374 of your text and the Section 5.4 lecture video.

The following circle graph shows the distribution of blood types in the United States. Use this graph to answer the following questions.

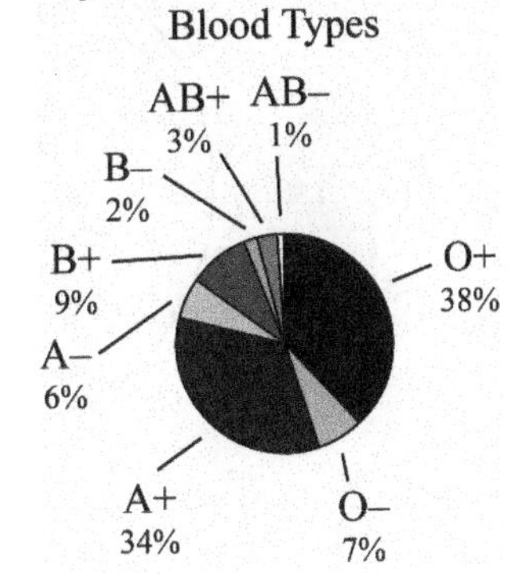

1. Together what percent of people in the U.S. have O+ or O– type blood?

2. If 120 people in a state need blood one day, how many would be expected to need B+ blood type?

3. What percent of people in the U.S. have blood type B+?

4. What percent of all people in the U.S. do not have A+ blood type?

5. If 500 people in a city donated a pint of blood at a blood drive, how many pints of A- blood would be expected to be collected?

Objective B Draw circle graphs.

For extra help, see Example 2 on page 770 of your text and the Section 5.4 lecture video.

Use the given information to complete a circle graph. Note that each circle is divided into 100 sections.

6. The table below lists the eye color of participants in a recent study.

Eye Color	Percent
Blue	21%
Brown	43%
Hazel	29%
Green	1%
Gray	3%
Amber	3%

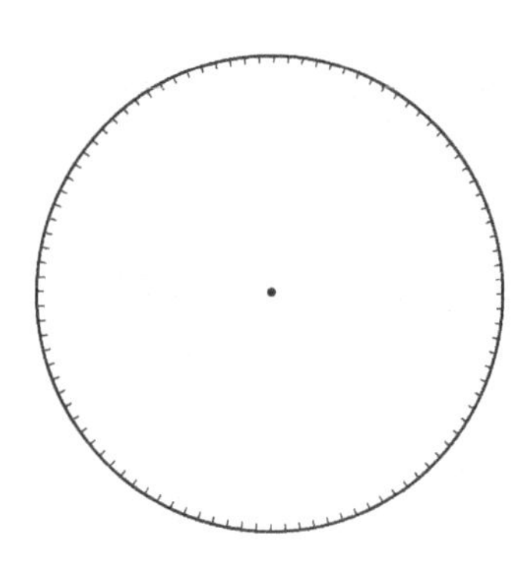

7. The table below lists the number of nights that guests stayed at a vacation lodge during a recent month.

Length of Stay (in nights)	Percent
1	20%
2	30%
3–5	15%
5–7	25%
more than 7	10%

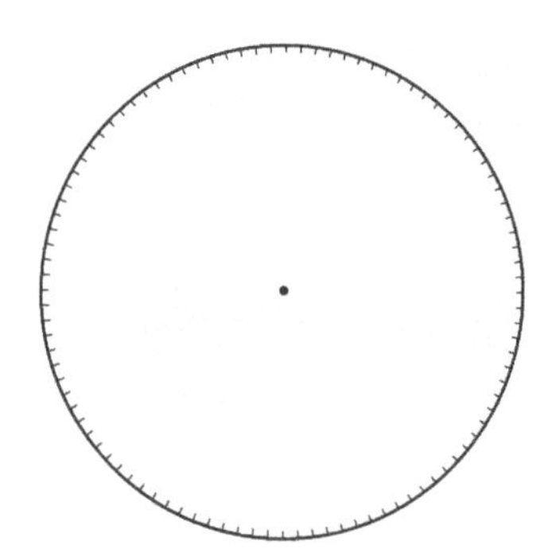

8. The table below lists the favorite desserts of the fifth-grade class at Johnson Academy.

Dessert	Percent
Ice cream	26%
Cake	8%
Cookies	28%
Pie	17%
Fruit	15%
Other	6%

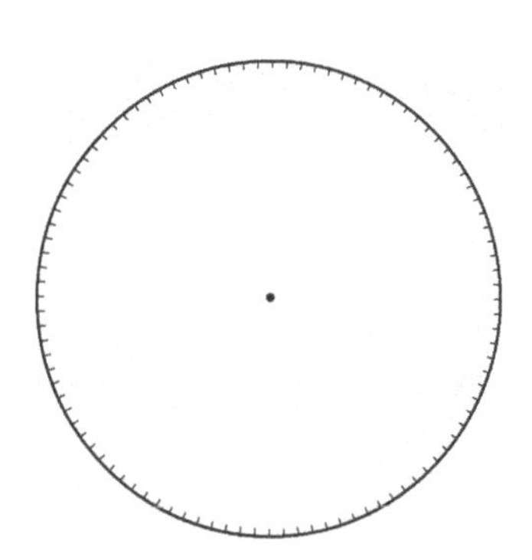

Name: Date:
Instructor: Section:

Chapter 6 GEOMETRY

6.1 Basic Geometric Figures

Learning Objectives

A Draw and name segments, rays, and lines. Also, identify endpoints, if they exist.
B Name an angle in five different ways, and given an angle, measure it with a protractor.
C Classify an angle as right, straight, acute, or obtuse.
D Identify perpendicular lines.
E Classify a triangle as equilateral, isosceles, or scalene and as right, obtuse, or acute. Given a polygon of twelve, ten, or fewer sides, classify it as a dodecagon, a decagon, and so on.
F Given a polygon of n sides, find the sum of its angle measures using the formula $(n-2)\cdot 180°$.

Key Terms

Use the vocabulary terms listed below to complete each statement in Exercises 1–5.

segment **ray** **parallel** **perpendicular** **coplanar**

1. Lines which lie in the same plane are called ________________ .

2. Two lines are ________________ if they intersect to form a right angle.

3. A ________________ has exactly one endpoint.

4. Coplanar lines that do not intersect are ________________ .

5. A ________________ is a geometric figure consisting of two endpoints and all points between them.

GUIDED EXAMPLES AND PRACTICE

Objective A Draw and name segments, rays, and lines. Also, identify endpoints, if they exist.

Review these examples for Objective A:

1. Draw the segment with endpoints A and B. Name the segment in two ways.

A B

The segment can be named $\overline{AB}$ or $\overline{BA}$.

Practice these exercises:

1. Give a name for the segment with endpoints M and N.

2. Draw and name the ray with endpoint T.

Since T is the endpoint we start at T and extend the ray to W and beyond.

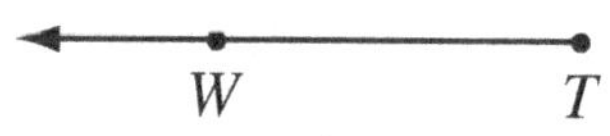

The ray is named $\overrightarrow{TW}$.

3. Name the line in seven different ways.

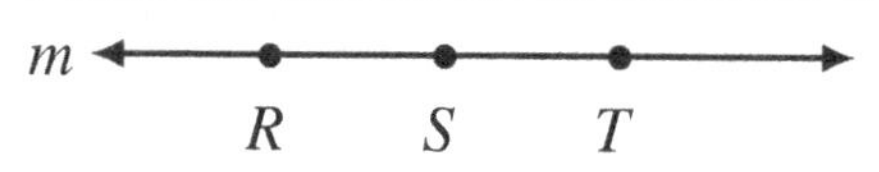

The line can be named m, $\overleftrightarrow{RS}$, $\overleftrightarrow{SR}$, $\overleftrightarrow{RT}$, $\overleftrightarrow{TR}$, $\overleftrightarrow{ST}$, or $\overleftrightarrow{TS}$.

2. Name the ray.

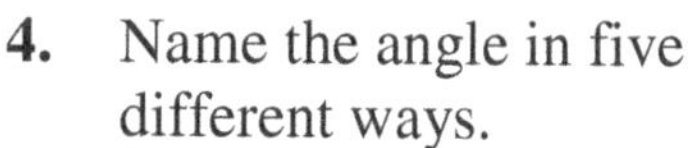

3. Name the line in seven different ways.

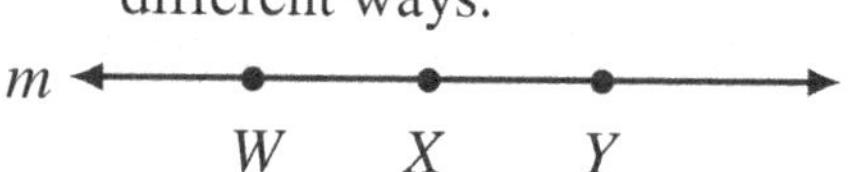

Objective B Name an angle in five different ways, and given an angle, measure it with a protractor.

Review these examples for Objective B:

4. Name the angle in five different ways.

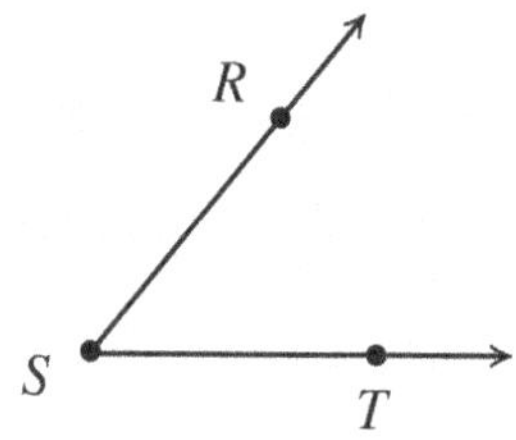

The angle can be named angle RST, angle TSR, $\angle RST$, $\angle TSR$, or $\angle S$.

5. Use a protractor to measure this angle.

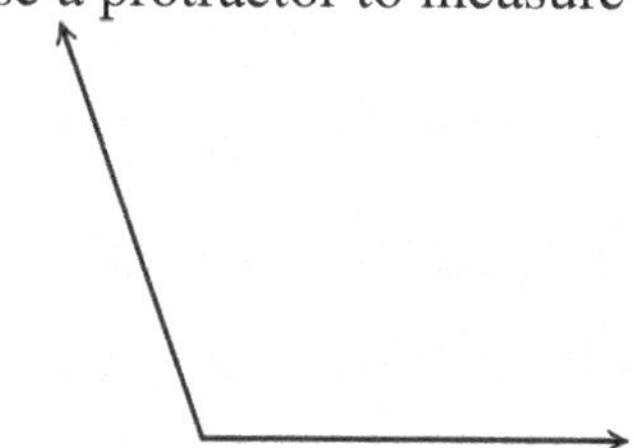

Place the $\triangle$ of the protractor at the vertex of the angle and line up one of the sides at 0°. Since 0° is on the inside scale, we check where the other side of the angle crosses the inside scale. It crosses at 110°. Thus, the measure of the angle is 110°.

Practice these exercises:

4. Name the angle in five different ways.

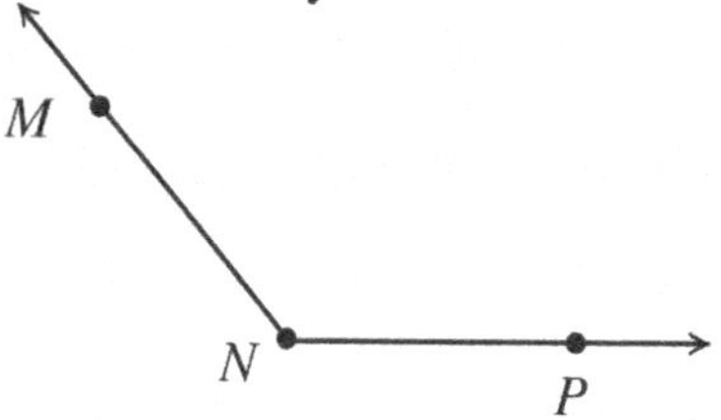

5. Use a protractor to measurc this angle.

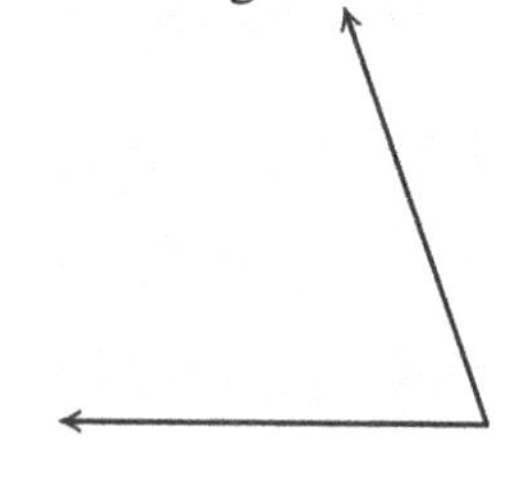

Name: Date:
Instructor: Section:

Objective C Classify an angle as right, straight, acute, or obtuse.

Review this example for Objective C:

6. Classify the angle as right, straight, acute, or obtuse.

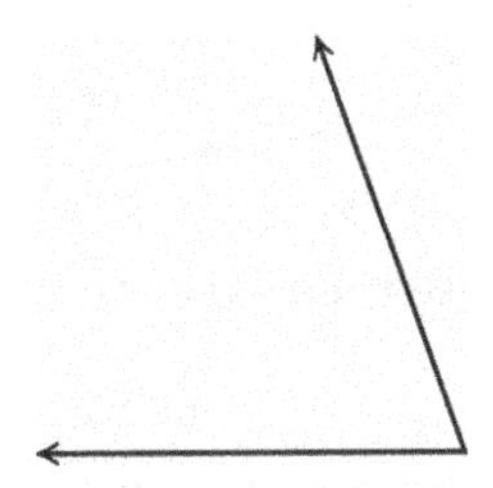

The measure of the angle is greater than 0° and less than 90°, so this is an acute angle.

Practice this exercise:

6. Classify the angle as right, straight, acute, or obtuse.

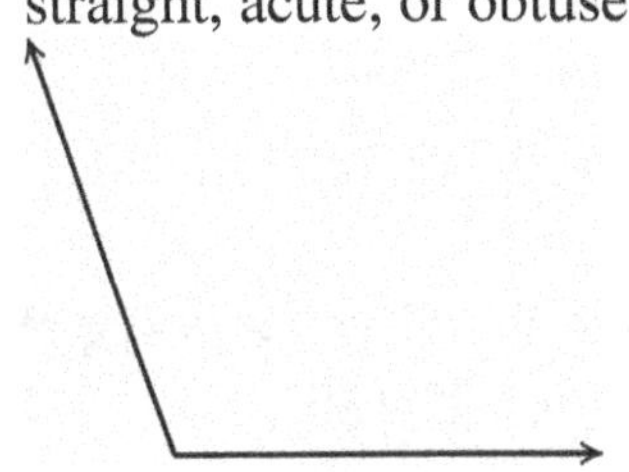

Objective D Identify perpendicular lines.

Review this example for Objective D:

7. Determine whether the pair of lines is perpendicular. Use a protractor.

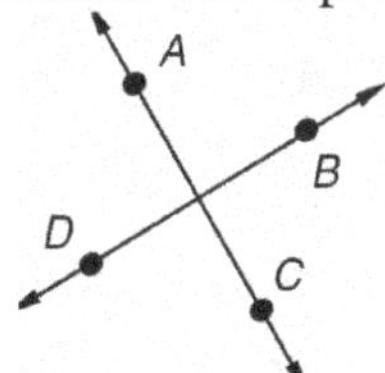

Measuring with a protractor, we find that the lines intersect to form a right angle. (In fact, they form four right angles.) Thus, the lines are perpendicular.

Practice this exercise:

7. Determine whether the pair of lines is perpendicular. Use a protractor.

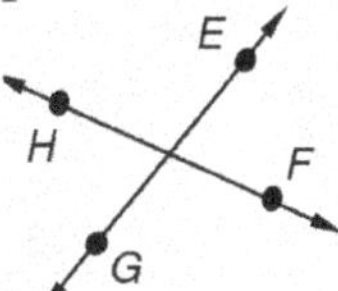

Objective E Classify a triangle as equilateral, isosceles, or scalene and as right, obtuse, or acute. Given a polygon of twelve, ten, or fewer sides, classify it as a dodecagon, a decagon, and so on.

Review these examples for Objective E:

8. Classify the triangle as equilateral, isosceles, or scalene. Then classify it as right, obtuse, or acute.

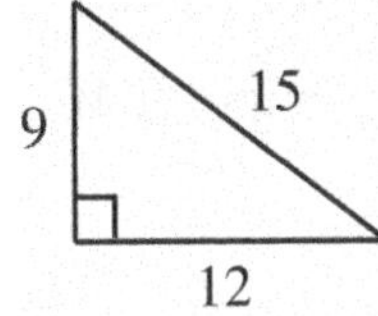

All the sides are different lengths, so this is a scalene triangle.

One angle is a right angle, so this is a right triangle.

Practice these exercises:

8. Classify the triangle as equilateral, isosceles, or scalene. Then classify it as right, obtuse, or acute.

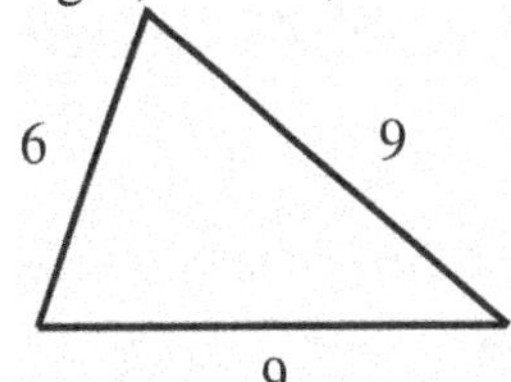

9. Classify the polygon by name.

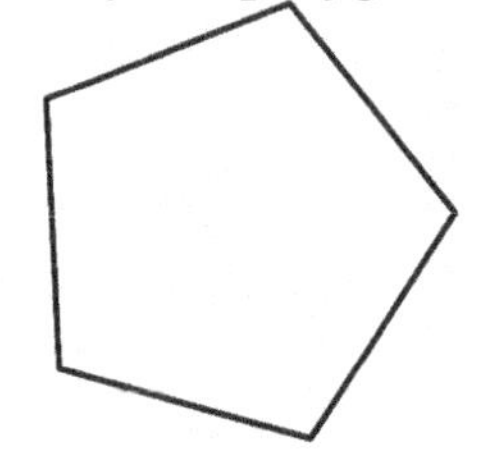

The polygon has five sides, so it is a pentagon.

9. Classify the polygon by name.

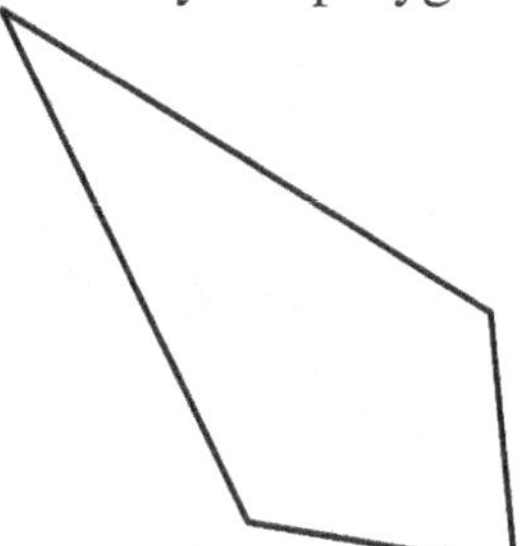

Objective F Given a polygon of n sides, find the sum of its angle measures using the formula $(n - 2)\cdot 180°$.

Review these examples for Objective F:

10. Find the sum of the angles of a nonagon.

A nonagon has 9 sides.

$$\begin{aligned}(n-2)\cdot 180° &= (9-2)\cdot 180°\\ &= 7\cdot 180°\\ &= 1260°\end{aligned}$$

11. Find the missing angle measure.

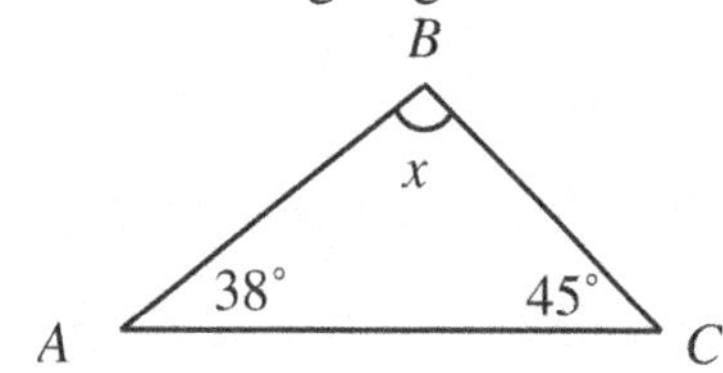

$$\begin{aligned}m\angle A + m\angle B + m\angle C &= 180°\\ 38° + x + 45° &= 180°\\ x + 83° &= 180°\\ x &= 180° - 83°\\ x &= 97°\end{aligned}$$

The missing angle measurc is 97°.

Practice these exercises:

10. Find the sum of the angles of a 16-sided polygon.

11. Find the missing angle measure.

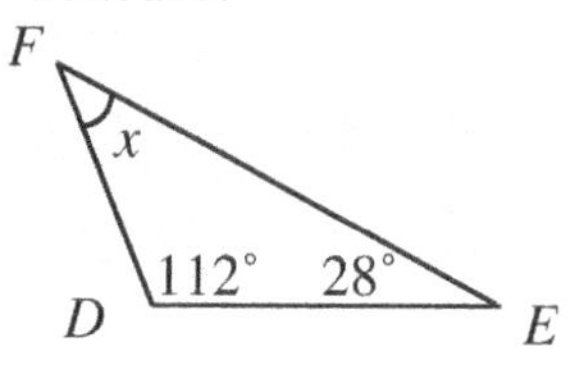

ADDITIONAL EXERCISES

Objective A Draw and name segments, rays, and lines. Also, identify endpoints, if they exist.

For extra help, see pages 390–391 of your text and the Section 6.1 lecture video.

1. Draw the segment whose endpoints are E and F. Name the segment in two ways.

E F

2. Draw the ray with endpoint M. Name the ray.

L  M

3. Name the line in seven different ways.

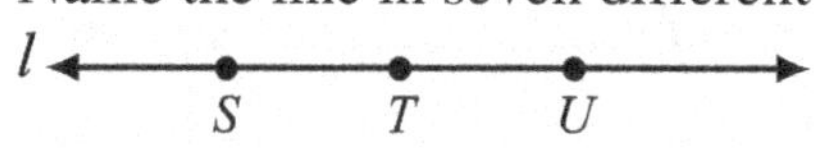

Name: Date:
Instructor: Section:

Objective B Name an angle in five different ways, and given an angle, measure it with a protractor.
For extra help, see pages 391–393 of your text and the Section 6.1 lecture video.

4. Name the angle in five different ways.

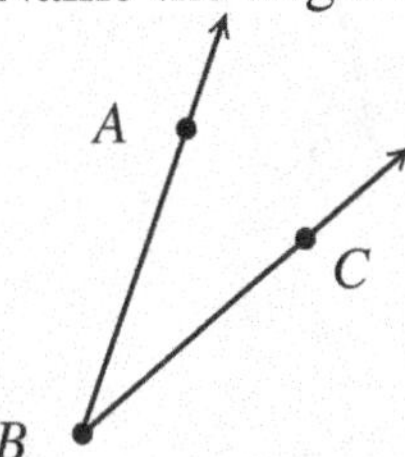

Use a protractor to measure each angle.

5.

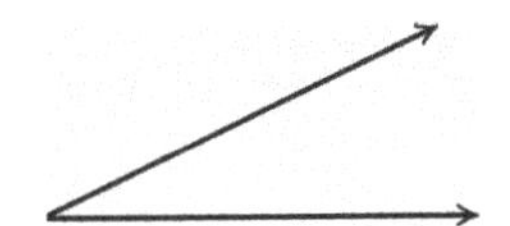

6.

Objective C Classify an angle as right, straight, acute, or obtuse.
For extra help, see page 393 of your text and the Section 6.1 lecture video.
Classify the angle as right, straight, acute, or obtuse.

7.

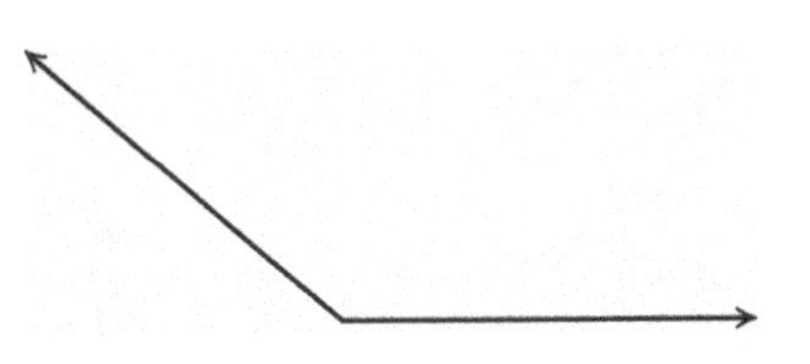

8.

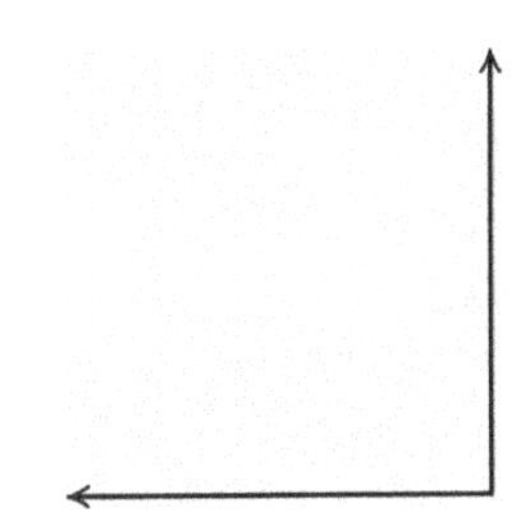

9.

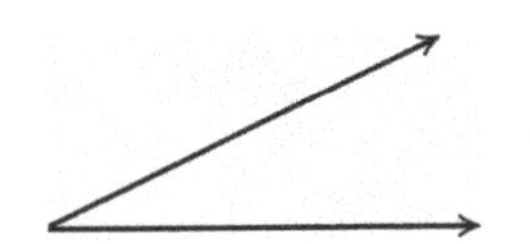

Objective D Identify perpendicular lines.
For extra help, see page 394 of your text and the Section 6.1 lecture video.
Determine whether the pair of lines is perpendicular. Use a protractor.

10.

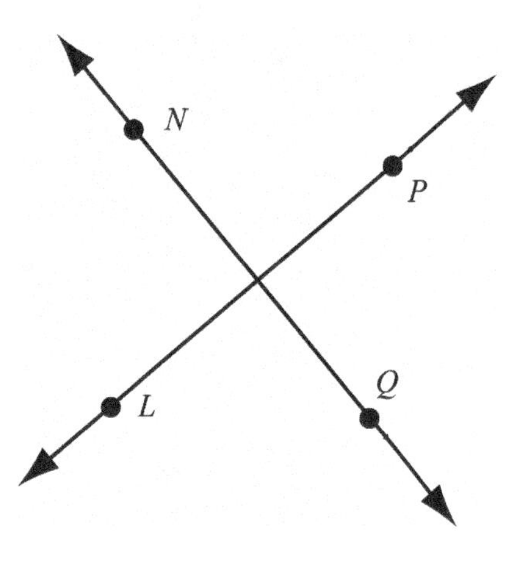

11.

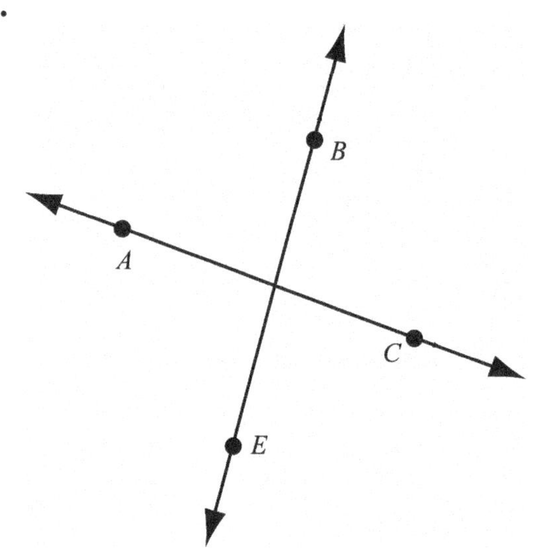

Objective E Classify a triangle as equilateral, isosceles, or scalene and as right, obtuse, or acute. Given a polygon of twelve, ten, or fewer sides, classify it as a dodecagon, a decagon, and so on.
For extra help, see pages 394–395 of your text and the Section 6.1 lecture video.
Classify the triangle as equilateral, isosceles, or scalene. Then classify it as right, obtuse, or acute.

12.

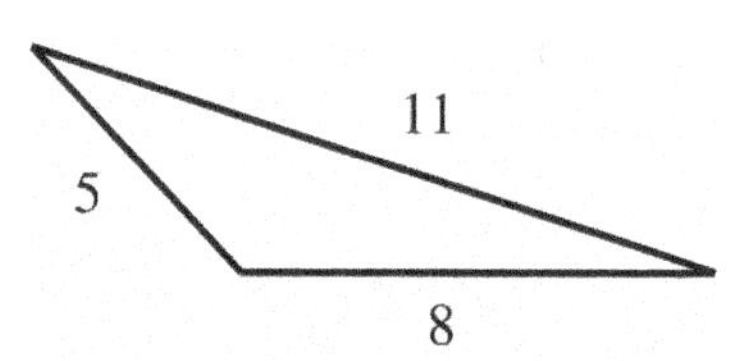

13.

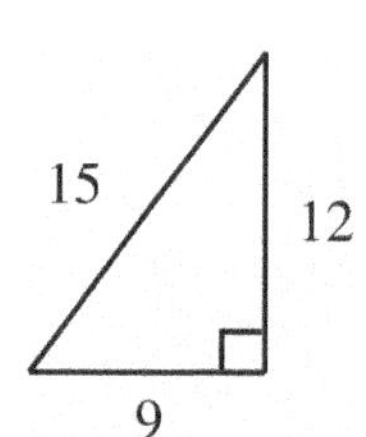

14.

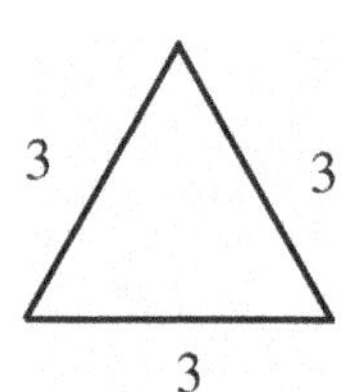

15.

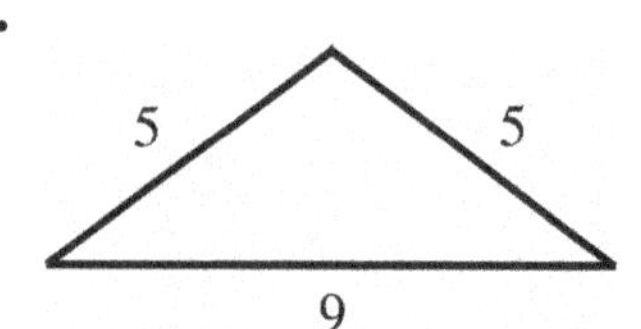

Name: Date:
Instructor: Section:

Classify the polygon by name.

16.

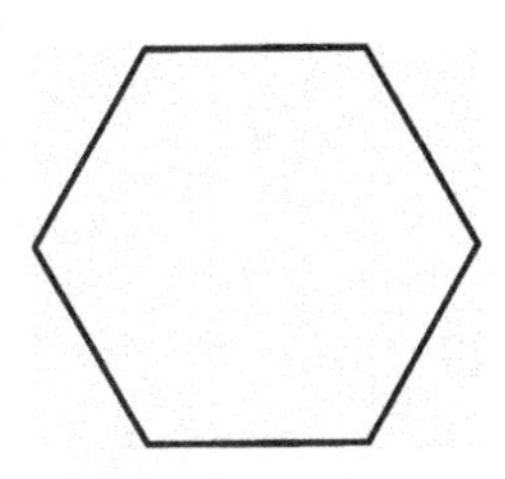

Objective F Given a polygon of n sides, find the sum of its angle measures using the formula $(n-2)\cdot 180°$.

For extra help, see Examples 1–2 on pages 395–396 of your text and the Section 6.1 lecture video.

Find the sum of the angle measures of each of the following.

17. An octagon

18. A 15-sided polygon

Find each missing angle measure.

19.

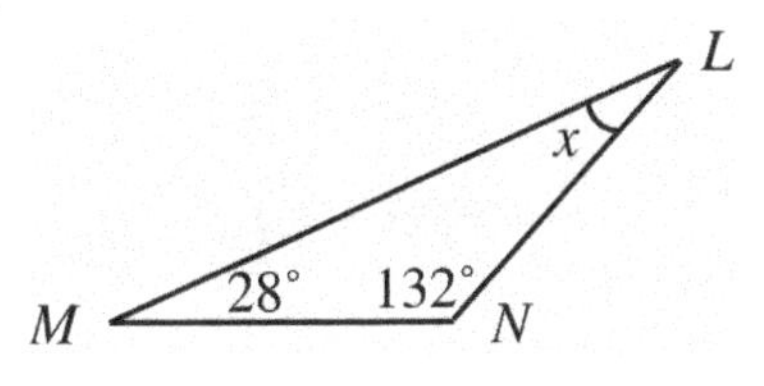

20.

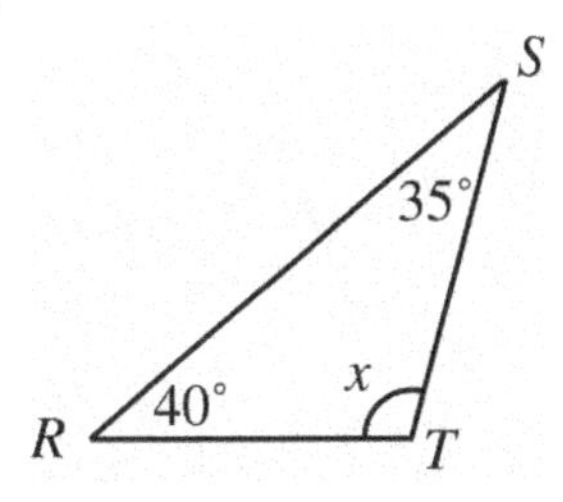

21.

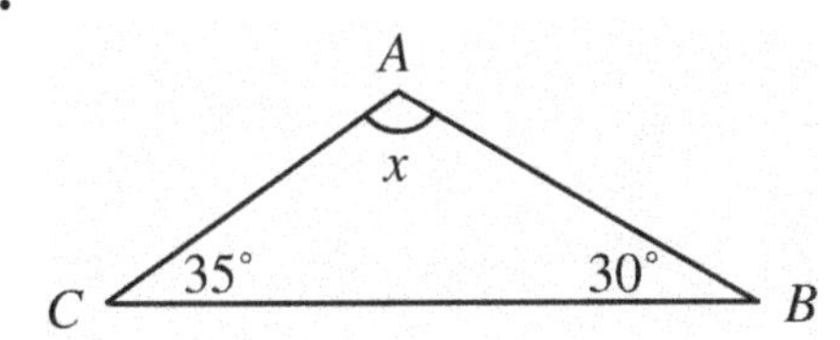

Name:
Instructor:

Date:
Section:

Chapter 6 GEOMETRY

6.2 Perimeter

Learning Objectives
A Find the perimeter of a polygon.
B Solve applied problems involving perimeter.

Key Terms
Use the vocabulary terms listed below to complete each statement in Exercises 1–3.

polygon **square** **rectangle**

1. A geometric figure with three or more sides is called a ________________ .
2. A ________________ is any 4-sided polygon which has four 90° angles.
3. A ________________ is any polygon which has four 90° angles and four sides of the same length.

GUIDED EXAMPLES AND PRACTICE

Objective A Find the perimeter of a polygon.

Review this example for Objective A:

1. Find the perimeter of a rectangle that is 4.5 ft by 3.5 ft.

$P = 2l + 2w$ We could also use $P = 2(l + w)$
$= 2 \cdot 4.5 \text{ ft} + 2 \cdot 3.5 \text{ ft}$
$= 9 \text{ ft} + 7 \text{ ft}$
$= 16 \text{ ft}$

Practice this exercise:

1. Find the perimeter of a square whose sides are 5 cm long.

Objective B Solve applied problems involving perimeter.

Review this example for Objective B:

2. A fence is to be built around a 25 ft by 20 ft play area. How many feet of fence will be needed? If fencing sells for $4.95 per foot, what will the fencing cost?

1. *Familiarize*. We make a drawing and let P = the perimeter.

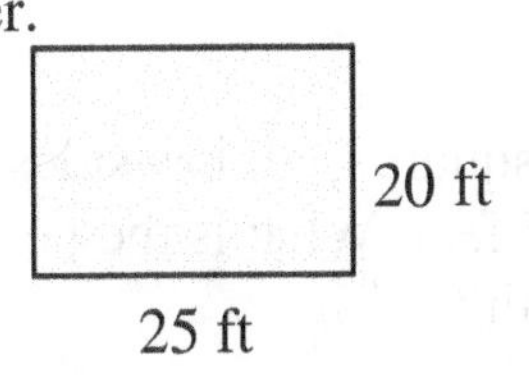

Practice this exercise:

2. A fence is to be built around a 200 m by 180 m field. If the fencing sells for $1.85 per meter, what will the fencing cost?

2. *Translate*. The perimeter of the play area is given by

$$P = 2(l + w) = 2 \cdot (25 \text{ ft} + 20 \text{ ft})$$

3. *Solve*. We calculate the perimeter.

$$P = 2 \cdot (25 \text{ ft} + 20 \text{ ft})$$
$$= 2 \cdot (45 \text{ ft}) = 90 \text{ ft}$$

The we multiply the perimeter by \$4.95 to find the cost of the fencing:

$$\text{Cost} = \$4.95 \times \text{Perimeter}$$
$$= \$4.95 \times 90 \text{ ft} = \$445.50$$

4. *Check*. We repeat the calculations. The answers check.

5. *State*. The 90 ft of fencing will cost \$445.50.

ADDITIONAL EXERCISES

Objective A Find the perimeter of a polygon.

For extra help, see Examples 1–5 on pages 401–403 of your text and the Section 6.2 lecture video.

Find the perimeter of the polygon.

1.

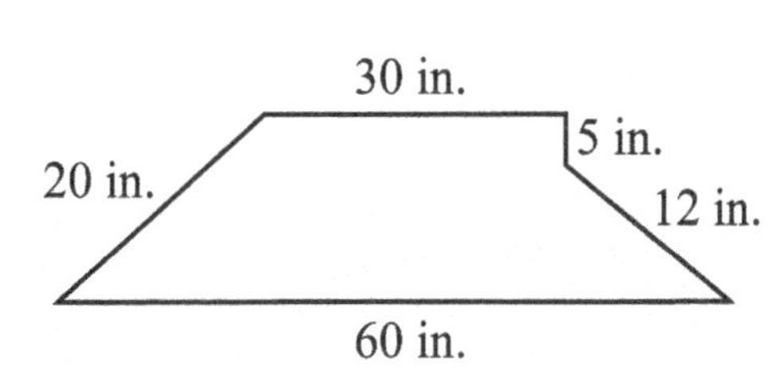

2.

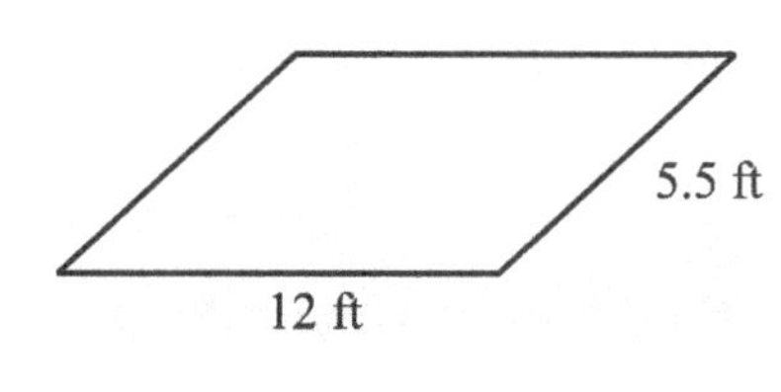

3. Find the perimeter of the rectangle that is 5.76 cm by 8.42 cm.

4. Find the perimeter of the square that is 7 mi on a side.

Objective B Solve applied problems involving perimeter.

For extra help, see Example 6 on page 403 of your text and the Section 6.2 lecture video.

Solve.

5. A homeowner decides to fence in a rectangular yard for his dog. The dimensions of the area to be fenced in are 20 ft by 35 ft. What is the perimeter of this area? If the fencing costs \$2.19 per foot, how much will it cost him to fence in this area?

6. A rectangular room is 12.5 ft by 14.5 ft. What is the perimeter of this room?

7. A rectangular picture is 8.89 cm by 12.7 cm. What is the perimeter of the picture?

8. A square quilt has sides of length 8.5 feet. What is the perimeter of the quilt?

Name: Date:
Instructor: Section:

Chapter 6 GEOMETRY

6.3 Area

Learning Objectives
A Find the area of a rectangle and a square.
B Find the area of a parallelogram, a triangle, and a trapezoid.
C Solve applied problems involving areas of rectangles, squares, parallelograms, triangles, and trapezoids.

Key Terms
Use the formulas listed below to complete each statement in Exercises 1–3.

$A=\frac{1}{2}\cdot b\cdot h$ $A=b\cdot h$ $A=\frac{1}{2}\cdot h\cdot(a+b)$

1. The formula for area of a parallelogram is ________________ .

2. The formula for area of a trapezoid is ________________ .

3. The formula for area of a triangle is ________________ .

GUIDED EXAMPLES AND PRACTICE

Objective A Find the area of a rectangle and a square.

Review these examples for Objective A:

1. Find the area of a rectangle that is 4.5 m by 2.3 m.

$$A=l\cdot w=4.5\text{ m}\times 2.3\text{ m}$$
$$=4.5\times 2.3\times \text{m}\times\text{m}$$
$$=10.35\text{ m}^2$$

2. Find the area of a square with sides of length 17 cm.

$$A=s\cdot s=17\text{ cm}\cdot 17\text{ cm}$$
$$=17\cdot 17\cdot\text{cm}\cdot\text{cm}$$
$$=289\text{ cm}^2$$

Practice these exercises:

1. Find the area of a rectangle that is 8 m by 6 m.

2. Find the area of a square with sides of length 6.4 yd.

Objective B Find the area of a parallelogram, a triangle, and a trapezoid.

Review these examples for Objective A:

3. Find the area.

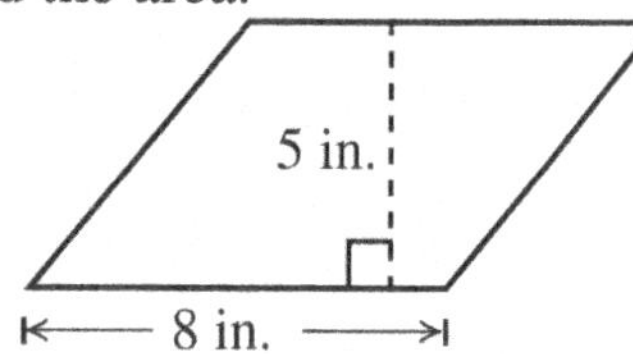

$A = b \cdot h = (8 \text{ in.}) \cdot (5 \text{ in.}) = 40 \text{ in}^2$

The area is 40 in^2.

4. Find the area.

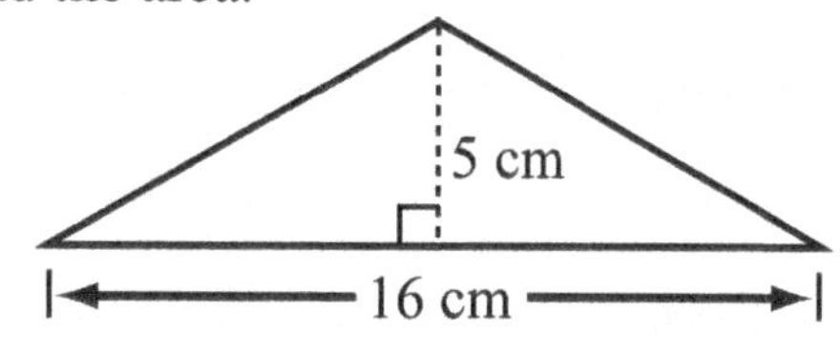

$$A = \frac{1}{2} \cdot b \cdot h$$
$$= \frac{1}{2} \cdot 16 \text{ cm} \cdot 5 \text{ cm}$$
$$= \frac{16 \cdot 5}{2} \text{ m}^2 = 40 \text{ cm}^2$$

The area is 40 m^2.

5. Find the area.

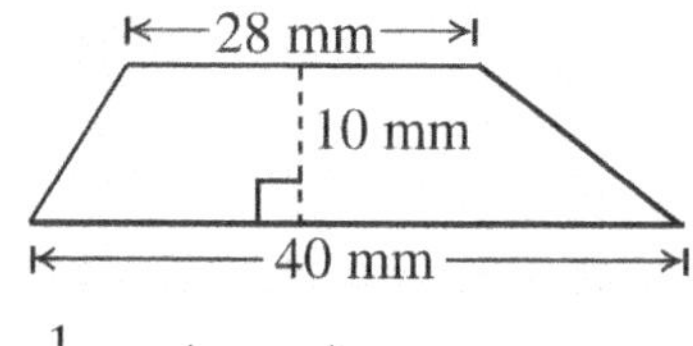

$$A = \frac{1}{2} \cdot h \cdot (a + b)$$
$$= \frac{1}{2} \cdot 10 \text{ mm} \cdot (28 + 40) \text{ mm}$$
$$= \frac{10 \cdot 68}{2} \text{ mm}^2 = \frac{2 \cdot 5 \cdot 68}{2 \cdot 1} \text{ mm}^2$$
$$= \frac{2}{2} \cdot \frac{5 \cdot 68}{1} \text{ mm}^2$$
$$= 340 \text{ mm}^2$$

The area is 340 mm^2.

Practice these exercises:

3. Find the area.

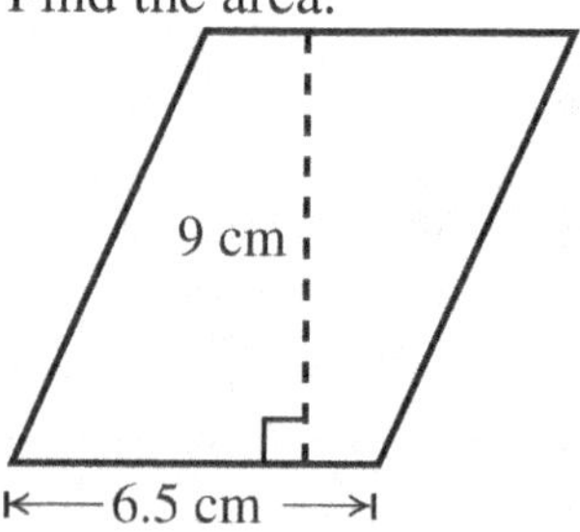

4. Find the area.

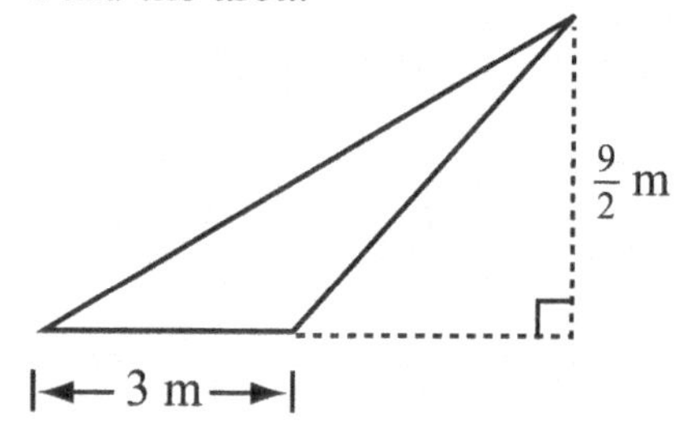

5. Find the area.

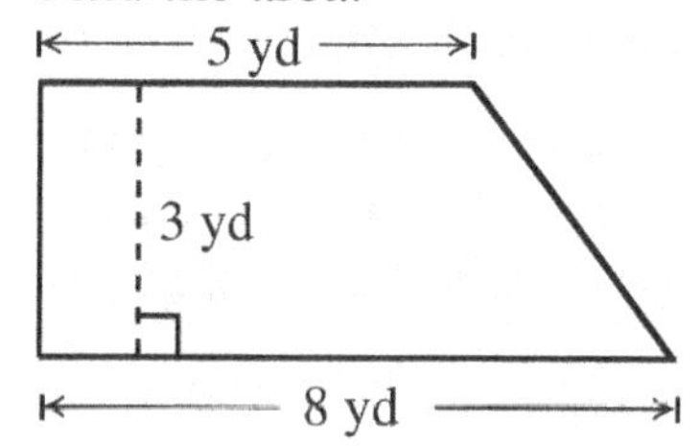

Name: Date:
Instructor: Section:

Objective C Solve applied problems involving areas of rectangles, squares, parallelograms, triangles, and trapezoids.

Review this example for Objective C:

6. A square flower garden 4 m on a side is dug in a 30 m by 20 m lawn. How much area is left over?

We make a drawing.

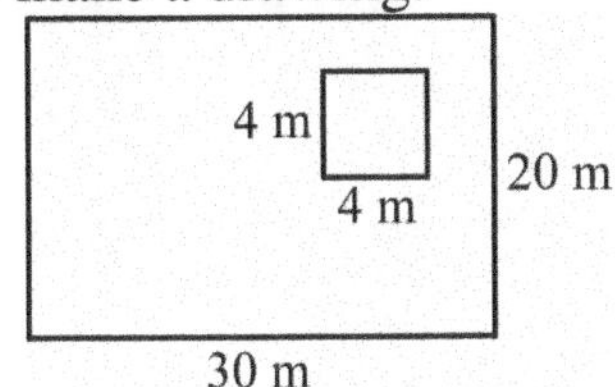

The area of the lawn is

$A = l \times w = (30\text{ m}) \times (20\text{ m}) = 600\text{ m}^2.$

The area of the garden is

$A = l \times w = (4\text{ m}) \times (4\text{ m}) = 16\text{ m}^2.$

The area left over is

$600\text{ m}^2 - 16\text{ m}^2 = 584\text{ m}^2.$

Practice this exercise:

6. A field measures 250 ft by 175 ft. A portion of the field measuring 180 ft by 150 ft is paved for a parking area. How much of the field is left unpaved?

ADDITIONAL EXERCISES

Objective A Find the area of a rectangle and a square.

For extra help, see Examples 1–4 on pages 406–407 of your text and the Section 6.3 lecture video.

Find the area.

1.

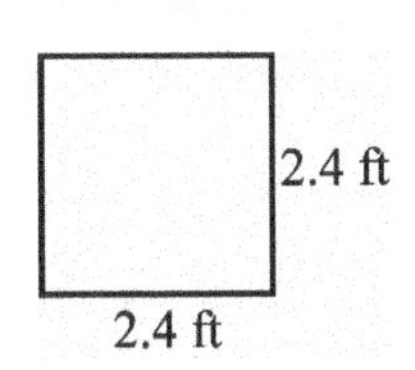

2.

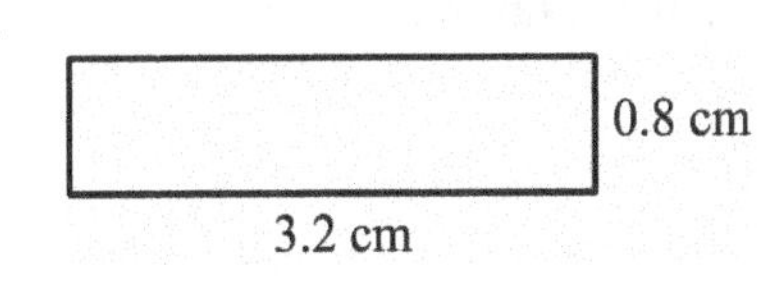

3. Find the area of the rectangle that is 12 ft by 8 ft.

4. Find the area of the square that is $\frac{3}{4}$ mi on a side.

Objective B Find the area of a parallelogram, a triangle, and a trapezoid.

For extra help, see Examples 5–9 on pages 408–410 of your text and the Section 6.3 lecture video.

Find the area.

5.

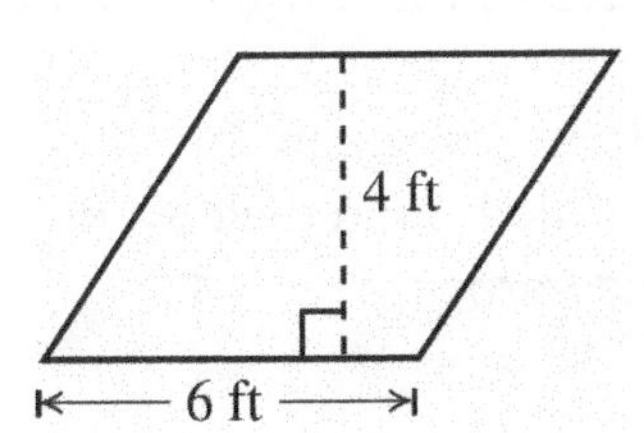

6.

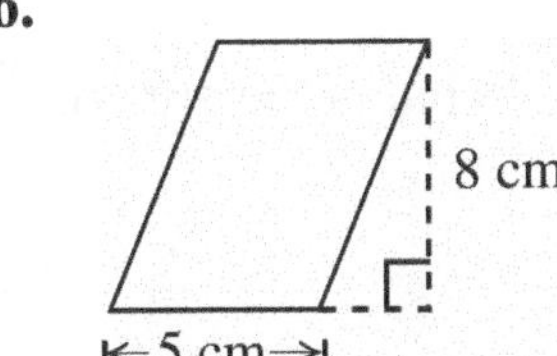

7.

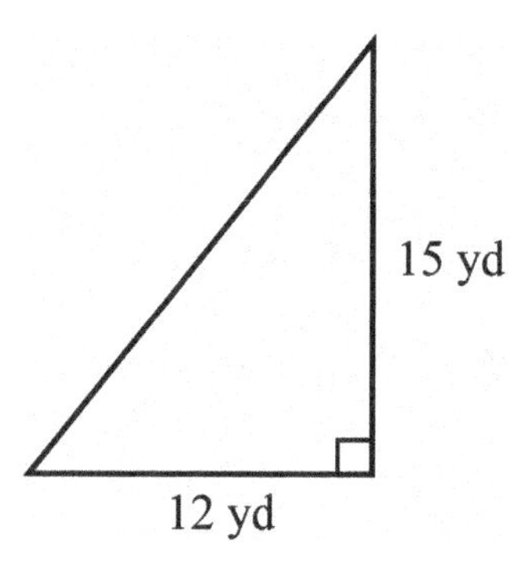

8.

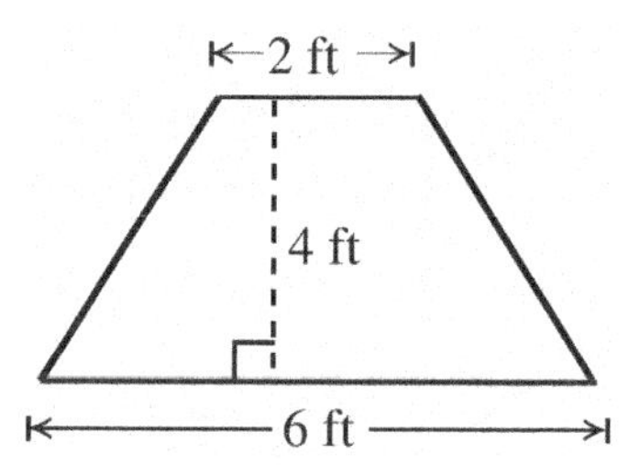

9.

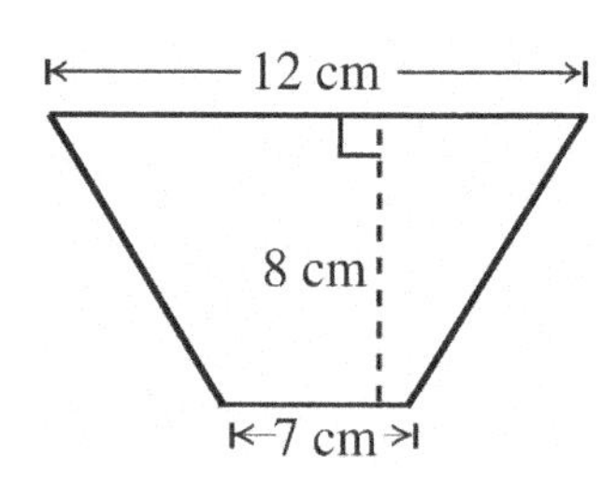

Objective C Solve applied problems involving areas of rectangles, squares, parallelograms, triangles, and trapezoids.

For extra help, see Examples 10–11 on pages 410–411 of your text and the Section 6.3 lecture video.

Solve.

10. A lot is 75 m by 50 m. A house 24 m by 10 m is built on the lot. How much area is left over for a lawn?

11. A city wants to put a sidewalk that is 2-ft wide around a community garden, as shown in the figure. The garden itself is 65 ft by 32 ft. Find the area of the sidewalk.

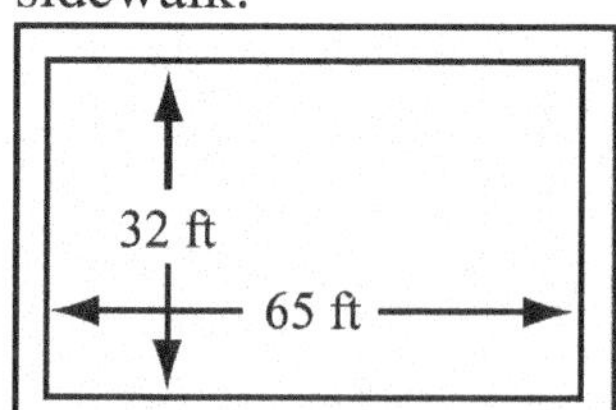

12. The page of a book is $10\frac{7}{8}$ in. by $8\frac{1}{2}$ in. The printed area of the page is $9\frac{1}{2}$ in. by $7\frac{1}{4}$ in. What is the area of the margin?

13. A square garden $12\frac{1}{2}$ ft on a side is placed in a lawn that is 48 ft by 32 ft.
a) Find the area of the lawn, excluding the garden.
b) It costs $0.35 per square foot to have a garden rototilled. Find the cost of rototilling the garden.

Name: Date:
Instructor: Section:

Chapter 6 GEOMETRY

6.4 Circles

Learning Objectives
A Find the length of a radius of a circle given the length of a diameter, and find the length of a diameter given the length of the radius.
B Find the circumference of a circle given the length of a diameter or a radius.
C Find the area of a circle given the length of a diameter or a radius.
D Solve applied problems involving circles.

Key Terms
Use the vocabulary terms listed below to complete each statement in Exercises 1–3. Terms may be used more than once.

radius **diameter** **circumference**

1. A ____________________ is a segment that passes through the center of a circle and has endpoints on the circle.
2. The ____________________ of a circle is the distance around it.
3. The length of the ____________________ of a circle is half the length of the ____________________ of the circle.

GUIDED EXAMPLES AND PRACTICE

Objective A Find the length of a radius of a circle given the length of a diameter, and find the length of a diameter given the length of the radius.

Review these examples for Objective A:

1. Find the length of a radius of this circle.

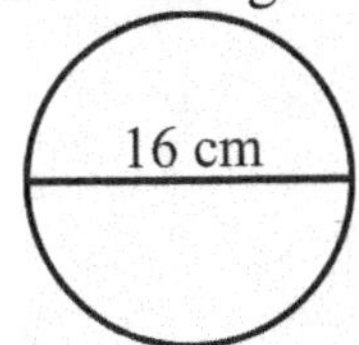

$r = \frac{d}{2} = \frac{16\text{ cm}}{2} = 8\text{ cm}$
The radius is 8 cm.

2. Find the length of a diameter of this circle.

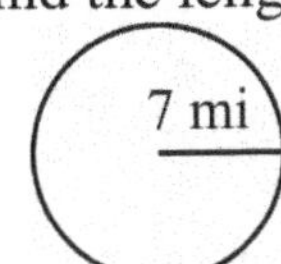

$d = 2 \cdot r = 2 \cdot 7\text{ mi} = 14\text{ mi}$
The diameter is 14 mi.

Practice these exercises:

1. Find the length of a radius of this circle.

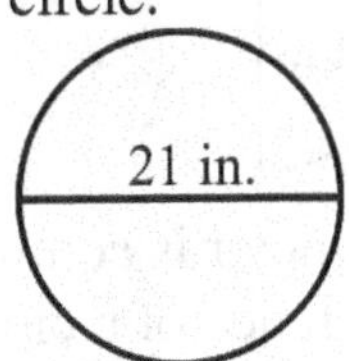

2. Find the length of a diameter of this circle.

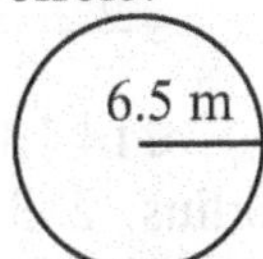

Objective B **Find the circumference of a circle given the length of a diameter or a radius.**

Review this example for Objective B:

3. Find the circumference of this circle. Use 3.14 for π.

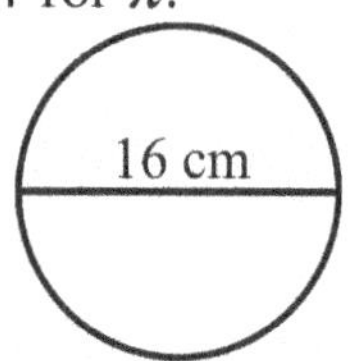

$C = \pi \cdot d \approx 3.14 \times 16 \text{ cm} \approx 50.24 \text{ cm}$

The circumference is about 50.24 cm.

Practice this exercise:

3. Find the circumference of this circle. Use 3.14 for π.

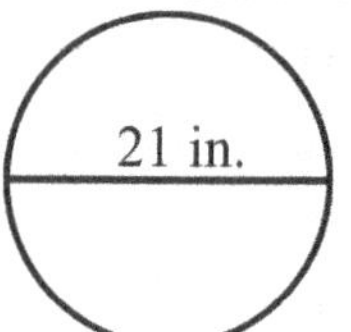

Objective C **Find the area of a circle given the length of a diameter or a radius.**

Review this example for Objective C:

4. Find the area of this circle. Use $\frac{22}{7}$ for π.

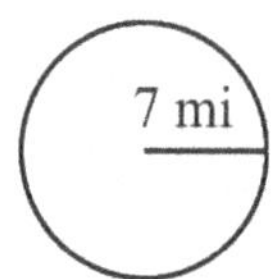

$A = \pi \cdot r \cdot r \approx \frac{22}{7} \cdot 7 \text{ mi} \cdot 7 \text{ mi} \approx 154 \text{ mi}^2$

The area is about 154 mi^2.

Practice this exercise:

4. Find the area of this circle. Use 3.14 for π.

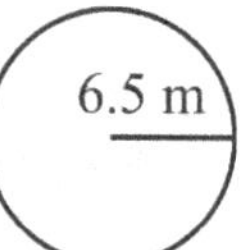

Objective D **Solve applied problems involving circles.**

Review this example for Objective D:

5. Find the perimeter. Use 3.14 for π.

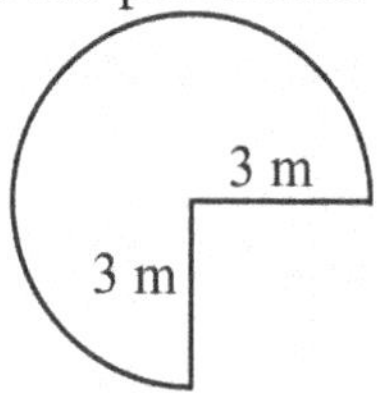

The perimeter is composed of three-fourths of the circumference of a circle with radius of 3 m, plus twice the radius of the circle.

Three-fourths of the circumference:

$$\frac{3}{4} \cdot 2 \cdot \pi \cdot r \approx \frac{3}{4} \cdot 2 \cdot 3.14 \cdot 3 \text{ m}$$
$$\approx \frac{3 \cdot 2 \cdot 3.14 \cdot 3}{4} \text{ m}$$
$$\approx 14.13 \text{ m}$$

Twice the radius: $2 \cdot 3 \text{ m} = 6 \text{ m}$

Perimeter: 14.13 m + 6 m = 20.13 m.

The perimeter is about 20.13 m.

Practice this exercise:

5. Find the area of the figure in the example at the left. Use 3.14 for π.

Name: Date:
Instructor: Section:

ADDITIONAL EXERCISES

Objective A Find the length of a radius of a circle given the length of a diameter, and find the length of a diameter given the length of the radius.

For extra help, see Examples 1–2 on page 417 of your text and the Section 6.4 lecture video.

1. Find the length of the diameter for the circle.

2. Find the length of the diameter for the circle.

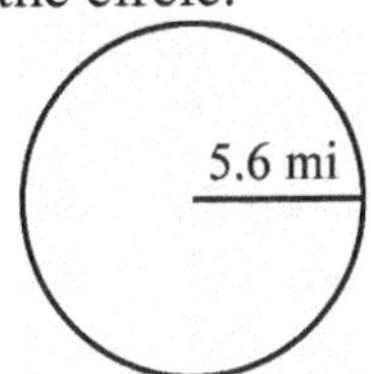

3. Find the length of the radius for the circle.

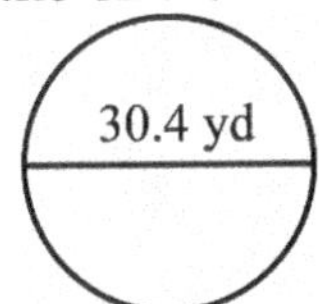

4. Find the length of the radius for the circle.

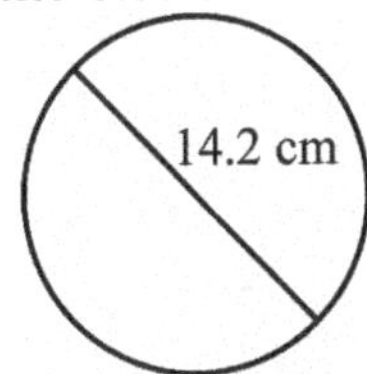

Objective B Find the circumference of a circle given the length of a diameter or a radius.

For extra help, see Examples 3–5 on pages 418–419 of your text and the Section 6.4 lecture video.

5. Find the circumference of the circle. Use $\frac{22}{7}$ for π.

6. Find the circumference of the circle. Use $\frac{22}{7}$ for π.

7. Find the circumference of the circle. Use 3.14 for π.

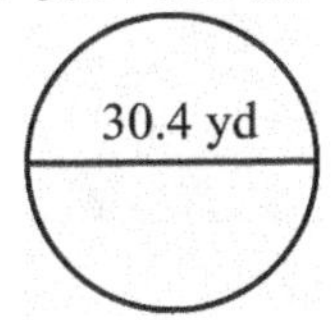

8. Find the circumference of the circle. Use 3.14 for π.

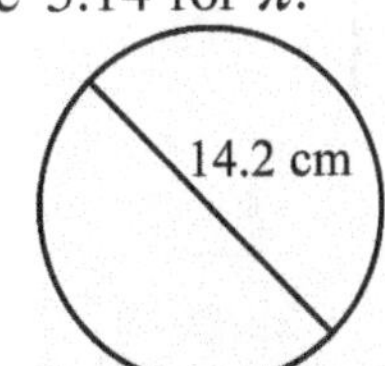

Objective C Find the area of a circle given the length of a diameter or a radius.
For extra help, see Examples 6–7 on pages 420–421 of your text and the Section 6.4 lecture video.

9. Find the area of the circle.
Use $\frac{22}{7}$ for π.

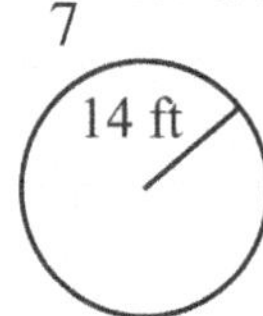

10. Find the area of the circle.
Use $\frac{22}{7}$ for π.

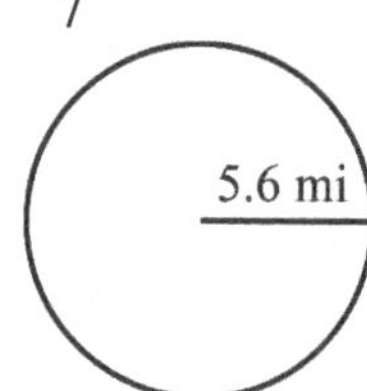

11. Find the area of the circle.
Use 3.14 for π.

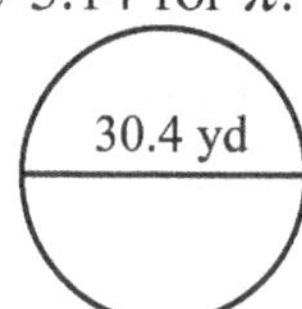

12. Find the area of the circle.
Use 3.14 for π.

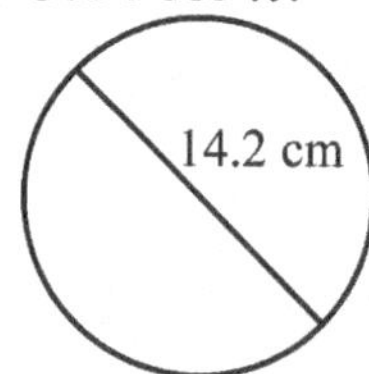

Objective D Solve applied problems involving circles.
For extra help, see Example 8 on page 421 of your text and the Section 6.4 lecture video.
Solve.

13. The top of a can of fruit has a radius of 4 cm. What is the diameter? the circumference? the area?
Use 3.14 for π.

14. The circumference of a tree is 4 ft. What is the tree's diameter?
Use 3.14 for π.

15. Find the perimeter of the figure.
Use 3.14 for π.

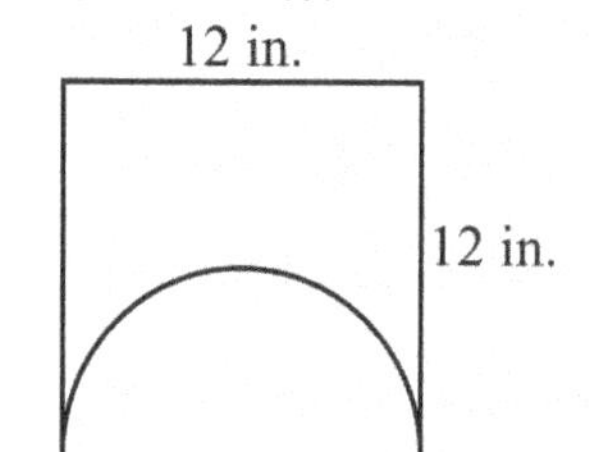

16. Find the area of the figure.
Use 3.14 for π.

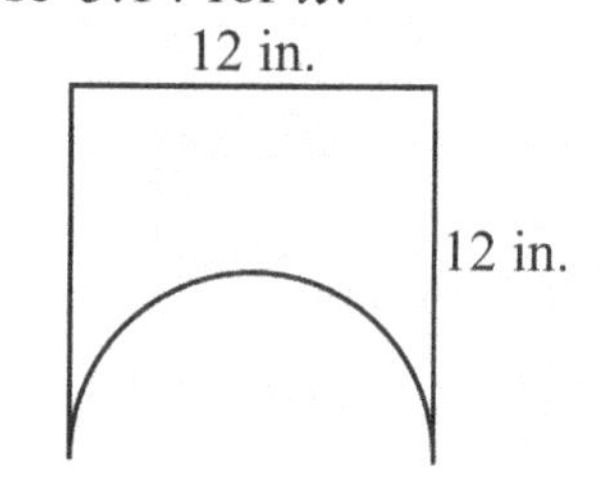

Name: Date:
Instructor: Section:

Chapter 6 GEOMETRY

6.5 Volume and Surface Area

Learning Objectives

A Find the volume and the surface area of a rectangular solid.
B Given the radius and the height, find the volume of a circular cylinder.
C Given the radius, find the volume of a sphere.
D Given the radius and the height, find the volume of a circular cone.
E Solve applied problems involving volumes of rectangular solids, circular cylinders, spheres and cones.

Key Terms

Use the vocabulary terms listed below to complete each statement in Exercises 1–4.

rectangular solid **circular cylinder** **circular cone** **sphere**

1. The volume of a ____________________ is given by $V = \frac{4}{3} \cdot \pi \cdot r^3$, where r is the radius.

2. The volume of a ____________________ is given by $V = l \cdot w \cdot h$, where l is the length, w is the width, and h is the height.

3. The volume of a ____________________ is given by $V = \pi \cdot r^2 \cdot h$, where r is the radius of the base and h is the height.

4. The volume of a ____________________ is given by $V = \frac{1}{3} \cdot \pi \cdot r^2 \cdot h$, where r is the radius of the base and h is the height.

GUIDED EXAMPLES AND PRACTICE

Objective A Find the volume and the surface area of a rectangular solid.

Review these examples for Objective A:

1. Find the volume.

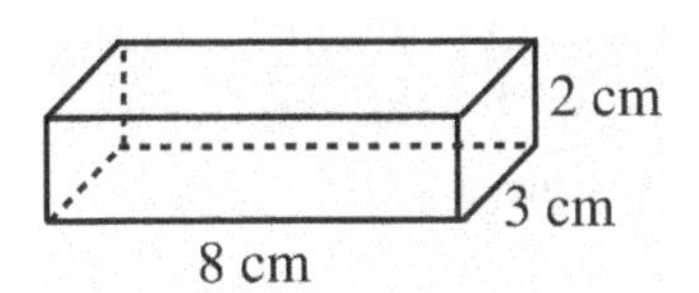

$$\begin{aligned} V &= l \cdot w \cdot h \\ &= 8 \text{ cm} \cdot 3 \text{ cm} \cdot 2 \text{ cm} \\ &= 24 \cdot 2 \text{ cm}^3 \\ &= 48 \text{ cm}^3 \end{aligned}$$

Practice these exercises:

1. Find the volume.

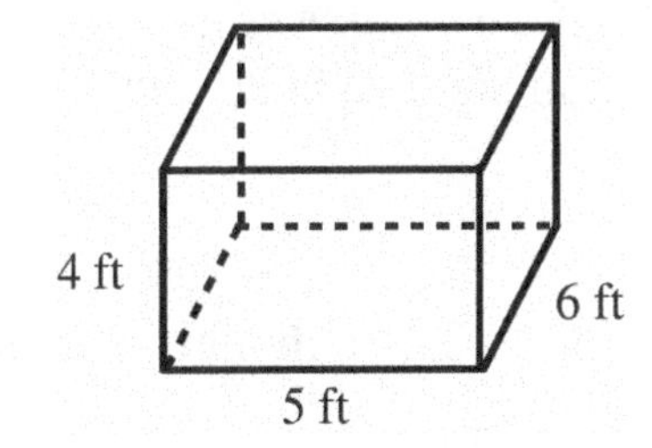

2. Find the surface area.

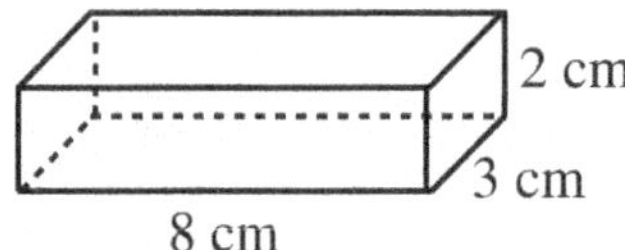

$SA = 2lw + 2lh + 2wh$
$= 2 \cdot 8 \text{ cm} \cdot 3 \text{ cm} + 2 \cdot 8 \text{ cm} \cdot 2 \text{ cm} + 2 \cdot 3 \text{ cm} \cdot 2 \text{ cm}$
$= 48 \text{ cm}^2 + 32 \text{ cm}^2 + 12 \text{ cm}^2$
$= 92 \text{ cm}^2$

2. Find the surface area.

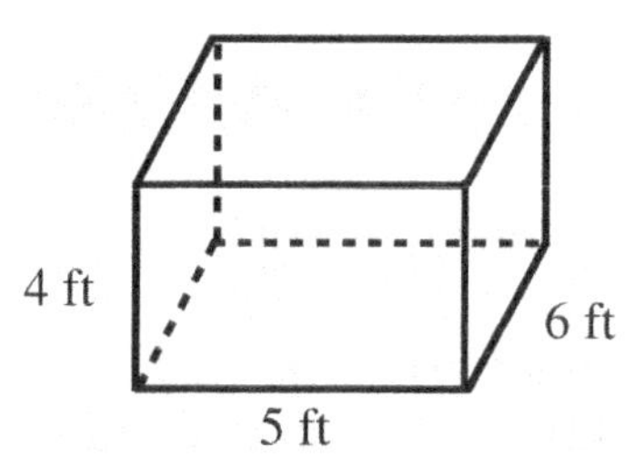

Objective B Given the radius and the height, find the volume of a circular cylinder.

Review this example for Objective B:

3. Find the volume of the circular cylinder. Use 3.14 for π.

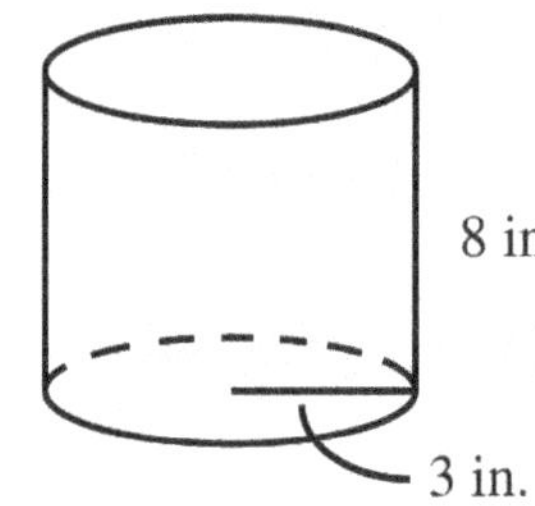

$V = \pi \cdot r^2 \cdot h$
$\approx 3.14 \times 3 \text{ in.} \times 3 \text{ in.} \times 8 \text{ in.}$
$\approx 226.08 \text{ in}^3$

Practice this exercise:

3. Find the volume of the circular cylinder. Use $\frac{22}{7}$ for π.

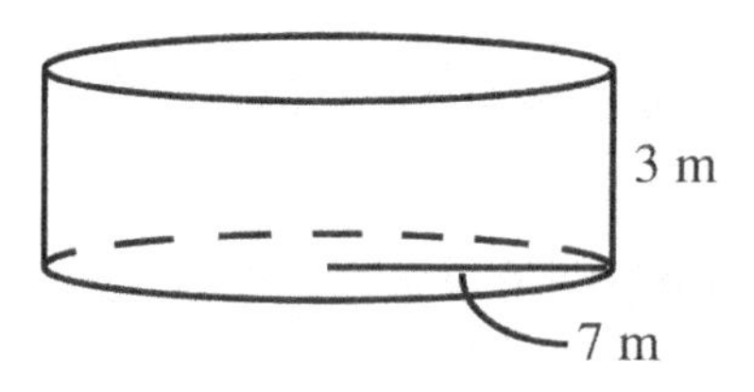

Objective C Given the radius, find the volume of a sphere.

Review this example for Objective C:

4. Find the volume of a sphere with radius 21 cm. Use $\frac{22}{7}$ for π.

$V = \frac{4}{3} \cdot \pi \cdot r^3$
$\approx \frac{4}{3} \times \frac{22}{7} \times (21 \text{ cm})^3 \approx \frac{4 \times 22 \times 9261}{3 \times 7} \text{ cm}^3$
$\approx 38{,}808 \text{ cm}^3$

Practice this exercise:

4. Find the volume of a sphere with radius 5 in. Use 3.14 for π.

Objective D Given the radius and the height, find the volume of a circular cone.

Review this example for Objective D:

5. Find the volume of a circular cone with radius 6 mm and height 20 mm. Use 3.14 for π.

$V = \frac{1}{3} \pi \cdot r^2 \cdot h$
$\approx \frac{1}{3} \times 3.14 \times (6 \text{ mm})^2 \times 20 \text{ mm}$
$\approx 753.6 \text{ mm}^2$

Practice this exercise:

5. Find the volume of a circular cone with radius 6 ft and height 4 ft. Use 3.14 for π.

Name: Date:
Instructor: Section:

Objective E Solve applied problems involving volumes of rectangular solids, circular cylinders, spheres and cones.

Review this example for Objective C:

6. A cylindrical log has a diameter of 12 in. and a height of 18 in. Find the volume. Use 3.14 for π.

1. *Familiarize.* We will use the formula for the volume of a circular cylinder, $V = \pi \cdot r^2 \cdot h$. Note that the radius of the log is 12 in./2 or 6 in.

2. *Translate.* We substitute in the formula.

$V \approx 3.14 \times (6 \text{ in.})^2 \times 18 \text{ in.}$

3. *Solve.* We carry out the calculation.

$$V \approx 3.14 \times (6 \text{ in.})^2 \times 18 \text{ in.}$$
$$\approx 3.14 \times 6 \text{ in.} \times 6 \text{ in.} \times 18 \text{ in.}$$
$$\approx 2034.72 \text{ in}^3$$

4. *Check.* We repeat the calculations. The answer checks.

5. *State.* The volume of the log is about 2034.72 in^3.

Practice this exercise:

6. The diameter of a beach ball is 27 cm. Find the volume. Use 3.14 for π.

ADDITIONAL EXERCISES

Objective A Find the volume and the surface area of a rectangular solid.

For extra help, see Examples 1–3 on pages 428–430 of your text and the Section 6.5 lecture video.

Find the volume and surface area of each figure.

1.

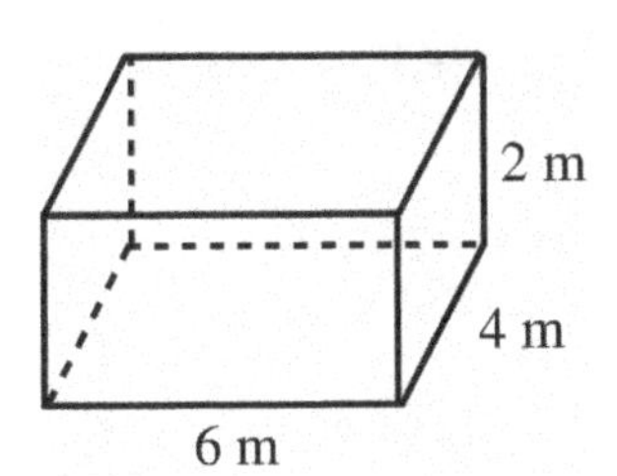

2.

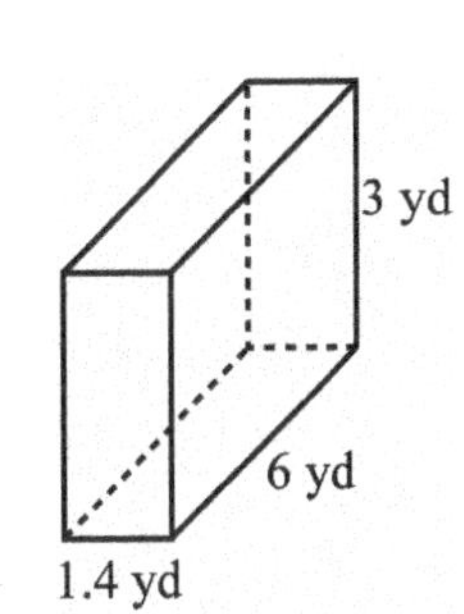

3.

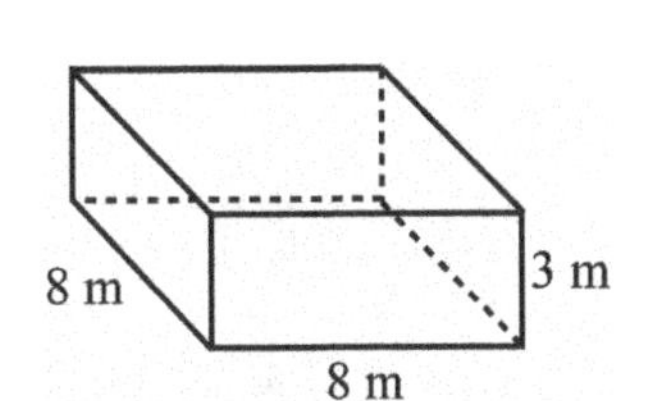

Objective B Given the radius and the height, find the volume of a circular cylinder.
For extra help, see Examples 4–5 page 431 of your text and the Section 6.5 lecture video.
Find the volume of each figure. Use 3.14 for π.

4.

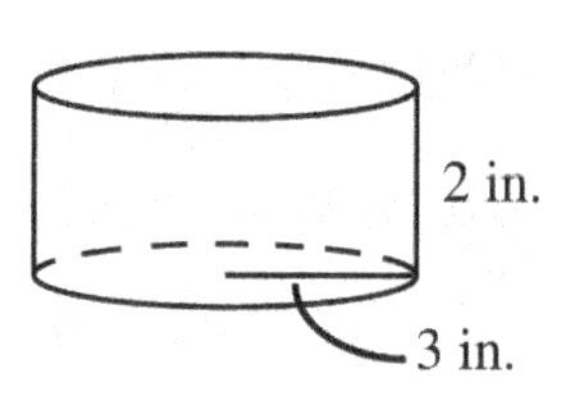

5.

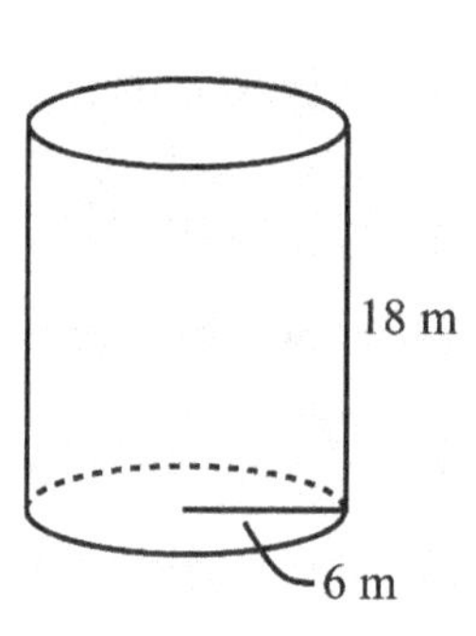

6.

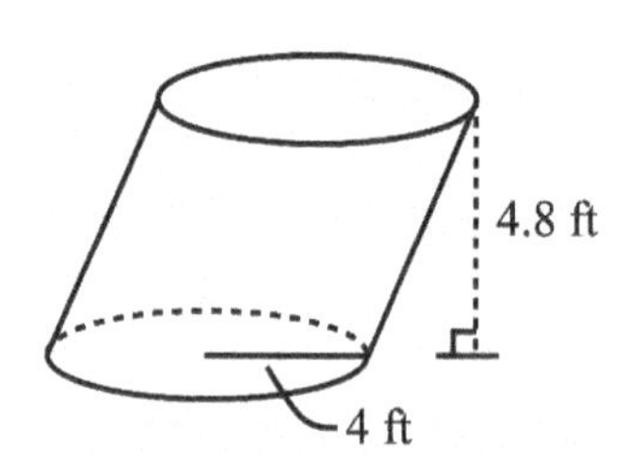

Objective C Given the radius, find the volume of a sphere.
For extra help, see Example 6 on page 432 of your text and the Section 6.5 lecture video.
Find the volume of each figure. Use 3.14 for π.

7.

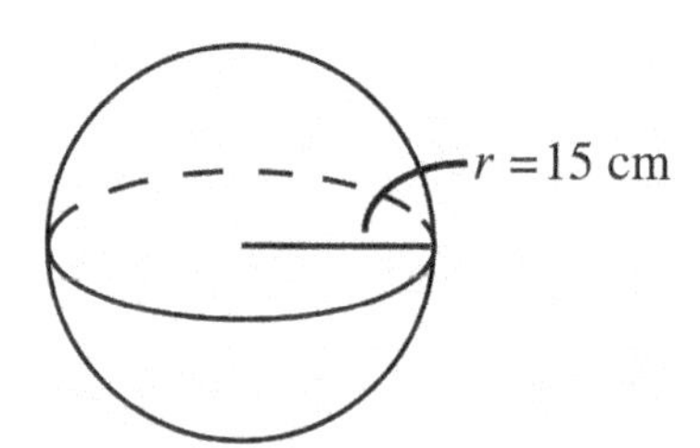

8.

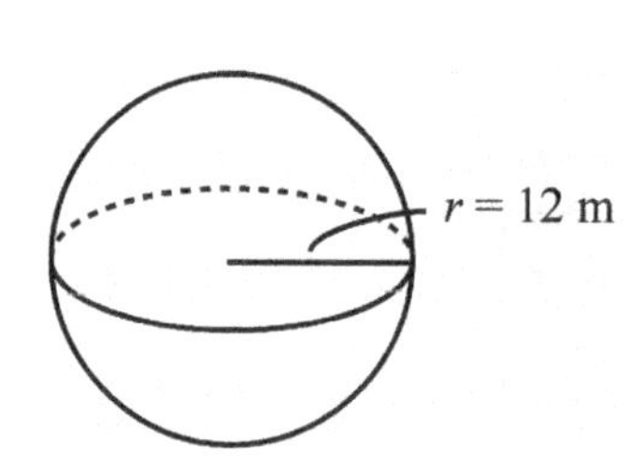

9.

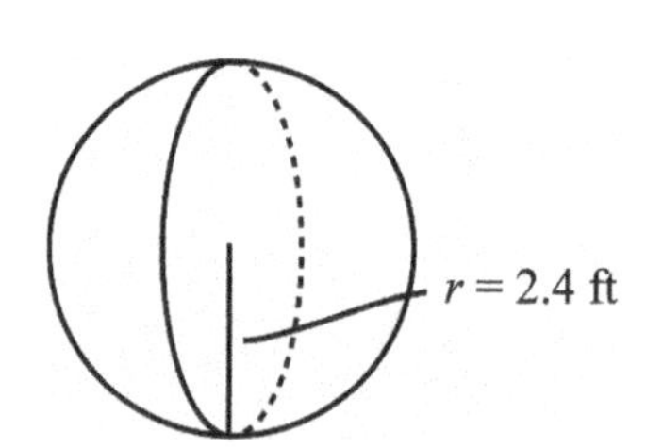

Name: Date:
Instructor: Section:

Objective D Given the radius and the height, find the volume of a circular cone.
For extra help, see Example 7 page 432 of your text and the Section 6.5 lecture video.
Find the volume of the circular cone. Use 3.14 for π.

10.

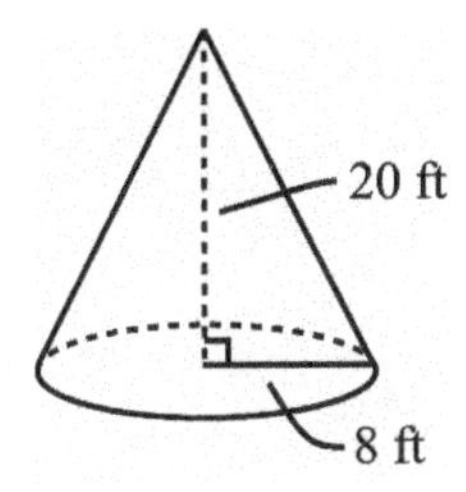

11.

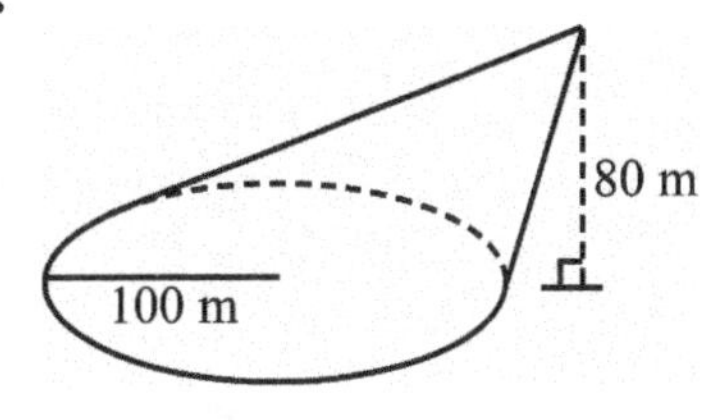

12.

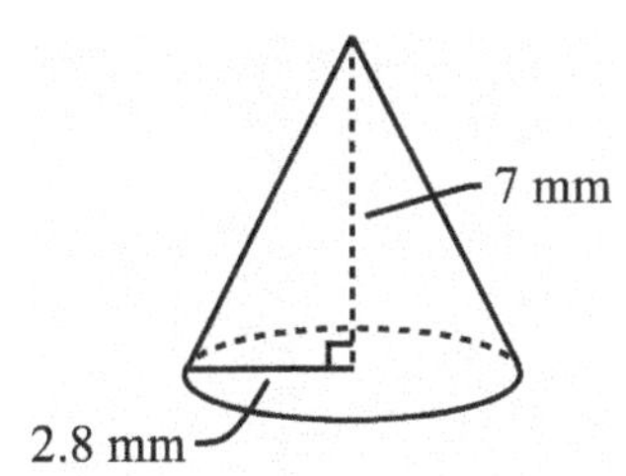

Objective E Solve applied problems involving volumes of rectangular solids, circular cylinders, spheres and cones.
For extra help, see Example 8 page 433 of your text and the Section 6.5 lecture video.
Solve.

13. The diameter of a child's spherical ball is 6 cm. Find the volume. Use 3.14 for π.

14. A cylindrical piece of copper pipe has a diameter of 4 in. and a length (height) of 16 in. Find the volume. Use 3.14 for π.

15. Javier stores photographs in a box that measures 4 in. by 6 in. by 4 in. Find the volume of the box.

16. A party hat shaped like a cone has a height of 8 in. and the radius of its base is 3 in. Find its volume. Use 3.14 for π.

Name: Date:
Instructor: Section:

Chapter 6 GEOMETRY

6.6 Relationships Between Angle Measures

Learning Objectives

A Identify complementary and supplementary angles and find the measure of a complement or a supplement of a given angle.
B Determine whether segments are congruent and whether angles are congruent.
C Use the Vertical Angle Property to find measures of angles.
D Identify pairs of corresponding angles, interior angles, and alternate interior angles and apply properties of transversals and parallel lines to find measures of angles.

Key Terms

Use the vocabulary terms listed below to complete each statement in Exercises 1–5.

complementary	**supplementary**	**congruent**
vertical angles	**transversal**	

1. A ____________________ is a line that intersects two or more coplanar lines in different points.

2. Two angles are ____________________ if the sum of their measures is 180°.

3. Two angles are ____________________ if and only if they have the same measure.

4. Two nonstraight angles are ____________________ if and only if their sides form two pairs of opposite rays.

5. Two angles are ____________________ if the sum of their measure is 90°.

GUIDED EXAMPLES AND PRACTICE

Objective A Identify complementary and supplementary angles and find the measure of a complement or a supplement of a given angle.

Review this example for Objective A:

1. Find the measure of a complement of an angle of 52°.

We subtract the measure of the given angle from 90° to find the measure of the complement.

$90° - 52° = 38°$

Practice this exercise:

1. Find the measure of a supplement of an angle of 74°.

Objective B Determine whether segments are congruent and whether angles are congruent.

Review this example for Objective B:

2. Determine if the pair of segments is congruent. Use a ruler.

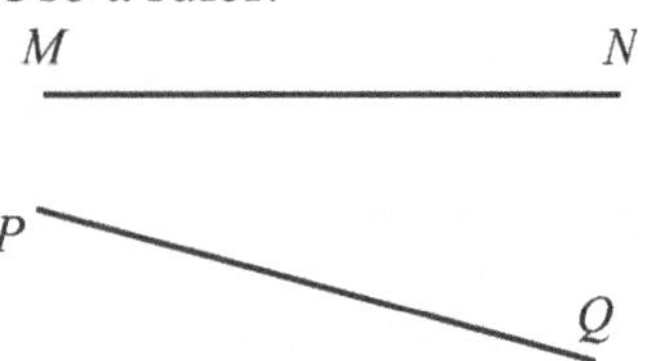

When we measure the segments with a ruler, we find that they have the same length. Thus, they are congruent.

Practice this exercise:

2. Determine if the angles are congruent. Use a protractor.

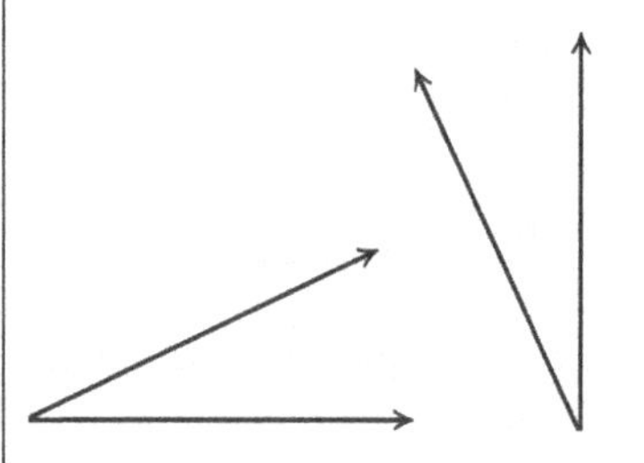

Objective C Use the Vertical Angle Property to find measures of angles.

Review this example for Objective C:

3. In the figure, $m\angle 2 = 60°$ and $m\angle 4 = 54°$. Find $m\angle 1$.

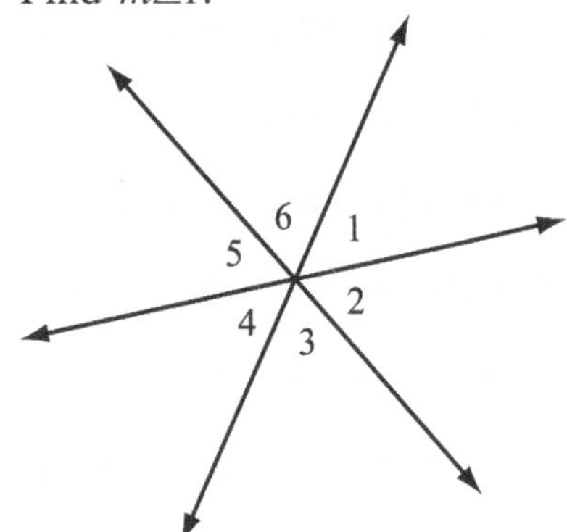

Since $\angle 1$ and $\angle 4$ are vertical angles, they have the same measure, so $m\angle 1 = 54°$.

Practice this exercise:

3. In the figure at the left, find $m\angle 3$.

Name: Date:
Instructor: Section:

Objective D Identify pairs of corresponding angles, interior angles, and alternate interior angles and apply properties of transversals and parallel lines to find measures of angles.

Review this example for Objective D:

4. If $p \parallel q$ and $m\angle 5 = 42°$, find $m\angle 6$, $m\angle 7$, and $m\angle 8$.

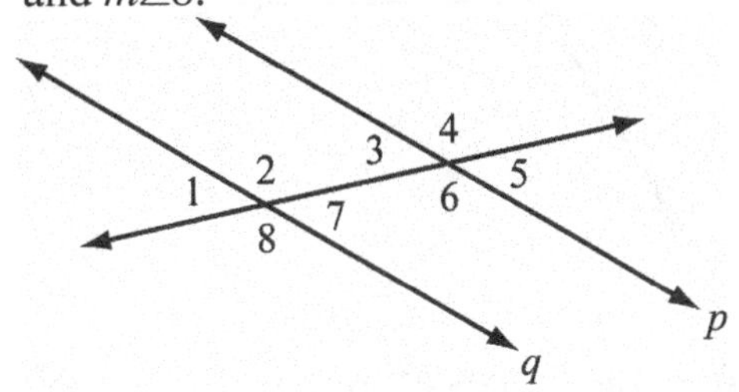

$m\angle 6 = 180° - 42° = 138°$
$m\angle 7 = 42°$ Corresponding angles with $\angle 5$
$m\angle 8 = 138°$ Corresponding angles with $\angle 6$

Practice this exercise:

4. Use the drawing at the left. If $p \parallel q$ and $m\angle 5 = 42°$, find the measures of $m\angle 1$, $m\angle 2$, $m\angle 3$, and $m\angle 4$.

ADDITIONAL EXERCISES

Objective A Identify complementary and supplementary angles and find the measure of a complement or a supplement of a given angle.

For extra help, see Examples 1–4 on pages 439–440 of your text and the Section 6.6 lecture video.

1. Find the complement of 49°.

2. Find the supplement of 24°.

3. Find the complement of 24°.

4. Find the supplement of 132°.

Objective B Determine whether segments are congruent and whether angles are congruent.

For extra help, see Examples 5–8 on page 441–442 of your text and the Section 6.6 lecture video.

Determine whether the pair of segments is congruent. Use a ruler.

5.

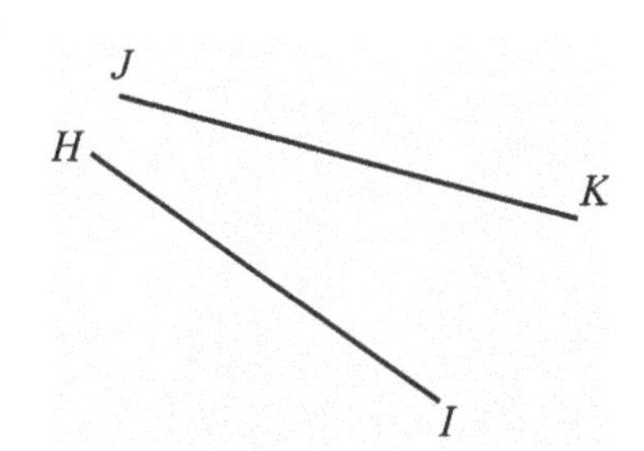

Determine whether the pair of angles is congruent. Use a protractor.

6.

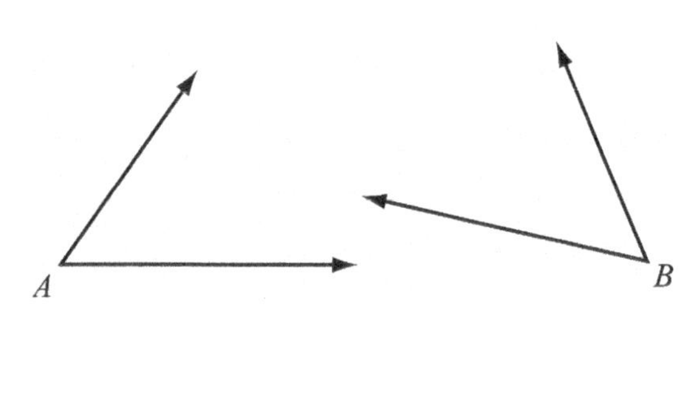

7.

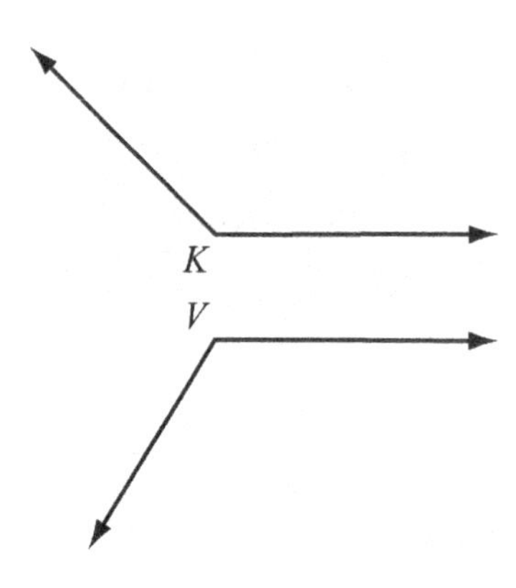

Objective C Use the Vertical Angle Property to find measures of angles.
For extra help, see Example 9 on page 443 of your text and the Section 6.6 lecture video.
In the figure, $m\angle 1 = 88°$ *and* $m\angle 5 = 32°$.

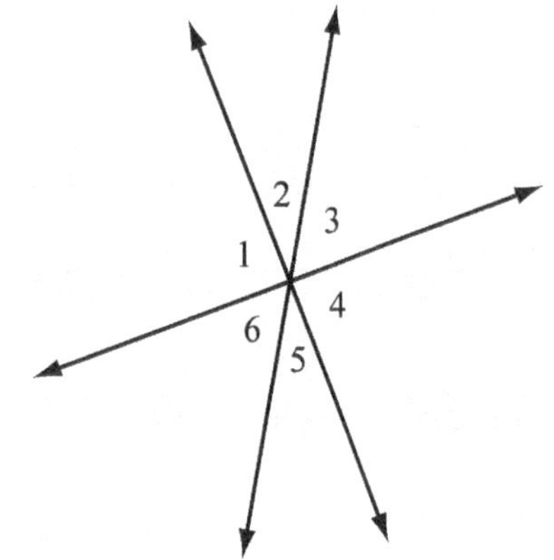

8. Find $m\angle 2$.

9. Find $m\angle 3$.

10. Find $m\angle 4$.

11. Find $m\angle 6$.

Name: Date:
Instructor: Section:

Objective D Identify pairs of corresponding angles, interior angles, and alternate interior angles and apply properties of transversals and parallel lines to find measures of angles.

For extra help, see Examples 10–13 on pages 444–446 of your text and the Section 6.6 lecture video.

12. Identify
a) all pairs of corresponding angles,
b) all interior angles,
c) all pairs of alternate interior angles.

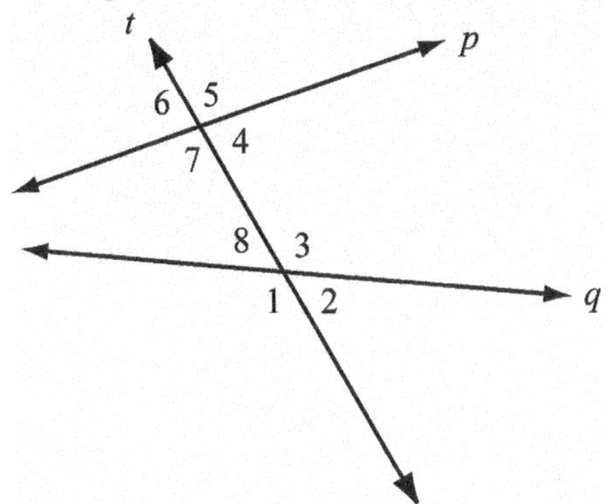

13. If $l \parallel n$ and $m\angle 4 = 113°$, what are the measures of the other angles?

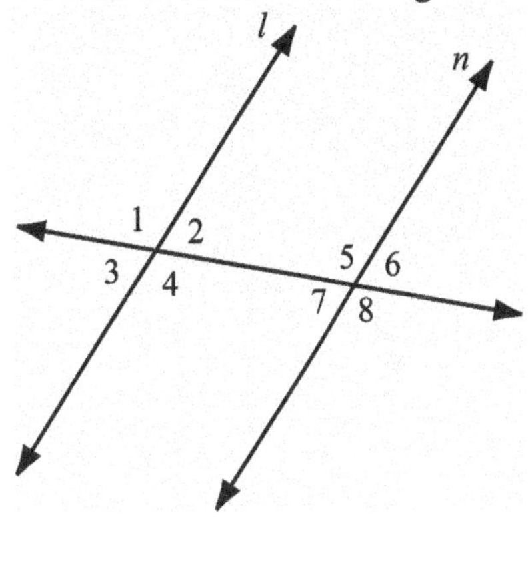

In each figure, $EF \parallel GH$. Identify pairs of congruent angles. Where possible, give the measure of the angles.

14.

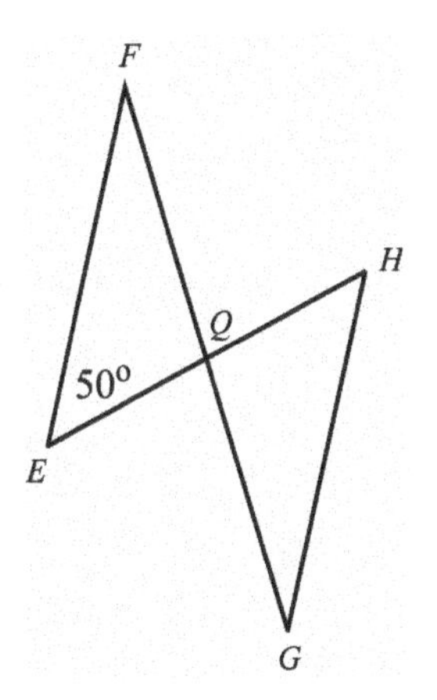

15.

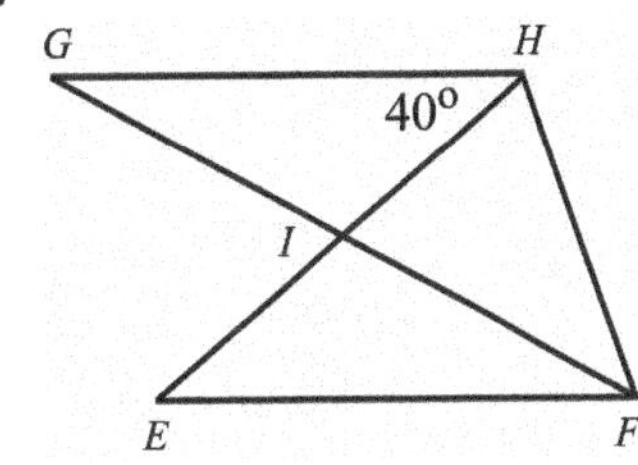

Name: Date:
Instructor: Section:

Chapter 6 GEOMETRY

6.7 Congruent Triangles and Properties of Parallelograms

Learning Objectives

A Identify the corresponding parts of congruent triangles and show why triangles are congruent using SAS, SSS, and ASA.

B Use properties of parallelograms to find lengths of sides and measures of angles of parallelograms.

Key Terms

Use the vocabulary terms listed below to complete each statement in Exercises 1–4. Terms may be used more than once.

diagonal **congruent** **180°** **360°**

1. A ________________ of a quadrilateral is a segment that joins two opposite vertices.

2. The sum of the measures of the angles in a triangle is ________________ .

3. The sum of the measures of the angles in a quadrilateral is ________________ .

4. Two triangles are ________________ if their corresponding angles are ________________ and their corresponding sides are ________________ .

GUIDED EXAMPLES AND PRACTICE

Objective A Identify the corresponding parts of congruent triangles and show why triangles are congruent using SAS, SSS, and ASA.

Review these examples for Objective A:

1. Suppose $\Delta MNP \cong \Delta RST$. What are the corresponding parts?

$\angle M \leftrightarrow \angle R$,
$\angle N \leftrightarrow \angle S$,
$\angle P \leftrightarrow \angle T$,
$\overline{MN} \leftrightarrow \overline{RS}$,
$\overline{MP} \leftrightarrow \overline{RT}$,
$\overline{NP} \leftrightarrow \overline{ST}$

Practice these exercises:

1. Suppose $\Delta WXY \cong \Delta DEF$. Which of the following is not true?

A. $\overline{XY} \cong \overline{DE}$
B. $\overline{WY} \cong \overline{DF}$
C. $\angle X \cong \angle E$
D. $\angle Y \cong \angle F$

2. Which property (if any) should be used to show that the pair of triangles is congruent?

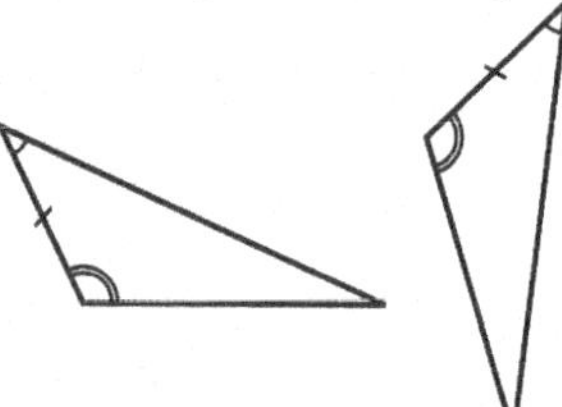

Two angles and the included side of one triangle are congruent to two angles and the included side of the other triangle, so they are congruent by the ASA Property.

2. Which property (if any) should be used to show that the pair of triangles is congruent?

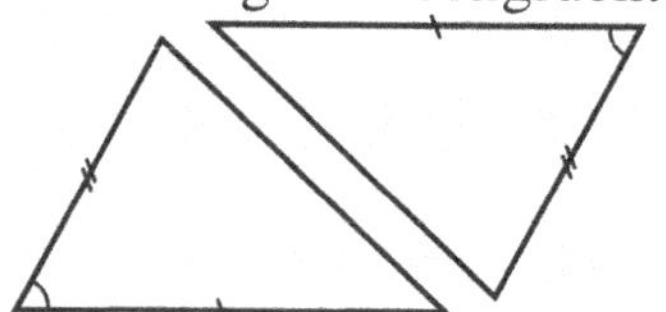

Objective B Use properties of parallelograms to find lengths of sides and measures of angles of parallelograms.

Review this example for Objective B:

3. For the parallelogram below, if $m\angle R = 46°$, $RS = 4$ and $UR = 8$, find $m\angle S$. Then find ST and UT.

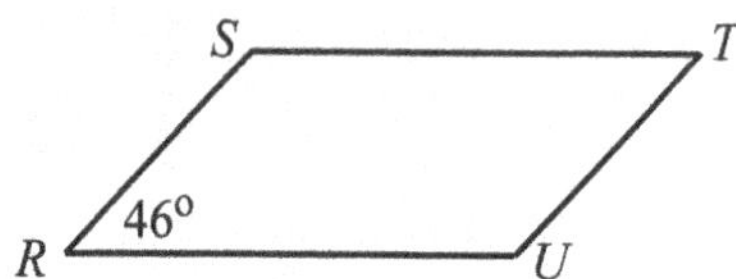

$m\angle S = 180° - 46° = 134°$
$ST = UR = 8$
$UT = RS = 4$

Practice this exercise:

3. For the parallelogram at left, find $m\angle T$ and $m\angle U$.

ADDITIONAL EXERCISES

Objective A Identify the corresponding parts of congruent triangles and show why triangles are congruent using SAS, SSS, and ASA.

For extra help, see Examples 1–15 on pages 449–454 of your text and the Section 6.7 lecture video.

Add, and if possible, simplify.

1. Name the corresponding parts of the congruent triangles. $\triangle NFS \cong \triangle ART$

Name: Date:
Instructor: Section:

Which property (if any) should be used to show that the pair of triangles is congruent?

2.

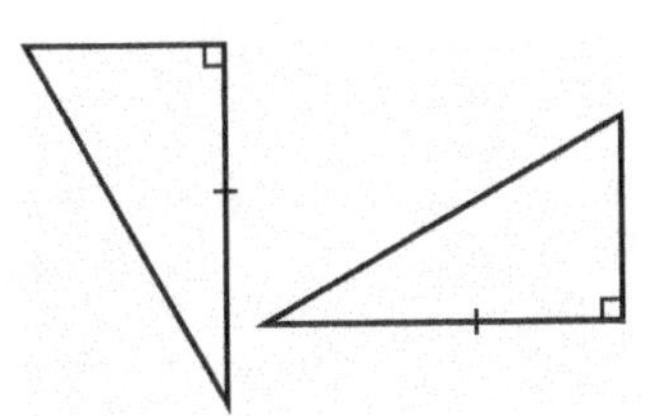

3.

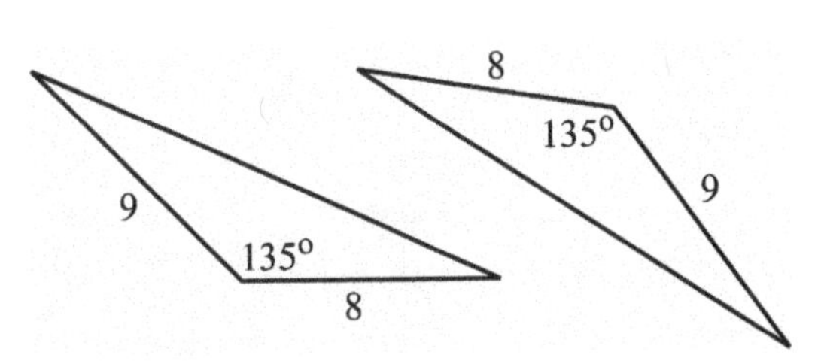

4.

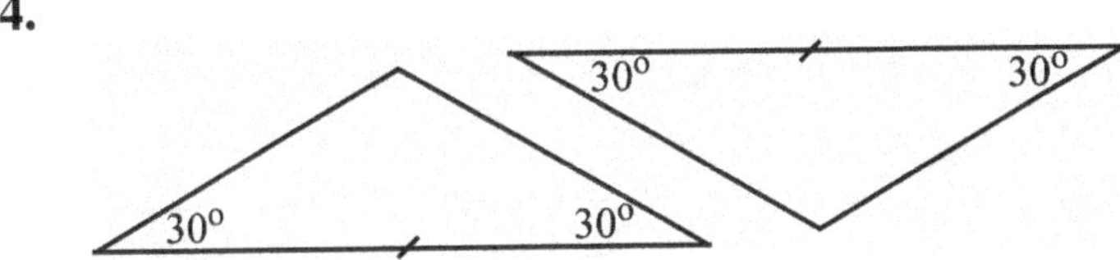

Objective B Use properties of parallelograms to find lengths of sides and measures of angles of parallelograms.
For extra help, see Examples 16–17 on page 455 of your text and the Section 6.7 lecture video.

5. Find the measures of the angles of the parallelogram.

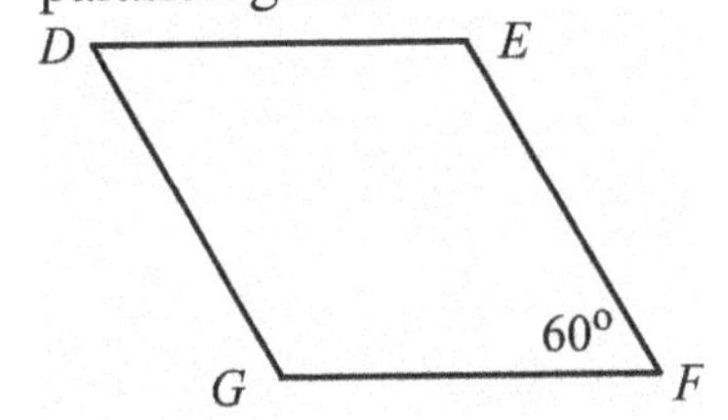

6. For the parallelogram below, if $DE = 7.5$ and $GD = 9.7$, find EF and FG. Then find the perimeter.

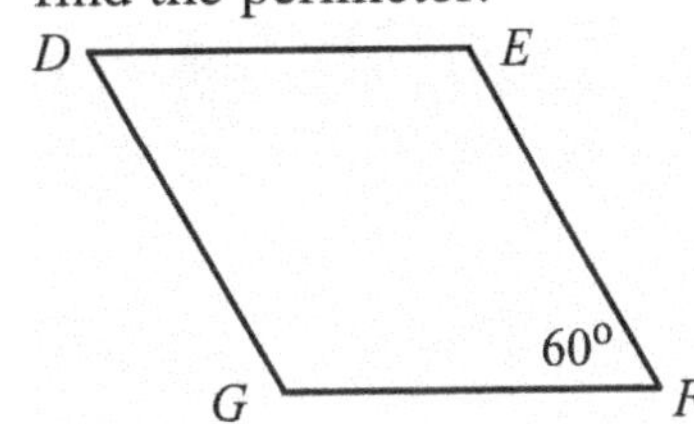

7. Find the lengths of the sides of the parallelogram.
The perimeter of $\square WXYZ$ is 30.

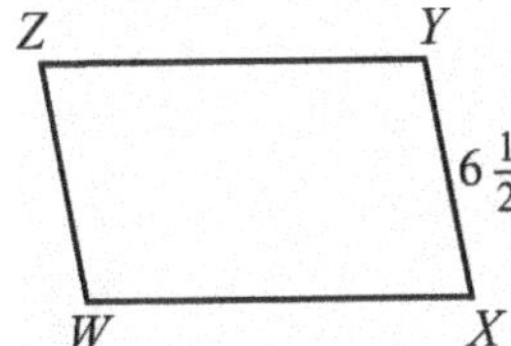

8. $EF = 12$ and $FG = 16$. Find the length of each diagonal.

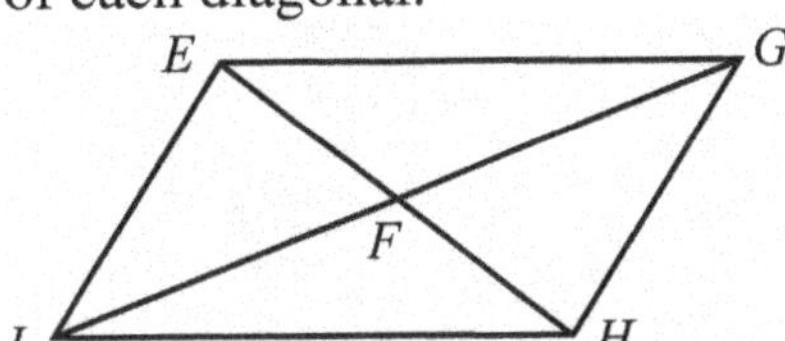

Name: Date:
Instructor: Section:

Chapter 6 GEOMETRY

6.8 Similar Triangles

Learning Objectives
A Identify the corresponding parts of similar triangles and determine which sides of a given pair of triangles have lengths that are proportional.
B Find lengths of sides of similar triangles using proportions.

Key Terms
Use the vocabulary terms listed below to complete each statement in Exercises 1–2. Terms may be used more than once.

proportional **congruent**

1. Two figures are similar if and only if their vertices can be matched so that corresponding angles are ____________________ and the lengths of corresponding sides are ____________________.

2. Two figures are congruent if and only if their vertices can be matched so that corresponding angles are ____________________ and the lengths of corresponding sides are ____________________.

GUIDED EXAMPLES AND PRACTICE

Objective A Identify the corresponding parts of similar triangles and determine which sides of a given pair of triangles have lengths that are proportional.

Review this example for Objective A:

1. ΔABC and ΔTSR are similar. Name their corresponding sides and angles and determine which sides are proportional.

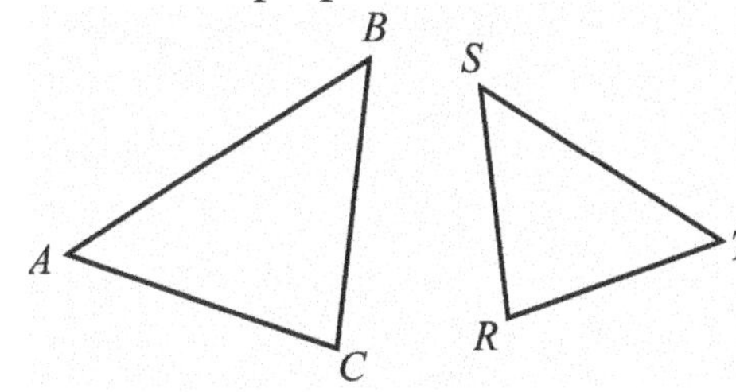

The corresponding vertices are A and T, B and S, and C and R. Then we have $\overline{AB} \leftrightarrow \overline{TS}$, $\overline{BC} \leftrightarrow \overline{SR}$, $\overline{CA} \leftrightarrow \overline{RT}$, $\angle A \leftrightarrow \angle T$, $\angle B \leftrightarrow \angle S$, $\angle C \leftrightarrow \angle R$

We also have $\frac{AB}{TS} = \frac{BC}{SR} = \frac{CA}{RT}$.

Practice this exercise:

1. For the similar triangles MNP and EDF below, determine which sides are proportional.

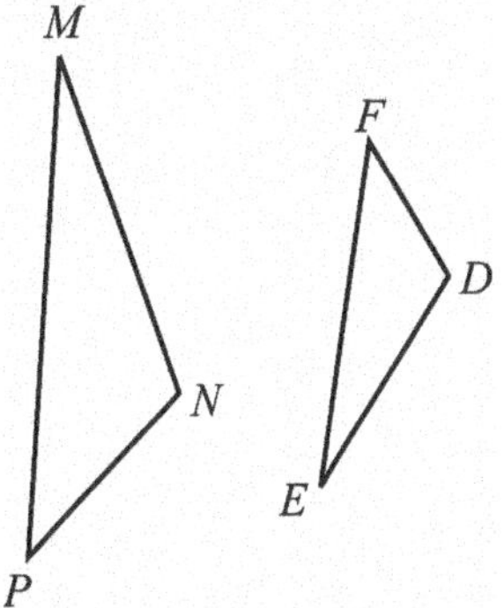

Objective B Find lengths of sides of similar triangles using proportions.

Review this example for Objective B:

2. The triangles below are similar. Find the missing length x.

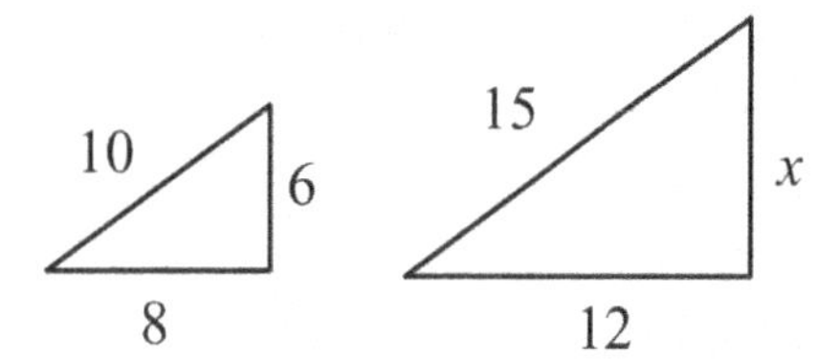

The ratio of 6 to x is the same as the ratio of 8 to 12 (and also as the ratio of 10 to 15). We write and solve a proportion.

$$\frac{6}{x}=\frac{8}{12}$$

$6\cdot 12 = x\cdot 8$ Equating cross products

$\frac{6\cdot 12}{8}=\frac{x\cdot 8}{8}$ Dividing both sides by 8

$9 = x$ Simplifying

The missing length is 9. (We could also have used the proportion $\frac{6}{x}=\frac{10}{15}$ to find x.)

Practice this exercise:

2. The triangles below are similar. Find the missing length y.

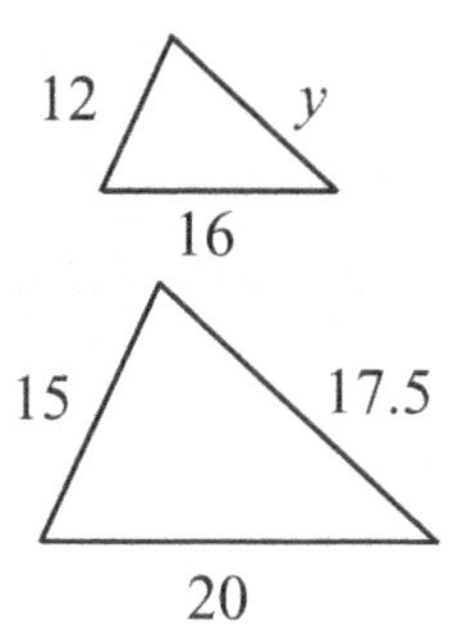

ADDITIONAL EXERCISES

Objective A Identify the corresponding parts of similar triangles and determine which sides of a given pair of triangles have lengths that are proportional.

For extra help, see Examples 1–4 on pages 461–462 of your text and the Section 6.8 lecture video.

For each pair of similar triangles, name the congruent angles and proportional sides.

1. $\Delta RST \sim \Delta UVW$

2. $\Delta KLS \sim \Delta BRN$

Name: Date:
Instructor: Section:

Name the proportional sides in these similar triangles.

3. Triangles XYZ and NQP.

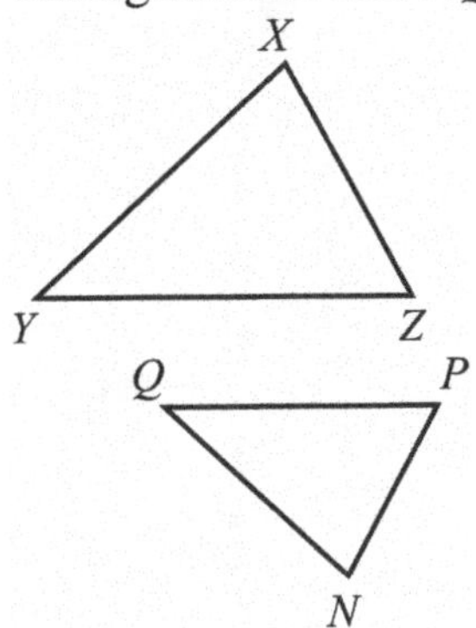

4. Triangles STU and KLJ.

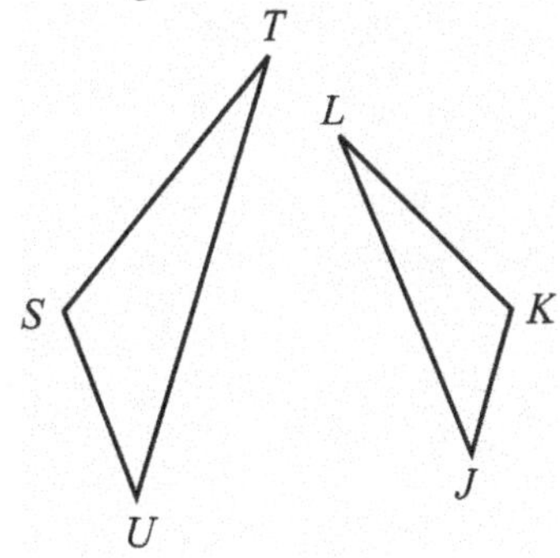

ADDITIONAL EXERCISES

Objective B Find lengths of sides of similar triangles using proportions.

For extra help, see Examples 5–8 on pages 463–465 of your text and the Section 6.8 lecture video.

5. The triangles are similar. Find the missing length, x.

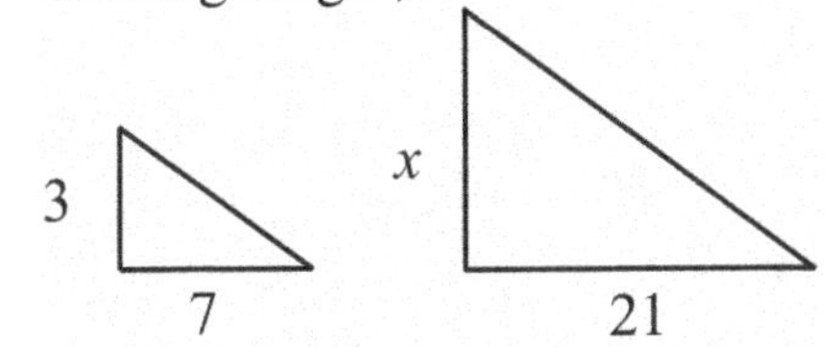

6. The triangles are similar. Find the missing lengths, x and y.

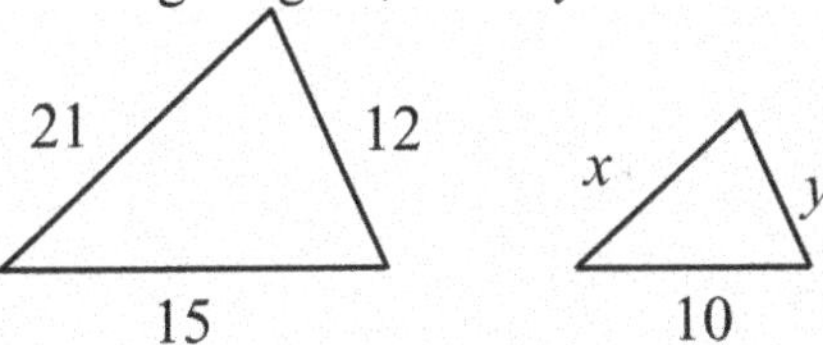

7. How tall is a tree that casts a 30-ft shadow at the same time that an 8-ft light pole casts a 15-ft shadow?

8. When a tree 15 ft tall casts a shadow 6 ft long, how long a shadow is cast by a child who is 4 ft tall?

Name: Date:
Instructor: Section:

Chapter 7 INTRODUCTION TO REAL NUMBERS AND ALGEBRAIC EXPRESSIONS

7.1 Introduction to Algebra

Learning Objectives
A Evaluate algebraic expressions by substitution.
B Translate phrases to algebraic expressions.

Key Terms
Use the vocabulary terms listed below to complete each statement in Exercises 1–6.

algebraic expression	**constant**	**evaluating**
substituting	**value**	**variable**

1. A combination of letters, numbers, and operation signs, such as $16x - 18y$ is called a(n) ____________________ .

2. A letter that can represent various numbers is a(n) ____________________ .

3. A letter than can stand for just one number is a(n) ____________________ .

4. When we replace a variable with a number, we are ____________________ for the variable.

5. When we replace all variables in an expression with numbers and carry out the operations, we are ____________________ the expression.

6. The result of evaluating an algebraic expression is called the ____________________ of the expression.

GUIDED EXAMPLES AND PRACTICE

Objective A Evaluate algebraic expressions by substitution.

Review this example for Objective A:

1. Evaluate $m - n$ for $m = 29$ and $n = 12$.

Substitute 29 for m and 12 for n and carry out the subtraction.

$m - n = 29 - 12 = 17$

Practice this exercise:

1. Evaluate $\frac{x}{y}$ for $x = 72$ and $y = 9$.

Objective B Translate phrases to algebraic expressions.

Review this example for Objective B:

2. Translate to an algebraic expression: 5 more than 7 times a number.

Let n = the number.
$7n + 5$

Practice this exercise:

2. Translate to an algebraic expression: 9 less the product of 6 and a number.

ADDITIONAL EXERCISES

Objective A Evaluate algebraic expressions by substitution.

For extra help, see Examples 1–6 on pages 486–488 of your text and the Section 7.1 lecture video.

Evaluate.

1. $10z$, when $z = 3$

2. $\frac{y}{x}$, when $x = 6$ and $y = 42$

3. $\frac{5p}{q}$, when $p = 8$ and $q = 10$

4. $\frac{a-b}{2}$, when $a = 25$ and $b = 13$

Objective B Translate phrases to algebraic expressions.

For extra help, see Examples 7–12 on pages 488–489 of your text and the Section 7.1 lecture video.

Translate to an algebraic expression. Choice of variables used may vary.

5. 18 less than w

6. The ratio of a number and 50

7. 40 increased by twice a number

8. One fifth of the sum of 3 and a number

9. The price of a treadmill after a 25% reduction if the price before the reduction was p

Name: Date:
Instructor: Section:

Chapter 7 INTRODUCTION TO REAL NUMBERS AND ALGEBRAIC EXPRESSIONS

7.2 The Real Numbers

Learning Objectives

A State the integer that corresponds to a real-world situation.
B Graph rational numbers on the number line.
C Convert from fraction notation for a rational number to decimal notation.
D Determine which of two real numbers is greater and indicate which, using < or >. Given an inequality like $a > b$, write another inequality with the same meaning. Determine whether an inequality like $-3 \leq 5$ is true or false.
E Find the absolute value of a real number.

Key Terms

Use the vocabulary terms listed below to complete each statement in Exercises 1–8.

absolute value	**graph**	**integers**	**natural numbers**
opposites	**set**	**whole numbers**	**rational numbers**

1. A(n) ____________________ is a collection of objects.

2. We call –1 and 1 ____________________ of each other.

3. To ____________________ a number means to find and mark its point on a number line.

4. The ____________________ of a number is its distance from zero on a number line.

5. The set of ____________________ = {1, 2, 3,...}.

6. The set of ____________________ = {0, 1, 2, 3,...}.

7. The set of ____________________ = {..., –4,–3,–2,–1, 0, 1, 2, 3, 4,...}.

8. The set of ____________________ = the set of numbers $\frac{a}{b}$, where a and b are integers and $b \neq 0$.

GUIDED EXAMPLES AND PRACTICE

Objective A State the integer that corresponds to a real-world situation.

Review this example for Objective A:

1. Tell what integers correspond to this situation: A student has \$106 in his checking account. The student owes \$248 on his credit card.

The integer 106 corresponds to having \$106 in a checking account. The integer –248 corresponds to a \$248 credit card debt.

Practice this exercise:

1. Tell which integer corresponds to this situation: A business lost \$1200 during a 30-day period.

Objective B Graph rational numbers on the number line.

Review this example for Objective B:

2. Graph –1.5.

The graph of –1.5 is halfway between –2 and –1.

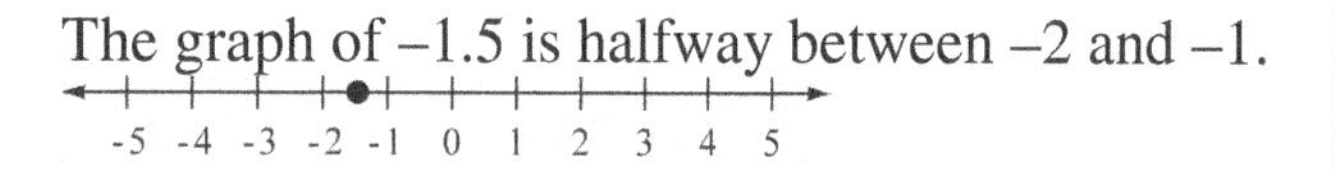

Practice this exercise:

2. Graph –2.1.

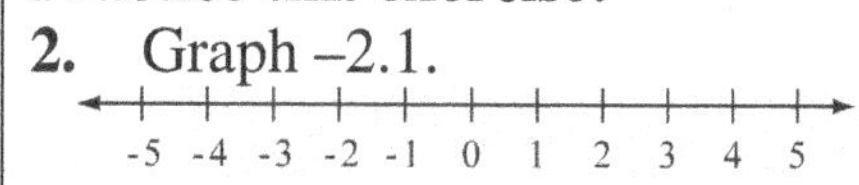

Objective C Convert from fraction notation for a rational number to decimal notation.

Review this example for Objective C:

3. Find decimal notation for $-\frac{7}{4}$.

We first find decimal notation for $\frac{7}{4}$. Since $\frac{7}{4}$ means $7 \div 4$, we divide.

$$\begin{array}{r} 1.75 \\ 4\overline{)7.00} \\ \underline{4} \\ 30 \\ \underline{28} \\ 20 \\ \underline{20} \\ 0 \end{array}$$

Thus $\frac{7}{4} = 1.75$, so $-\frac{7}{4} = -1.75$.

Practice this exercise:

3. Find decimal notation for $-\frac{1}{8}$.

Name: Date:
Instructor: Section:

Objective D Determine which of two real numbers is greater and indicate which, using < or >. Given an inequality like $a > b$, write another inequality with the same meaning. Determine whether an inequality like $-3 \leq 5$ is true or false.

Review these examples for Objective D:

4. Use either < or > for $\square$ to form a true sentence.

$-7 \square -10$

Since -7 is to the right of -10 on the number line, we have $-7 > -10$.

5. Write another inequality with the same meaning as $x > 8$.

The inequality $8 < x$ has the same meaning.

6. Determine whether each inequality is true or false.
a) $-4 \leq 1$ b) $6 \geq 6$ c) $-10 \geq 2$

a) $-4 \leq 1$ is true since $-4 < 1$ is true.
b) $6 \geq 6$ is true since $6 = 6$ is true.
c) $-10 \geq 2$ is false since neither $-10 > 2$ nor $-10 = 2$ is true.

Practice these exercises:

4. Use either < or > for $\square$ to form a true sentence.

$-8 \square 1$

5. Write another inequality with the same meaning as $-3 < t$.

6. Determine whether the inequality $-1 \geq -8$ is true or false.

Objective E Find the absolute value of a real number.

Review this example for Objective E:

7. Find $|-4.3|$.

The distance from -4.3 to 0 is 4.3, so $|-4.3| = 4.3$.

Practice this exercise:

7. Find $|59|$.

ADDITIONAL EXERCISES

Objective A State the integer that corresponds to a real-world situation.

For extra help, see Examples 1–4 on pages 494–495 of your text and the Section 7.2 lecture video.

State the integers that correspond to the situation.

1. On March 12, the temperature was 3° below zero. On March 21, it was 55° above zero.

2. Will withdrew \$480 from his savings account to buy textbooks. The next day, he deposited his paycheck of \$325.

3. April deposited her refund of \$525 and paid a fine of \$426 for running a red light.

Objective B Graph rational numbers on the number line.
For extra help, see Examples 5–7 on page 496 of your text and the Section 7.2 lecture video.
Graph the number on the number line.

4. -4.2

-5 -4 -3 -2 -1 0 1 2 3 4 5

5. 2.67

-5 -4 -3 -2 -1 0 1 2 3 4 5

6. $-\frac{7}{4}$

-5 -4 -3 -2 -1 0 1 2 3 4 5

Objective C Convert from fraction notation for a rational number to decimal notation.
For extra help, see Examples 8–9 on pages 496–497 of your text and the Section 7.2 lecture video.
Convert to decimal notation.

7. $-\frac{3}{8}$

8. $-\frac{8}{9}$

9. $-\frac{5}{4}$

10. $\frac{3}{20}$

Objective D Determine which of two real numbers is greater and indicate which, using < or >. Given an inequality like $a > b$, write another inequality with the same meaning. Determine whether an inequality like $-3 \leq 5$ is true or false.
For extra help, see Examples 10–27 on pages 499–500 of your text and the Section 7.2 lecture video.
Use < or > for $\square$ *to write a true sentence.*

11. $0 \square -6$

12. $-4.6 \square -1.2$

13. $\frac{2}{5} \square \frac{3}{8}$

14. Determine whether the statement $-3 \leq -4$ is true or false.

15. Write an inequality with the same meaning as $-3 \geq x$.

Objective E Find the absolute value of a real number.
For extra help, see Examples 28–32 on page 501 of your text and the Section 7.2 lecture video.
Find the absolute value.

16. $|-16|$

17. $|12|$

18. $\left|-\frac{1}{4}\right|$

19. $\left|\frac{0}{5}\right|$

Name: Date:
Instructor: Section:

Chapter 7 INTRODUCTION TO REAL NUMBERS AND ALGEBRAIC EXPRESSIONS

7.3 Addition of Real Numbers

Learning Objectives
A Add real numbers without using the number line.
B Find the opposite, or additive inverse, of a real number.
C Solve applied problems involving addition of real numbers.

Key Terms
Use the vocabulary terms listed below to complete each statement in Exercises 1–4.

additive inverses **negative** **positive** **zero**

1. The sum of two positive real numbers is ________________ .
2. The sum of two negative real numbers is ________________ .
3. If $a + b = a$, then b must be ________________ .
4. Two numbers whose sum is 0 are called ________________ .

GUIDED EXAMPLES AND PRACTICE

Objective A Add real numbers without using the number line.

Review this example for Objective A:

1. Add without using a number line: –15 + 9.

   We have a negative and a positive number. The absolute values are 15 and 9. The difference is 6. The negative number has the larger absolute value, so the answer is negative.
   $-15 + 9 = -6$

Practice this exercise:

1. Add without using a number line: –1.2 + (–3.4).

Objective B Find the opposite, or additive inverse, of a real number.

Review this example for Objective B:

2. Find the opposite of $\frac{5}{3}$.

   The opposite of $\frac{5}{3}$ is $-\frac{5}{3}$ because $\frac{5}{3}+\left(-\frac{5}{3}\right)=0$.

Practice this exercise:

2. Find the opposite of –20.

Objective C Solve applied problems involving addition of real numbers.

Review this example for Objective C:

3. The temperature in Monroe was 46° at 7AM and it rose 18° by noon. What was the temperature at noon?

Let T = the temperature at noon. Then we add 18° to 46° to find T.

$T = 46° + 18° = 64°$

The temperature was 64° at noon.

Practice this exercise:

3. Corey has $278 in his checking account. He writes a check for $154 to pay for a textbook. What is the balance in his checking account?

ADDITIONAL EXERCISES

Objective A Add real numbers without using the number line.

For extra help, see Examples 5–13 on pages 506–507 of your text and the Section 7.3 lecture video.

Add. Do not use the number line except as a check.

1. $27+(-27)$

2. $-3+12$

3. $-1.3+(-3.8)$

4. $-\frac{3}{20}+\frac{7}{15}$

5. $36+(-45)+657+(-36)+(-657)$

Objective B Find the opposite, or additive inverse, of a real number.

For extra help, see Examples 14–23 on pages 508–509 of your text and the Section 7.3 lecture video.

6. Find the opposite, or additive inverse of 16.3.

7. Evaluate $-x$ then $x=-54$.

8. Evaluate $-(-x)$ when $x=-19$.

9. Find the opposite of $-\frac{7}{3}$.

Objective C Solve applied problems involving addition of real numbers.

For extra help, see Example 24 on page 509 of your text and the Section 7.3 lecture video.

Solve.

10. In a football game, the home team lost 5 yd on the first play and gained 8 yd on the second play. Find the total gain or loss.

11. Joe plays golf. For the first hole, he scores 1 under par. At the second hole, he is 2 over par. For the third hole, he is 2 under par. What is his current score?

12. On December 1, LaKenya's credit card bill shows that she owes $125. During the month of December, LaKenya sends a check for $95 to the credit card company, charges another $860 for gifts, and then pays off another $500 of her bill. What is the new balance of LaKenya's account at the end of December?

Name: Date:
Instructor: Section:

Chapter 7 INTRODUCTION TO REAL NUMBERS AND ALGEBRAIC EXPRESSIONS

7.4 Subtraction of Real Numbers

Learning Objectives
A Subtract real numbers and simplify combinations of additions and subtractions.
B Solve applied problems involving subtraction of real numbers.

Key Terms
Use the vocabulary terms listed below to complete each statement in Exercises 1–2.

difference **opposite**

1. The ____________________ $a - b$ is the number c for which $a = b + c$.

2. To subtract, add the ____________________ of the number being subtracted.

GUIDED EXAMPLES AND PRACTICE

Objective A Subtract real numbers and simplify combinations of additions and subtractions.

Review these examples for Objective A:

1. Subtract $6-(-7)$.

The opposite of –7 is 7. We change the subtraction to addition and add the opposite.

$$6-(-7)=6+7=13$$

2. Simplify: $5-(-1)-3+7$.

$$5-(-1)-3+7=5+1+(-3)+7$$
$$=10$$

Practice these exercises:

1. Subtract 2 – 12.

2. Simplify: $-8-4+12-(-9)$.

Objective B Solve applied problems involving subtraction of real numbers.

Review this example for Objective B:

3. The temperature in Milton ranged from a high of 22° to a low of –5° one day. Find the difference between the high temperature and the low temperature.

Let D = the difference in temperatures. We subtract the low temperature from the high temperature to find D.

$$D = 22° - (-5)° = 22° + 5° = 27°$$

The difference between the high and the low temperatures was 27°.

Practice this exercise:

3. The highest temperature ever recorded in Bogstown was 105°. The lowest recorded temperature was –11°. How much higher was the highest temperature than the lowest?

ADDITIONAL EXERCISES

Objective A Subtract real numbers and simplify combinations of additions and subtractions.

For extra help, see Examples 1–12 on pages 513–515 of your text and the Section 7.4 lecture video.

Subtract.

1. $18-(-40)$

2. $-2-(-17)$

3. $8-4.63$

4. $13-20$

5. Simplify.
$-1.7-(-15.7)+(-12.3)-(-16)$

Objective B Solve applied problems involving subtraction of real numbers.

For extra help, see Example 13 on page 515 of your text and the Section 7.4 lecture video.

Solve.

6. From an elevation of 38 ft below sea level, Devin climbed to an elevation of 92 ft above sea level. How much higher was Devin at the end of his climb than at the beginning?

7. On January 10, the temperature fell from 12 °F to –8°F. By how much did the temperature drop?

8. Skylar has $385 in a checking account. He writes a check for $425 to pay for Christmas gifts. What is the balance in his checking account?

Name: Date:
Instructor: Section:

Chapter 7 INTRODUCTION TO REAL NUMBERS AND ALGEBRAIC EXPRESSIONS

7.5 Multiplication of Real Numbers

Learning Objectives
A Multiply real numbers.
B Solve applied problems involving multiplication of real numbers.

Key Terms
Use the vocabulary terms listed below to complete each statement in Exercises 1–4. Each term will be used more than once.

negative **positive**

1. The product of a positive number and a negative number is ________________ .
2. The product of two negative numbers is ________________ .
3. The product of an even number of negative numbers is ________________ .
4. The product of an odd number of negative numbers is ________________ .

GUIDED EXAMPLES AND PRACTICE

Objective A Multiply real numbers.

Review this example for Objective A:

1. Multiply: $-2.4(3)$.

   The signs are different, so the answer is negative.

   $-2.4(3) = -7.2$

Practice this exercise:

1. Multiply: $-7(-9)$.

Objective B Solve applied problems involving multiplication of real numbers.

Review this example for Objective B:

2. Emily lost 1.5 lb each week for a period of 6 weeks. Express her total weight change as an integer.

   Emily's weight changes –1.5 lb each week, so the total weight loss l is given by

   $l = 6(-1.5) = -9$.

   The total weight change is –9 lb.

Practice this exercise:

2. Ivan owns a stock that fell in price $2 each day for a period of 6 days. Express the total loss as an integer.

ADDITIONAL EXERCISES

Objective A Multiply real numbers.

For extra help, see Examples 1–18 on pages 522–525 of your text and the Section 7.5 lecture video.

Multiply.

1. $6\cdot(-5)$

2. $-84\cdot(-4.5)$

3. $-\frac{4}{5}\cdot\left(\frac{7}{8}\right)$

4. $(-2)(-3)(-4)(-5)(-6)$

Objective B Solve applied problems involving multiplication of real numbers.

For extra help, see Example 19 page 525 of your text and the Section 7.5 lecture video.

Solve.

5. Vivian lost 1.5 lb each week for a period of 8 weeks. Express her total weight change as an integer.

6. The temperature of a chemical compound was –10°C at 4:30 PM. During a reaction, it dropped 2°C per minute until 4:38 PM. What was the temperature at 4:38 PM?

7. The price of a stock began the day at $42.75 per share and dropped $0.38 per hour for 8 hours. What was the price of the stock after 8 hr?

8. After diving 55 m below sea level, a diver rises at a rate of 6 m per minute for 8 min. Where is the diver in relation to the surface?

Name: Date:
Instructor: Section:

Chapter 7 INTRODUCTION TO REAL NUMBERS AND ALGEBRAIC EXPRESSIONS

7.6 Division of Real Numbers

Learning Objectives
A Divide integers.
B Find the reciprocal of a real number.
C Divide real numbers.
D Solve applied problems involving division of real numbers.

Key Terms
Use the vocabulary terms listed below to complete each statement in Exercises 1–4.

negative **not defined** **positive** **reciprocals**

1. Division by 0 is ________________ .

2. The quotient of two numbers with the same sign is ________________ .

3. The quotient of two numbers with different signs is ________________ .

4. Two numbers whose product is 1 are called ________________ .

GUIDED EXAMPLES AND PRACTICE

Objective A Divide integers.

Review this example for Objective A:

1. Divide: $-36 \div (-4)$.

The signs are the same, so the answer is positive.
$-36 \div (-4) = 9$

Practice this exercise:

1. Divide: $\frac{56}{-8}$.

Objective B Find the reciprocal of a real number.

Review this example for Objective B:

2. Find the reciprocal of $-\frac{4}{5}$.

The reciprocal of $-\frac{4}{5}$ is $-\frac{5}{4}$, because
$-\frac{4}{5}\left(-\frac{5}{4}\right) = 1$.

Practice this exercise:

2. Find the reciprocal of 2.

Objective C Divide real numbers.

Review this example for Objective C:

3. Divide: $-\frac{1}{3}\div\frac{2}{7}$.

To divide, we multiply by the reciprocal of the divisor.

$-\frac{1}{3}\div\frac{2}{7}=-\frac{1}{3}\cdot\frac{7}{2}=-\frac{7}{6}$

Practice this exercise:

3. Divide: $-\frac{3}{4}\div\left(-\frac{5}{11}\right)$.

Objective D Solve applied problems involving division of real numbers.

Review this example for Objective D:

4. During a chemical reaction, the temperature of a compound fell by the same number of degrees every minute. The temperature was 42°F at 8:05 AM. By 8:30 AM it had dropped to –3°F. What was the temperature change each minute?

We first find the number of degrees, d, the temperature changed altogether.

$d=42-(-3)=42+3=45$

The temperature changed 45°. Since the temperature dropped, we can express this as –45°.

The amount of time, t, that passed was $t=30-5$, or 25 min.

We divide to find the number of degrees, T, the temperature dropped each minute:

$T=\frac{d}{t}=\frac{-45}{25}=-1.8$

The temperature changed –1.8°F each minute.

Practice this exercise:

4. The price of a stock fell the same amount each day for 7 days. The price on the first day was \$98 and the price on the seventh day was \$80.50. What was the change in price each day?

ADDITIONAL EXERCISES

Objective A Divide integers.

For extra help, see Examples 1–7 on pages 529–530 of your text and the Section 7.6 lecture video.

Divide, if possible. Check each answer..

1. $\frac{26}{-2}$

2. $-66\div(-6)$

3. $\frac{66}{0}$

4. $\frac{0}{-3}$

Name: Date:
Instructor: Section:

Objective B Find the reciprocal of a real number.

For extra help, see Examples 8–13 on page 530 of your text and the Section 7.6 lecture video.

Find the reciprocal.

5. $\frac{4}{9}$

6. -12

7. $\frac{-1}{3t}$

8. $\frac{-2x}{3y}$

Objective C Divide real numbers.

For extra help, see Examples 14–23 on pages 532–533 of your text and the Section 7.6 lecture video.

Rewrite each division as a multiplication.

9. $-\frac{6.8}{1.3}$

10. $\frac{2x-1}{4}$

Divide.

11. $-\frac{3}{8}\div\left(-\frac{6}{7}\right)$

12. $-10.4\div 2.6$

Objective D   Solve applied problems involving division of real numbers.
For extra help, see Example 24 on page 534 of your text and the Section 7.6 lecture video.
A percent of increase is generally positive and a percent of decrease is generally negative. Use the information to find the percent increase or decrease. Round to the nearest tenth of a percent.

13.

2000	2010	Change	Percent of Increase or Decrease
59	98	39	

14.

2000	2010	Change	Percent of Increase or Decrease
135	110	−25	

15.

2000	2010	Change	Percent of Increase or Decrease
83	66	−17	

Name: Date:
Instructor: Section:

Chapter 7 INTRODUCTION TO REAL NUMBERS AND ALGEBRAIC EXPRESSIONS

7.7 Properties of Real Numbers

Learning Objectives

A Find equivalent fraction expressions and simplify fraction expressions.
B Use the commutative and associative laws to find equivalent expressions.
C Use the distributive laws to multiply expressions like 8 and $x - y$.
D Use the distributive laws to factor expressions like $4x - 12 + 24y$.
E Collect like terms.

Key Terms

Use the vocabulary terms listed below to complete each statement in Exercises 1–7.

associative	**commutative**	**distributive**	**terms**
equivalent	**factor**	**like**	

1. Two expressions that have the same value for all allowable replacements are called ____________________ expressions.
2. The statement $3 + x = x + 3$ illustrates a(n) ____________________ law.
3. The statement $(2p)q = 2(pq)$ illustrates a(n) ____________________ law.
4. The statement $5(a+b) = 5a + 5b$ illustrates a(n) ____________________ law.
5. The ____________________ of an expression are separated by addition signs.
6. To ____________________ an expression, we find an equivalent expression that is a product.
7. ____________________ terms have exactly the same variable factors.

GUIDED EXAMPLES AND PRACTICE

Objective A Find equivalent fraction expressions and simplify fraction expressions.

Review these examples for Objective A:

1. Write a fractional expression equivalent to $\frac{2}{5}$ with a denominator of $5y$.

   Note that $5y = 5 \cdot y$. The denominator, 5, is missing a factor of y. Thus we multiply by 1 using y/y.

   $$\frac{2}{5} = \frac{2}{5} \cdot 1 = \frac{2}{5} \cdot \frac{y}{y} = \frac{2y}{5y}$$

Practice these exercises:

1. Write a fractional expression equivalent to $\frac{3}{7}$ with a denominator of $7t$.

2. Simplify: $-\frac{24y}{15y}$

$$-\frac{24y}{15y}=-\frac{8\cdot 3y}{5\cdot 3y}=-\frac{8}{5}\cdot\frac{3y}{3y}=-\frac{8}{5}\cdot 1=-\frac{8}{5}$$

2. Simplify: $\frac{27x}{36x}$

Objective B Use the commutative and associative laws to find equivalent expressions.

Review these examples for Objective B:

3. Use a commutative law to write an equivalent expression.

a) $n+6$ b) xy

a) An equivalent expression is $6+n$, by the commutative law of addition.

b) An equivalent expression is yx, by the commutative law of multiplication.

4. Use an associative law to write an equivalent expression.

a) $(m+n)+1$ b) $5(st)$

a) An equivalent expression is $m+(n+1)$, by the associative law of addition.

b) An equivalent expression is $(5s)t$, by the associative law of multiplication.

Practice these exercises:

3. Use a commutative law to write an equivalent expression for $8+a$.

4. Use an associative law to write an equivalent expression for $(4x)y$.

Objective C Use the distributive laws to multiply expressions like 8 and $x - y$.

Review this example for Objective C:

5. Multiply: $5(2x-3y+z)$.

$$\begin{aligned}5(2x-3y+z)&=5\cdot 2x-5\cdot 3y+5\cdot z\\&=10x-15y+5z\end{aligned}$$

Practice this exercise:

5. Multiply: $3(x+4y-2z)$.

Objective D Use the distributive laws to factor expressions like $4x - 12 + 24y$.

Review this example for Objective D:

6. Factor: $8a+4b-12c$.

$$\begin{aligned}8a+4b-12c&=4\cdot 2a+4\cdot b-4\cdot 3c\\&=4(2a+b-3c)\end{aligned}$$

Practice this exercise:

6. Factor: $36m-27n+9p$.

Name: Date:
Instructor: Section:

Objective E Collect like terms.

Review this example for Objective E:

7. Collect like terms: $3x-5y+8x+y$.

$$\begin{aligned}3x-5y+8x+y &= 3x+8x-5y+y\\ &= 3x+8x-5y+1\cdot y\\ &= (3+8)x+(-5+1)y\\ &= 11x-4y\end{aligned}$$

Practice this exercise:

7. Collect like terms: $6a-4b-a+2b$.

ADDITIONAL EXERCISES

Objective A Find equivalent fraction expressions and simplify fraction expressions.

For extra help, see Examples 1–3 on pages 538–539 of your text and the Section 7.7 lecture video.

Find an equivalent expression with the given denominator.

1. $\frac{2}{7}=\frac{\square}{14x}$

2. $\frac{3}{10y}=\frac{\square}{10xy}$

Simplify.

3. $-\frac{35xy}{25xy}$

4. $\frac{48ab}{40a}$

Objective B Use the commutative and associative laws to find equivalent expressions.

For extra help, see Examples 4–11 on pages 539–542 of your text and the Section 7.7 lecture video.

Write an equivalent expression. Use a commutative law.

5. Write an equivalent expression for $18+xy$. Use a commutative law.

6. Write an equivalent expression for $3+(x+y)$. Use an associative law.

Use the commutative and associative laws to write three equivalent expressions.

7. $4+(m+n)$

8. $(xy)9$

Objective C Use the distributive laws to multiply expressions like 8 and $x - y$.
For extra help, see Examples 12–24 on pages 542–544 of your text and the Section 7.7 lecture video.
Multiply.

9. $3(2x+8)$

10. $\frac{3}{4}(y-12)$

11. $-4(-6x-y+9)$

12. $-1.4(-2.3a-1.6b+3.4)$

13. List the terms of the expression:
$2a-1.4b+10c$

Objective D Use the distributive laws to factor expressions like $4x - 12 + 24y$.
For extra help, see Examples 25–31 on page 545 of your text and the Section 7.7 lecture video.
Factor. Check by multiplying.

14. $9m-45$

15. $-5g+20$

16. $-21c+35d+7$

17. $\frac{3}{4}x-\frac{7}{4}y+\frac{1}{4}$

Objective E Collect like terms.
For extra help, see Examples 32–38 on page 546 of your text and the Section 7.7 lecture video.
Collect like terms.

18. $14x-x$

19. $8g-11g$

20. $16-40c-20-5c+6+8c$

21. $-4+8a-5b+36a-17b+12$

Name: Date:
Instructor: Section:

Chapter 7 INTRODUCTION TO REAL NUMBERS AND ALGEBRAIC EXPRESSIONS

7.8 Simplifying Expressions; Order of Operations

Learning Objectives

A Find an equivalent expression for an opposite without parentheses, where an expression has several terms.
B Simplify expressions by removing parentheses and collecting like terms.
C Simplify expressions with parentheses inside parentheses.
D Simplify expressions using the rules for order of operations.

Key Terms

Exercises 1–4 list the rules for order of operations. Use the vocabulary terms listed below to complete each statement.

additions **divisions** **exponential** **grouping**

1. Do all calculations within ________________ symbols before operations outside.
2. Evaluate all ________________ expressions.
3. Do all multiplications and ________________ in order from left to right.
4. Do all ________________ and subtractions in order from left to right.

GUIDED EXAMPLES AND PRACTICE

Objective A Find an equivalent expression for an opposite without parentheses, where an expression has several terms.

Review this example for Objective A:

1. Find an equivalent expression without parentheses for $-(2x-3y+7z)$.

   We change the sign of each term.
   $-(2x-3y+7z)=-2x+3y-7z$

Practice this exercise:

1. Find an equivalent expression without parentheses for $-(-3a+6b-c)$.

Objective B Simplify expressions by removing parentheses and collecting like terms.

Review this example for Objective B:

2. Remove parentheses and simplify: $6x-2(x-3y)$.

   $6x-2(x-3y)=6x-2x+6y=4x+6y$

Practice this exercise:

2. Remove parentheses and simplify: $3m-n-(2m+5n)$.

Objective C **Simplify expressions with parentheses inside parentheses.**

Review this example for Objective C:

3. Simplify: $3[5-(8-4)]$.

$$\begin{aligned}3[5-(8-4)] &= 3[5-4]\\ &= 3[1]\\ &= 3\end{aligned}$$

Practice this exercise:

3. Simplify: $3[-16\div(4\cdot 2)]$.

Objective D **Simplify expressions using the rules for order of operations.**

Review this example for Objective D:

4. Simplify: $10-(6-4\cdot 5)$.

$$\begin{aligned}10-(6-4\cdot 5) &= 10-(6-20)\\ &= 10-(-14)\\ &= 10+14\\ &= 24\end{aligned}$$

Practice this exercise:

4. Simplify:
$100\div(-25)+12\div 3$.

ADDITIONAL EXERCISES

Objective A **Find an equivalent expression for an opposite without parentheses, where an expression has several terms.**

For extra help, see Examples 1–5 on pages 551–552 of your text and the Section 7.8 lecture video.

Find an equivalent expression without parentheses.

1. $-(3x+y-10z)$

2. $-(a-4b+6c)$

3. $-(-10m-3n-18)$

4. $-(-2p+5q-8t)$

Objective B **Simplify expressions by removing parentheses and collecting like terms.**

For extra help, see Examples 6–14 on pages 552–553 of your text and the Section 7.8 lecture video.

Remove parentheses and simplify.

5. $5y+2y-(3y+1)$

6. $4p-5q-2(3p-9q)$

7. $12m-n-3(5m-2n+6p)$

8. $(15x+7y-8z)-4(-4x-2y+10z)$

Name: Date:
Instructor: Section:

Objective C Simplify expressions with parentheses inside parentheses.
For extra help, see Examples 15–19 on pages 553–554 of your text and the Section 7.8 lecture video.
Simplify.

9. $6[3-2(7-4)]$

10. $[5(11-2)+7]-[9-(3+5)]$

11. $[4(x+5)-10]+[3(x-6)+7]$

12. $5\{[3(x-4)+8]-6(2x-1)+9\}$

Objective D Simplify expressions using the rules for order of operations.
For extra help, see Examples 20–23 on pages 554–555 of your text and the Section 7.8 lecture video.
Simplify.

13. $35-4(-5)+8$

14. $2^4+13\cdot16-(12+32\cdot3)$

15. $6\cdot7-2\cdot4+10$

16. $\dfrac{2(8-9)-3\cdot4}{5\cdot6-3(8-1)}$

Name: Date:
Instructor: Section:

Chapter 8 SOLVING EQUATIONS AND INEQUALITIES

8.1 Solving Equations: The Addition Principle

Learning Objectives
A Determine whether a given number is a solution of a given equation.
B Solve equations using the addition principle.

Key Terms
Use the vocabulary terms listed below to complete each statement in Exercises 1–4.

addition principle **equation** **equivalent equations** **solution**

1. A(n) ________________ is a number sentence that says that the expressions on either side of the equals sign represent the same number.

2. Any replacement for the variable that makes an equation true is called a(n) ________________ of the equation.

3. Equations with the same solution are called ________________ .

4. The ________________ states that for any real numbers a, b, and c, $a = b$ is equivalent to $a + c = b + c$.

GUIDED EXAMPLES AND PRACTICE

Objective A Determine whether a given number is a solution of a given equation.

Review this example for Objective A:

1. Determine whether 23 is a solution of $a-19=4$.

$a-19=4$		Writing the equation
$23-19$	4	Substituting 23 for a
4		TRUE

Yes, 23 is a solution.

Practice this exercise:

1. Determine whether 6 is a solution of $x+7=13$.

Objective B Solve equations using the addition principle.

Review this example for Objective B:

2. Solve: $-8+y=-15$.

$-8+y=-15$

$8-8+y=8-15$ Using the addition princple adding 8 on both sides.

$y=-7$

Practice this exercise:

2. Solve: $-3+x=-8$.

Check: $\frac{-8+y=-15}{\begin{array}{c|c} -8+(-7) & -15 \\ -15 & \end{array}}$ TRUE

The solution is –7.

ADDITIONAL EXERCISES

Objective A Determine whether a given number is a solution of a given equation.

For extra help, see Examples 1–5 on pages 570–571 of your text and the Section 8.1 lecture video.

Determine whether the given number is a solution of the given equation.

1. $3;\ x+19=22$

2. $-6;\ 9t=-54$

3. $56;\ \frac{y}{8}=9$

4. $14;\ 5x+3=73$

5. $-9;\ 3(x-1)=24$

Objective B Solve equations using the addition principle.

For extra help, see Examples 6–9 on pages 572–573 of your text and the Section 8.1 lecture video.

Solve using the addition principle. Don't forget to check!

6. $x+8=20$

7. $y-5=21$

8. $-6+x=17$

9. $10.8=x+4.9$

10. $6\frac{2}{3}=4\frac{1}{2}+x$

Name: Date:
Instructor: Section:

Chapter 8 SOLVING EQUATIONS AND INEQUALITIES

8.2 Solving Equations: The Multiplication Principle

Learning Objectives
A Solve equations using the multiplication principle.

Key Terms
Use the vocabulary terms listed below to complete each statement in Exercises 1–4.

coefficient **identity** **inverse** **principle**

1. The multiplication ________________ states that for any real numbers a, b, and c, $c \neq 0$, $a = b$ is equivalent to $a \cdot c = b \cdot c$.

2. The multiplicative ________________ of 3 is $\frac{1}{3}$.

3. The multiplicative ________________ is 1 since $1 \cdot x = x$.

4. The ________________ in $8x$ is 8.

GUIDED EXAMPLES AND PRACTICE

Objective A Solve equations using the multiplication principle.

Review these examples for Objective A:

1. Solve: $9x = -108$.

$9x = -108$

$\frac{9x}{9} = \frac{-108}{9}$ Dividing by 9 on both sides

$1 \cdot x = -12$ Simplifying

$x = -12$ Identity property of 1.

Check: $9x = -108$

$9(-12) \mid -108$

$-108 \mid$ TRUE

The solution is –12.

Practice these exercises:

1. Solve: $36y = -216$.

2. Solve: $-\frac{3}{4}w=\frac{5}{8}$.

The reciprocal of $-\frac{3}{4}$ is $-\frac{4}{3}$.

$$-\frac{3}{4}w=\frac{5}{8}$$

$$-\frac{4}{3}\left(-\frac{3}{4}\right)w=-\frac{4}{3}\left(\frac{5}{8}\right) \quad \text{Multiplying by } -\frac{4}{3} \text{ on both sides}$$

$$1\cdot w=-\frac{20}{24} \quad \text{Simplifying}$$

$$w=-\frac{5}{6} \quad \text{Identity property of 1.}$$

Check: $-\frac{3}{4}w=\frac{5}{8}$

$$-\frac{3}{4}\left(-\frac{5}{6}\right) \;\Big|\; \frac{5}{8}$$

$$\frac{5}{8} \;\Big|\; \quad \text{TRUE}$$

The solution is $-\frac{5}{6}$.

2. Solve: $\frac{2}{5}v=\frac{1}{3}$.

ADDITIONAL EXERCISES

Objective A Solve equations using the multiplication principle.

For extra help, see Examples 1–8 on pages 576–579 of your text and the Section 8.2 lecture video.

Solve using the multiplication principle. Don't forget to check!

1. $11x=-66$

2. $-32x=-96$

3. $-\frac{4}{5}a=-\frac{16}{15}$

4. $5.9x=17.7$

Name: Date:
Instructor: Section:

Chapter 8 SOLVING EQUATIONS AND INEQUALITIES

8.3 Using the Principles Together

Learning Objectives
A Solve equations using both the addition principle and the multiplication principle.
B Solve equations in which like terms may need to be collected.
C Solve equations by first removing parentheses and collecting like terms; solve equations with an infinite number of solutions and equations with no solutions.

Key Terms
Use the vocabulary terms listed below to complete each statement in Exercises 1–4.

clear fractions **distributive law** **infinitely many solutions** **no solution**

1. We remove parentheses in an equation by multiplying using the ________________ .

2. We multiply every term on both sides of an equation by the least common multiple of all denominators in order to ________________ .

3. When solving an equation, if we end with a true equation, the equation has ________________ .

4. When solving an equation, if we end with a false equation, the equation has ________________

GUIDED EXAMPLES AND PRACTICE

Objective A Solve equations using both the addition principle and the multiplication principle.

Review this example for Objective A:

1. Solve: $4x-5=-41$.

$$4x-5=-41$$
$$4x-5+5=-41+5 \quad \text{Adding 5 on both sides}$$
$$4x=-36$$
$$\frac{4x}{4}=\frac{-36}{4} \quad \text{Dividing by 4 on both sides}$$
$$x=-9 \quad \text{Simplifying}$$

Check: $4x-5=-41$

$$\begin{array}{r|l} 4(-9)-5 & -41 \\ -36-5 & \\ -41 & \quad \text{TRUE} \end{array}$$

The solution is –9.

Practice this exercise:

1. Solve: $-6x+11=29$.

Objective B Solve equations in which like terms may need to be collected.

Review these examples for Objective B:

2. Solve: $7y+2y=-72$.

$$7y+2y=-72$$
$$9y=-72 \quad \text{Collecting like terms}$$
$$\frac{9y}{9}=\frac{-72}{9} \quad \text{Dividing by 9 on both sides}$$
$$y=-8 \quad \text{Simplifying}$$

The number –8 checks, so the solution is –8.

3. Solve: $\frac{2}{5}+\frac{3}{5}x=\frac{18}{15}+\frac{1}{5}x$.

The denominators are 5 and 15. The number 15 is the least common multiple of the denominators. We multiply by 15 on both sides of the equation.

$$15\left(\frac{2}{5}+\frac{3}{5}x\right)=15\left(\frac{18}{15}+\frac{1}{5}x\right)$$
$$15\cdot\frac{2}{5}+15\cdot\frac{3}{5}x=15\cdot\frac{18}{15}+15\cdot\frac{1}{5}x$$
$$6+9x=18+3x \quad \text{Simplifying}$$
$$6+9x-3x=18+3x-3x \quad \text{Adding } -3x$$
$$6+6x=18 \quad \text{Simplifying}$$
$$6-6+6x=18-6 \quad \text{Subtracting 6}$$
$$6x=12 \quad \text{Simplifying}$$
$$\frac{6x}{6}=\frac{12}{24} \quad \text{Dividing both sides by 6}$$
$$x=2$$

The number 2 checks, so the solution is 2.

4. Solve: $3.6x+14.7=0.3-1.2x$.

The greatest number of decimal places in any one number is *one*. Multiplying by 10, which has *one* 0, will clear all decimals.

$$10(3.6x+14.7)=10(0.3-1.2x)$$
$$10(3.6x)+10(14.7)=10(0.3)-10(1.2x)$$
$$36x+147=3-12x$$
$$36x+147-147=3-12x-147$$
$$36x=-144-12x$$
$$36x+12x=-144-12x+12x$$
$$48x=-144$$
$$\frac{48x}{48}=\frac{-144}{48}$$
$$x=-3$$

The number –3 checks, so the solution is –3.

Practice these exercises:

2. Solve: $3x+5x=-32$.

3. Solve: $\frac{1}{6}+2y=5y-\frac{7}{12}$.

4. Solve:
$0.86x-0.67=0.11x+3.08$.

Name: Date:
Instructor: Section:

Objective C Solve equations by first removing parentheses and collecting like terms; solve equations with an infinite number of solutions and equations with no solutions.

Review these examples for Objective C:

5. Solve: $3(4x+3)=6(2x-4)$.

First, we use the distributive law to multiply and remove parentheses.

$$3(4x+3)=6(2x-4)$$
$$12x+9=12x-24$$
$$-12x+12x+9=-12x+12x-24$$
$$9=-24 \quad \text{False}$$

There are no solutions.

6. Solve: $2x-4x+1=1-2x$.

$$2x-4x+1=1-2x$$
$$-2x+1=1-2x \quad \text{Collecting like terms}$$
$$2x-2x+1=1-2x+2x \quad \text{Adding } 2x$$
$$1=1 \quad \text{True for all real numbers.}$$

There are infinitely many solutions.

Practice these exercises:

5. Solve: $4(5x-2)=2(10x+1)$.

6. Solve: $y+3y-4=4y-4$.

ADDITIONAL EXERCISES

Objective A Solve equations using both the addition principle and the multiplication principle.

For extra help, see Examples 1–4 on pages 582–583 of your text and the Section 8.3 lecture video.

Solve. Don't forget to check!

1. $3x+5=29$

2. $6x-5=37$

3. $-23=7+5y$

4. $-7x-18=-28\frac{1}{2}$

Objective B Solve equations in which like terms may need to be collected.
For extra help, see Examples 5–9 on pages 583–586 of your text and the Section 8.3 lecture video.
Solve.

5. $-3y-4y=28$

6. $4y-3+2y=6y+8-y$

Solve. Clear fractions or decimals first.

7. $\frac{2}{3}x-\frac{1}{4}x=\frac{2}{5}x+1$

8. $4.07-0.61x=0.82x-8.8$

Objective C Solve equations by first removing parentheses and collecting like terms; solve equations with an infinite number of solutions and equations with no solutions.
For extra help, see Examples 10–13 on pages 586–588 of your text and the Section 8.3 lecture video.
Solve.

9. $3(2+5y)-10=11$

10. $4x+3-5x-18=6-8x+7x-21$

11. $13x-2-7x=2(3x-1)+5$

Name: Date:
Instructor: Section:

Chapter 8 SOLVING EQUATIONS AND INEQUALITIES

8.4 Formulas

Learning Objectives
A Evaluate a formula.
B Solve a formula for a specified letter.

Key Terms
Use the vocabulary terms listed below to complete each statement in Exercises 1–2.

evaluating **formula**

1. A(n) ________________ is an equation relating several quantities.

2. When we replace the variables in an expression with numbers and calculate the result, we are ________________ the expression.

GUIDED EXAMPLES AND PRACTICE

Objective A Evaluate a formula.

Review this example for Objective A:

1. The formula $d = 65t$ gives the distance traveled, in miles, by a vehicle traveling 65 mph for t hr. A car travels 65 mph for 3 hr. How many miles did the car travel?

   We substitute 3 for t and calculate d.

   $$d = 65t = 65(3) = 195$$

   The car traveled 195 mi.

Practice this exercise:

1. The formula $d = 55t$ gives the distance traveled, in miles, by a vehicle traveling 55 mph for t hr. A car travels 55 mph for 4 hr. How many miles did the car travel?

Objective B Solve a formula for a specified letter.

Review this example for Objective B:

2. Solve $y = 3x - 6$ for x.

   $y = 3x - 6$ We want x alone.
   $y + 6 = 3x - 6 + 6$ Adding 6
   $y + 6 = 3x$ Simplifying
   $\dfrac{y+6}{3} = \dfrac{3x}{3}$ Dividing both sides by 3
   $\dfrac{y+6}{3} = x$ Simplifying

Practice this exercise:

2. Solve $y = 2x + 5$ for x.

ADDITIONAL EXERCISES

Objective A Evaluate a formula.

For extra help, see Examples 1–3 on pages 593–594 of your text and the Section 8.4 lecture video.

Solve.

1. The formula $A = lw$ gives the area of a rectangle with length l and width w. A rectangle has a length of 4 m and a width of $\frac{1}{2}$ m. What is the area of the rectangle?

2. The formula $I = Prt$ gives the simple interest earned by an investment with principal P, interest rate r, and time t. How much simple interest is earned on \$2000 invested at 6% for 2 yr? (Use 0.06 for 6%.)

3. The cost for one month of Camden's cell phone, in dollars, is given by the formula $c = 45 + 0.1m$, where m is the number of text messages sent or received that month. How much was his cell phone bill for a month in which he sent or received 80 text messages?

Objective B Solve a formula for a specified letter.

For extra help, see Examples 4–12 on pages 594–596 of your text and the Section 8.4 lecture video.

Solve for the indicated letter.

4. $P = ax + b$, for x

5. $A = \dfrac{p+q+r}{3}$, for q

6. $A = 4\pi r^2$, for r^2

7. $y = 18 - x$, for x

Name: Date:
Instructor: Section:

Chapter 8 SOLVING EQUATIONS AND INEQUALITIES

8.5 Applications of Percent

Learning Objectives
A Solve applied problems involving percent.

Key Terms
Use the vocabulary terms listed below to complete each statement in Exercises 1–4.

%	is	of	what number

1. ________________ translates to "$\cdot$" or "$\times$".
2. ________________ translates to "=".
3. ________________ translates to any letter.
4. ________________ translates to "$\times\frac{1}{100}$" or "$\times 0.01$".

GUIDED EXAMPLES AND PRACTICE

Objective A Solve applied problems involving percent.

Review these examples for Objective A:

1. What percent of 240 is 156?

$$\underbrace{\text{What percent}}_{\downarrow} \ \underbrace{\text{of}}_{\downarrow} \ \underbrace{240}_{\downarrow} \ \underbrace{\text{is}}_{\downarrow} \ \underbrace{156}_{\downarrow}$$

Translate: $p \times 240 = 156$

Solve $p \cdot 240 = 156$.

$$\frac{p \cdot 240}{240} = \frac{156}{240}$$
$$p = 0.65$$
$$p = 65\%$$

2. 4.5 is 25% of what number?

$$\underbrace{4.5}_{\downarrow} \ \underbrace{\text{is}}_{\downarrow} \ \underbrace{25\%}_{\downarrow} \ \underbrace{\text{of}}_{\downarrow} \ \underbrace{\text{what number}}_{\downarrow}$$

Translate: $4.5 = 0.25 \times b$

Solve $4.5 = 0.25 \times b$.

$$\frac{4.5}{0.25} = \frac{0.25 \times b}{0.25}$$
$$18 = b$$

Practice these exercises:

1. What percent of 60 is 9?

2. 72 is 45% of what number?

3. What number is 55% of 280?

What number	is	55%	of	280
↓	↓	↓	↓	↓
Translate: a	$=$	0.55	$\times$	280

$a = 0.55 \times 280 = 154$

3. What number is 40% of 670?

ADDITIONAL EXERCISES

Objective A Solve applied problems involving percent.

For extra help, see Examples 1–9 on pages 603–606 of your text and the Section 8.5 lecture video.

Solve.

1. What number is 24% of 50?

2. What percent of 450 is 72?

3. 35 is 140% of what number?

4. In 2005, shoppers spent \$9 billion in Father's Day gifts and \$14 billion on Mother's Day gifts. What percent of the amount spent for Mother's Day was spent for Father's Day?
Source: National Retail Federation.

5. Chad left a \$6 tip for a meal that cost \$30.
a) What percent of the cost of the meal was the tip?
b) What was the total cost of the meal including the tip?

Name: Date:
Instructor: Section:

Chapter 8 SOLVING EQUATIONS AND INEQUALITIES

8.6 Applications and Problem Solving

Learning Objectives
A Solve applied problems by translating to equations.

Key Terms
Use the vocabulary terms listed below to complete each statement in Exercises 1–5.

check **familiarize** **solve** **state** **translate**

1. To ________________ yourself with a problem, read it carefully, choose a variable to represent the unknown, and make a drawing.

2. To ________________ a problem into mathematical language, write an equation.

3. To ________________ an equation, find all replacements that make the equation true.

4. Always ________________ the answer in the original problem.

5. As a final problem-solving step, ________________ the answer to the problem clearly.

GUIDED EXAMPLES AND PRACTICE

Objective A Solve applied problems by translating to equations.

Review this example for Objective A:

1. The sum of three consecutive even integers is 198. What are the integers?

1. *Familiarize.* We let x = the smallest integer, then $x + 2$ = the next integer and $x + 4$ = the largest integer.
2. *Translate.* Rephrase the question and translate.

$$\underbrace{\text{First integer}}_{\downarrow\; x}\;\underbrace{\text{plus}}_{\downarrow\; +}\;\underbrace{\text{Second integer}}_{\downarrow\; (x+2)}\;\underbrace{\text{plus}}_{\downarrow\; +}\;\underbrace{\text{Third integer}}_{\downarrow\; (x+4)}\;\underbrace{\text{is}}_{\downarrow\; =}\;\underbrace{198}_{\downarrow\; 198}$$

$$x + (x+2) + (x+4) = 198$$

Practice this exercise:

1. A rectangle has a perimeter of 88 ft. The length is 2 ft more than twice the width. Find the dimensions of the rectangle.

3. *Solve*. We solve the equation.

$$x+(x+2)+(x+4)=198$$
$$3x+6=198$$
$$3x+6-6=198-6$$
$$3x=192$$
$$\frac{3x}{3}=\frac{192}{3}$$
$$x=64$$

If $x = 64$, then $x + 2 = 66$, and $x + 4 = 68$.

4. *Check*. The sum of 64, 66 and 68 is 198. The answer checks.

5. *State*. The integers are 64, 66, and 68.

ADDITIONAL EXERCISES

Objective A Solve applied problems by translating to equations.

For extra help, see Examples 1–9 on pages 611–621 of your text and the Section 8.6 lecture video.

Solve.

1. A 60-in. board is cut into two pieces. One piece is four times the length of the other. Find the lengths of the pieces.

2. The balance in Clayton's charge card account grew 3%, to \$669.50, in one month. What was his balance at the beginning of the month?

3. The sum of three consecutive integers is 69. What are the numbers?

4. Craig left an 18% tip for a meal. The total cost of the meal, including the tip, was \$51.92. What was the cost of the meal before the tip was added?

Name: Date:
Instructor: Section:

Chapter 8 SOLVING EQUATIONS AND INEQUALITIES

8.7 Solving Inequalities

Learning Objectives
A Determine whether a given number is a solution of an inequality.
B Graph an inequality on the number line.
C Solve inequalities using the addition principle.
D Solve inequalities using the multiplication principle.
E Solve inequalities using the addition principle and the multiplication principle together.

Key Terms
Use the vocabulary terms listed below to complete each statement in Exercises 1–4.

equivalent **graph** **inequality** **set-builder notation**

1. A(n) ________________ is a number sentence with $<$, $>$, $\leq$, or $\geq$ as its verb.

2. A(n) ________________ of an inequality is a drawing that represents its solutions.

3. The sentences $x + 4 < 10$ and $x < 6$ are ________________ since they have the same solution set.

4. The solution set $\{x|x > 2\}$ is written using ________________ .

GUIDED EXAMPLES AND PRACTICE

Objective A Determine whether a given number is a solution of an inequality.

Review this example for Objective A:

1. Determine whether –4 is a solution of $x < 1$.

   Since $-4 < 1$ is true, –4 is a solution.

Practice this exercise:

1. Determine whether 5 is a solution of $x > 3$.

Objective B Graph an inequality on the number line.

Review this example for Objective B:

2. Graph the inequality $x > -3$.

   The solutions of $x > -3$ are all numbers greater than –3. We shade all points to the right of –3. Use an open circle at –3 to indicate that –3 is not part of the graph.

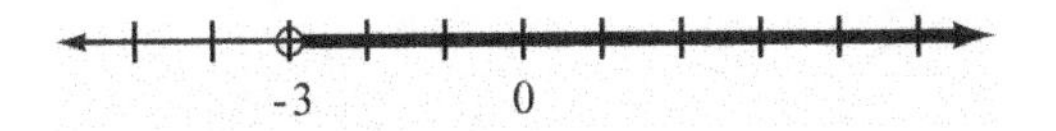

Practice this exercise:

2. Graph $x > 2$.

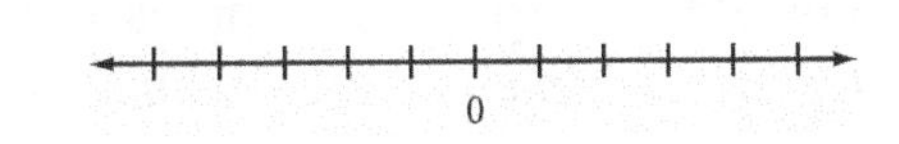

Objective C Solve inequalities using the addition principle.

Review this example for Objective C:

3. Solve $x+3\geq -2$.

$$x+3\geq -2$$
$$x+3-3\geq -2-3 \quad \text{Subtracting 3 from both sides}$$
$$x\geq -5 \quad \text{Simplifying}$$

The solution set is $\{x|x\geq -5\}$.

Practice this exercise:

3. Solve $x+5>4$.

Objective D Solve inequalities using the multiplication principle.

Review this example for Objective D:

4. Solve $-3x>3$. Then graph.

$$-3x>3$$
$$\frac{-3x}{-3}<\frac{3}{-3} \quad \text{The symbol must be reversed because we are dividing by a negative number, } -3.$$
$$x<-1$$

The solution set is $\{x|x<-1\}$.

-1 0

Practice this exercise:

4. Solve $-8x>32$.

Objective E Solve inequalities using the addition principle and the multiplication principle together.

Review this example for Objective E:

5. Solve $8x+8-6x>12$.

$$8x+8-6x>12$$
$$2x+8>12 \quad \text{Collecting like terms}$$
$$2x+8-8>12-8 \quad \text{Subtracting 8 on both sides}$$
$$2x>4 \quad \text{Simplifying}$$
$$\frac{2x}{2}>\frac{4}{2} \quad \text{Dividing both sides by 2}$$
$$x>2$$

The solution set is $\{x|x>2\}$.

Practice this exercise:

5. Solve $13-8y-y\leq 40$.

ADDITIONAL EXERCISES

Objective A Determine whether a given number is a solution of an inequality.

For extra help, see Examples 1–4 on page 628 of your text and the Section 8.7 lecture video.

Determine whether each number is a solution of $x\leq -6$.

1. 0

2. –3

3. –6

4. –9

5. –5.4

Name: Date:
Instructor: Section:

Objective B Graph an inequality on the number line.
For extra help, see Examples 5–7 on page 629 of your text and the Section 8.7 lecture video.
Graph on a number line.

6. $m < 2$

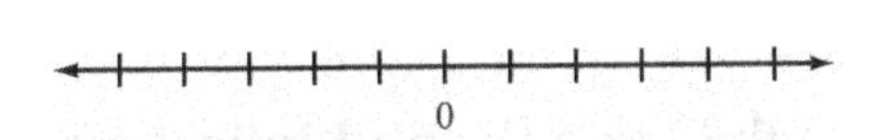

7. $t \geq -2$

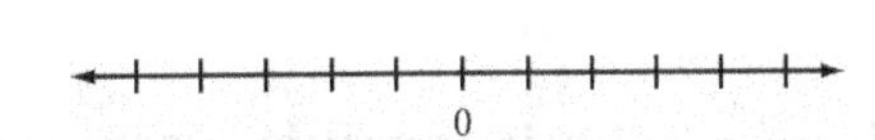

8. $-2 \leq x < 5$

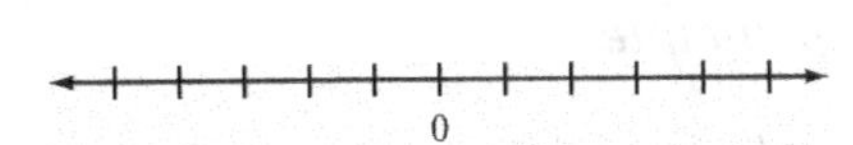

9. $-4 < x < 0$

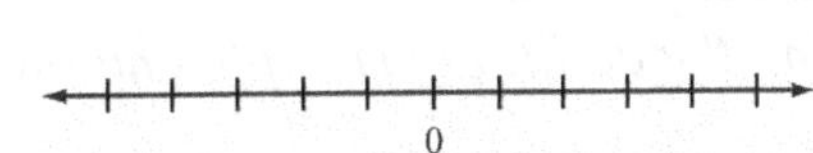

Objective C Solve inequalities using the addition principle.
For extra help, see Examples 8–10 on pages 630–631 of your text and the Section 8.7 lecture video.
Solve using the addition principle. Then graph.

10. $x + 4 < 3$

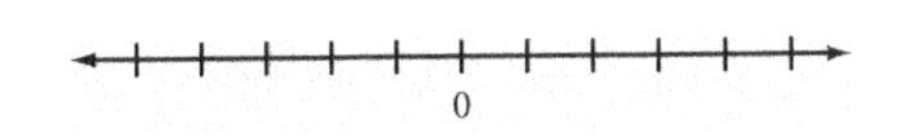

11. $x + 3 \geq -2$

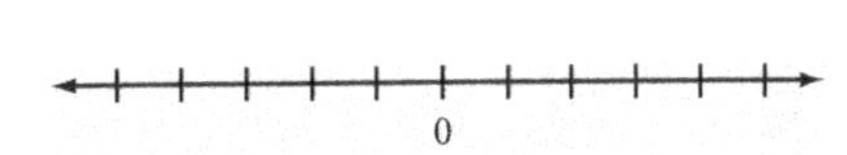

Solve using the addition principle.

12. $2x + 7 \leq 3x + 2$

13. $-7 + y \geq 14$

Objective D Solve inequalities using the multiplication principle.
For extra help, see Examples 11–12 on pages 632–633 of your text and the Section 8.7 lecture video.
Solve using the multiplication principle. Then graph.

14. $-3x \geq 6$

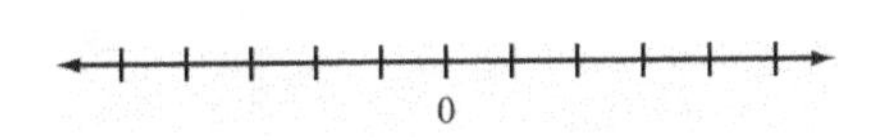

15. $-11x < -44$

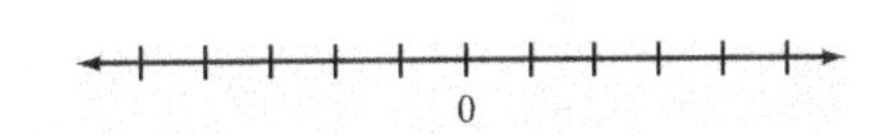

Solve using the multiplication principle.

16. $3x \le -18$

17. $-4p < 8$

Objective E Solve inequalities using the addition principle and the multiplication principle together.

For extra help, see Examples 13–18 on pages 633–635 of your text and the Section 8.7 lecture video.

Solve using the addition principle and the multiplication principle.

18. $2+3y > 17$

19. $12-10x \le 3-9x$

20. $4(3y-5) \le 4$

21. $2(r-4)+5 \ge 4(r+8)-16$

Name: Date:
Instructor: Section:

Chapter 8 SOLVING EQUATIONS AND INEQUALITIES

8.8 Applications and Problem Solving with Inequalities

Learning Objectives
A Translate number sentences to inequalities.
B Solve applied problems using inequalities.

Key Terms
In Exercises 1–6, match each phrase with its translation from the column on the right. Letters may be used more than once or not at all.

1. ________________ x is at least 5.
2. ________________ x is at most 5.
3. ________________ x cannot exceed 5.
4. ________________ x must exceed 5.
5. ________________ x is no more than 5.
6. ________________ x is no less than 5.

A. $x < 5$
B. $x \leq 5$
C. $x > 5$
D. $x \geq 5$

GUIDED EXAMPLES AND PRACTICE

Objective A Translate number sentences to inequalities.

Review this example for Objective A:

1. Translate to an inequality: a number is greater than 10.

   Let x = the number.
   $x > 10$

Practice this exercise:

1. Translate to an inequality: a number is less than or equal to 50.

Objective B Solve applied problems using inequalities.

Review this example for Objective B:

2. David has saved \$4500 for college tuition. If his local community college charges a \$65 registration fee plus \$395 per credit hour, what is the greatest number of credit hours for which David can register?

   1. *Familiarize.* Suppose that David registered for 9 credit hours. The cost would then be 9(\$395) + \$65, or \$3620. This shows that David can register for more than 9 credit hours without exceeding \$4500. Let n = the number of credit hours.

Practice this exercise:

2. As part of the requirements for a class, Hannah must volunteer at a community shelter for 4 weeks with an average of at least 12 hr per week. For the first 3 weeks, Hannah volunteered 15 hr, 6 hr, and 11 hr. How many hr must she volunteer in the fourth week in order to meet the requirement?

2. *Translate.* Registration fee plus cost per credit hour cannot exceed $4500.

$$65 + 395n \le 4500$$

3. *Solve.*

$$\begin{aligned} 65 + 395n &\le 4500 \\ 65 + 395n - 65 &\le 4500 - 65 \\ 395n &\le 4435 \\ \frac{395n}{395} &\le \frac{4435}{395} \\ n &\le 11.2 \quad \text{Round to nearest tenth.} \end{aligned}$$

4. *Check.* Although the solution set of the inequality is all numbers less than or equal to about 11.2, since n = the number of credit hours, we round *down* to 11 credit hours. If David registers for 11 credit hours, the cost will be \$65 + \$395(11), or \$4410. If David registers for 12 credit hours, the cost will exceed \$4500.

5. *State.* At most, David can register for 11 credit hours.

ADDITIONAL EXERCISES

Objective A Translate number sentences to inequalities.

For extra help, see page 640 of your text and the Section 8.8 lecture video.

Translate to an inequality.

1. A number is at most 13.

2. The coffee weighs more than 1.5 lb.

3. The speed of the car was between 45 and 65 mph.

4. The amount spent on advertising is not to exceed \$1500.

Objective B Solve applied problems using inequalities.

For extra help, see Examples 1–2 on pages 640–642 of your text and the Section 8.8 lecture video.

Solve.

5. A student is taking a history course in which five tests are given. To get a C, he must average at least 70 on the five tests. The student got scores of 74, 82, 60 and 68 on the first four tests. Determine (in terms of an inequality) what scores on the last test will allow him to get at least a C.

6. Tin stays solid at Fahrenheit temperatures below 449.7°. Use the formula $F = \frac{9}{5}C + 32$ to determine (in terms of an inequality) those Celsius temperatures for which tin stays solid.

7. To print business cards, Timeless Printing charges a \$25 design fee plus \$0.20 per card. Damon can spend no more than \$85 for business cards. What numbers of cards will allow him to stay within budget?

Name: Date:
Instructor: Section:

Chapter 9 GRAPHS OF LINEAR EQUATIONS

9.1 Graphs and Applications of Linear Equations

Learning Objectives

A Plot points associated with ordered pairs of numbers; determine the quadrant in which a point lies.
B Find the coordinates of a point on a graph.
C Determine whether an ordered pair is a solution of an equation with two variables.
D Graph linear equations of the type $y = mx + b$ and $Ax + By = C$, identifying the y-intercept.
E Solve applied problems involving graphs of linear equations.

Key Terms

Use the vocabulary terms listed below to complete each statement in Exercises 1–6.

axes **coordinates** **graph** **ordered pair** **origin** ***y*-intercept**

1. We graph number pairs on a plane using two perpendicular number lines called ____________________ .

2. On the plane, the perpendicular number lines cross at a point called the ____________________ .

3. The numbers in an ordered pair are called ____________________ .

4. The notation (3, –2) is an example of a(n) ____________________ .

5. The ____________________ of an equation is a drawing that represents all its solutions.

6. A graph crosses the y-axis at the ____________________ .

GUIDED EXAMPLES AND PRACTICE

Objective A Plot points associated with ordered pairs of numbers; determine the quadrant in which a point lies.

Review these examples for Objective A:

1. Plot the point (3, –2).

   The first coordinate is positive, so starting at the origin, move 3 units to the right. The second coordinate is negative, so we then move down 2 units.

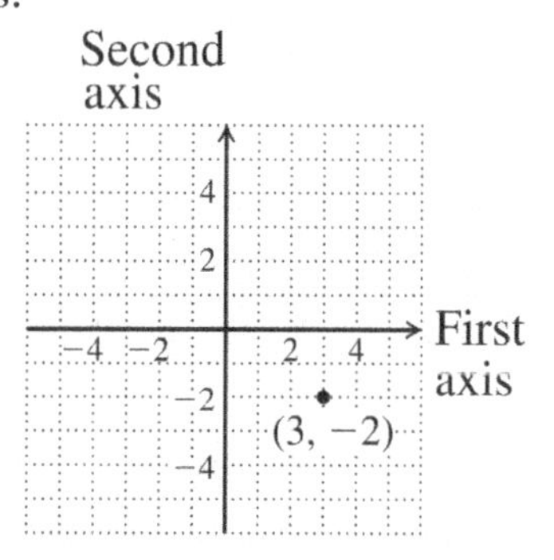

2. In which quadrant is the point (–3, –5) located?

   Both coordinates are negative, so (–3, –5) is in quadrant III.

Practice these exercises:

1. Which point is (–2, 4)?

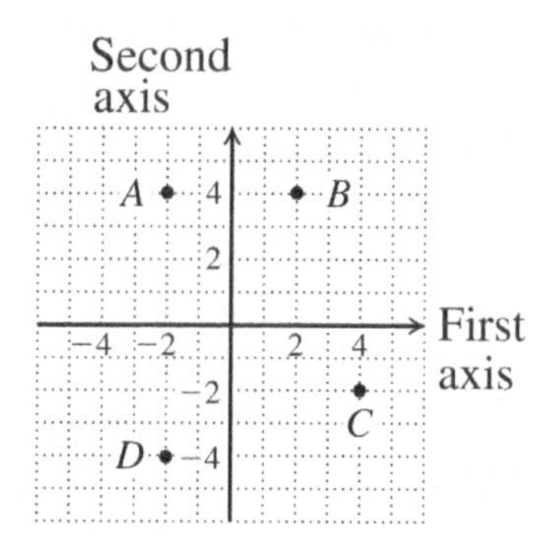

2. In which quadrant is the point (2, –1) located?

Objective B Find the coordinates of a point on a graph.

Review this example for Objective B:

3. Find the coordinates of point *P*.

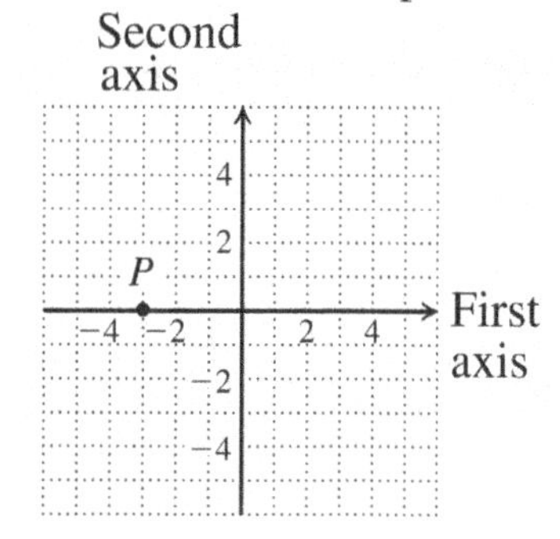

   Point *P* is 3 units to the left of the origin and 0 units up or down. Its coordinates are (–3, 0).

Practice this exercise:

3. Find the coordinates of point *M*.

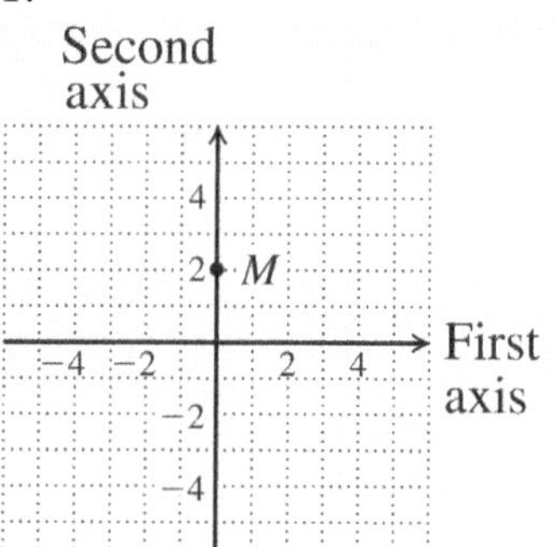

Name: Date:
Instructor: Section:

Objective C Determine whether an ordered pair is a solution of an equation with two variables.

Review this example for Objective C:

4. Determine whether (–2, 2) is a solution of $2b - a = 6$.

$$\begin{array}{c|l} 2b - a & = 6 \\ \hline 2\cdot 2-(-2) & 6 \\ 4+2 & \\ 6 & \end{array} \quad \text{TRUE}$$

Practice this exercise:

4. Determine whether (–4, 1) is a solution of $n - m = -5$.

Objective D Graph linear equations of the type $y = mx + b$ and $Ax + By = C$, identifying the y-intercept.

Review this example for Objective D:

5. Graph $y = -\frac{3}{2}x + 2$ and identify the y-intercept.

From the equation, we see that the y-intercept is (0, 2). We find two other pairs that are solutions, using multiples of 2 to avoid fractions. Then we complete and label the graph.

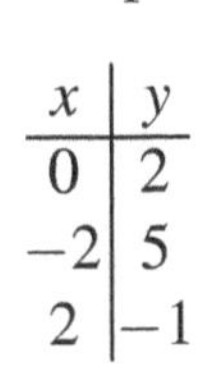

x	y
0	2
–2	5
2	–1

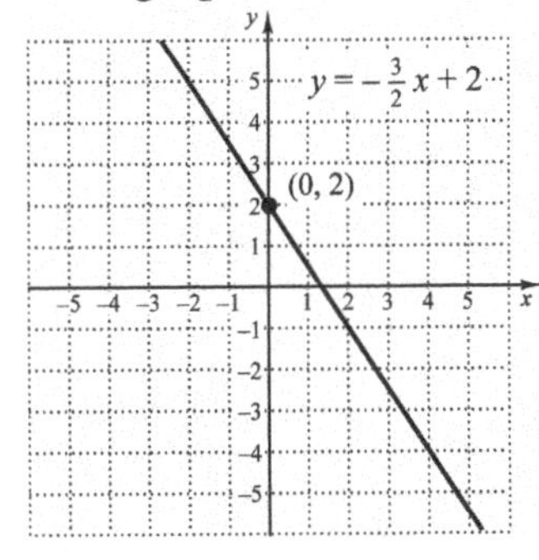

Practice this exercise:

5. Graph $x + 4y = 4$ and identify the y-intercept.

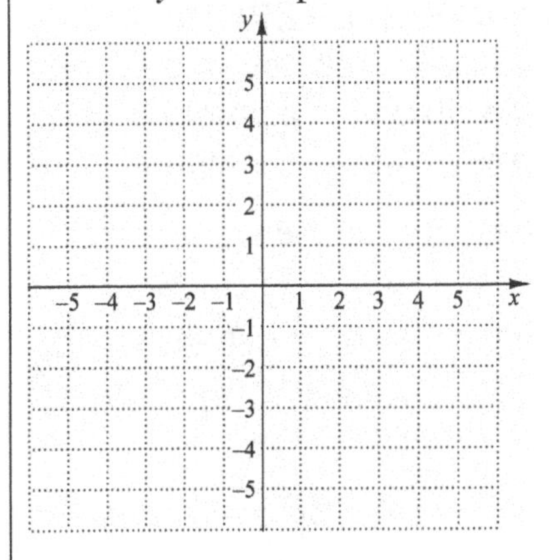

Objective E Solve applied problems involving graphs of linear equations.

Review this example for Objective E:

6. The weekly salary of a salesperson at Shoe City is given by the equation $w = 200 + 0.04s$, where s is that person's sales for the week. Graph the equation and then use the graph to estimate the salesperson's sales when the week's pay is \$375.

We choose some values for s and find the corresponding w-values.

When $s = 1000$, $w = 200 + 0.04(1000) = 240$.

When $s = 3000$, $w = 200 + 0.04(3000) = 320$.

Practice this exercise:

6. The cost c, in dollars of renting a 20-ft moving van at Rent King is given by the equation $c = 0.45m + 59.95$, where m is the number of miles the truck is driven. Graph the equation and then use the graph to estimate how far a van can be driven on a budget of \$150.

When $s = 5000$, $w = 200 + 0.04(5000) = 400$.
Plot these points and draw the graph.

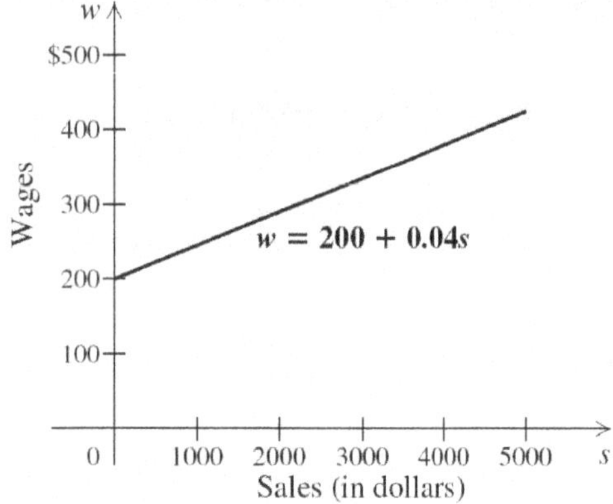

To estimate the sales when the week's pay is \$375, locate 375 on the w-axis, go across horizontally to the graph, and then go down vertically to the s-axis. We find that sales are about \$4400 when a week's pay is \$375.

ADDITIONAL EXERCISES

Objective A Plot points associated with ordered pairs of numbers; determine the quadrant in which a point lies.

For extra help, see Examples 1–2 on page 657 of your text and the Section 9.1 lecture video.

1. Plot and label the set of points.
$(3, 2), (-1, 4), (2, -2), (0, 3),$
$(-3, -4), (0, -5), (-4, 0), (1, 0)$

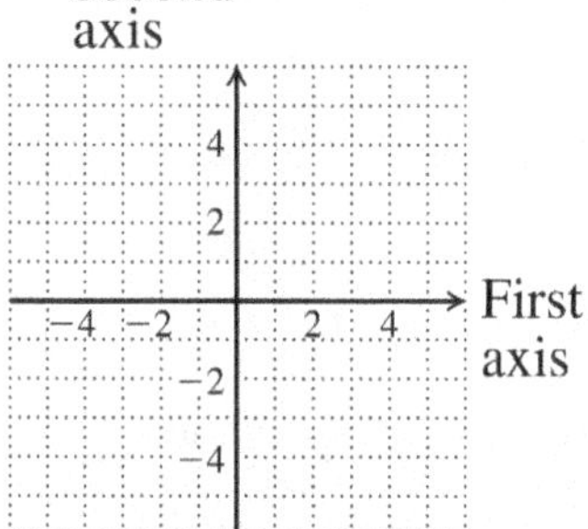

In which quadrant is each point located?

2. $(-8, 10)$

3. $(-9, -12)$

Name: Date:
Instructor: Section:

Complete each sentence using the words positive *or* negative *or the numerals* I, II, III, or IV.

4. In quadrants III and _________, the first coordinate is always _________ .

5. In quadrants IV and _________, the second coordinate is always _________ .

Objective B Find the coordinates of a point on a graph.

For extra help, see Example 3 on page 658 of your text and the Section 9.1 lecture video.

Determine the coordinates of the points from the given graph.

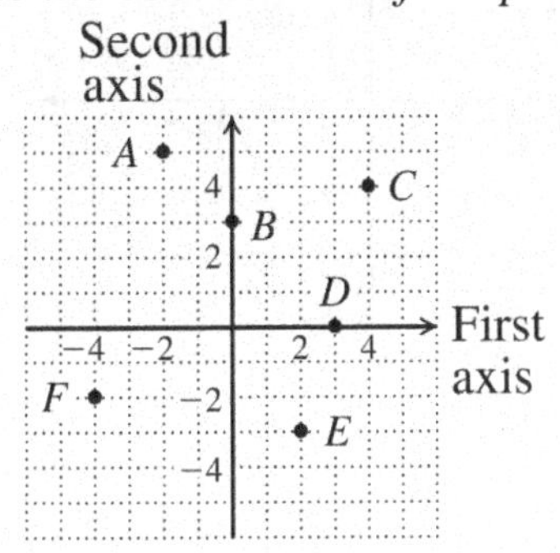

6. *A* and *B*

7. *C* and *D*

8. *E* and *F*

Objective C Determine whether an ordered pair is a solution of an equation with two variables.

For extra help, see Examples 4–5 on pages 658–659 of your text and the Section 9.1 lecture video.

Determine whether each ordered pair is a solution of the given equation.

9. $(-2, 1)$; $x + 4y = 2$

10. $(3, 7)$; $y = 2x - 1$

11. $(-4, -5)$; $p - q = 1$

12. $(5, -2)$; $2d + c = 8$

Objective D Graph linear equations of the type $y = mx + b$ and $Ax + By = C$, identifying the y-intercept.

For extra help, see Examples 6–10 on pages 660–664 of your text and the Section 9.1 lecture video.

Graph the equation and identify the y-intercept.

13. $y = \frac{2}{3}x$

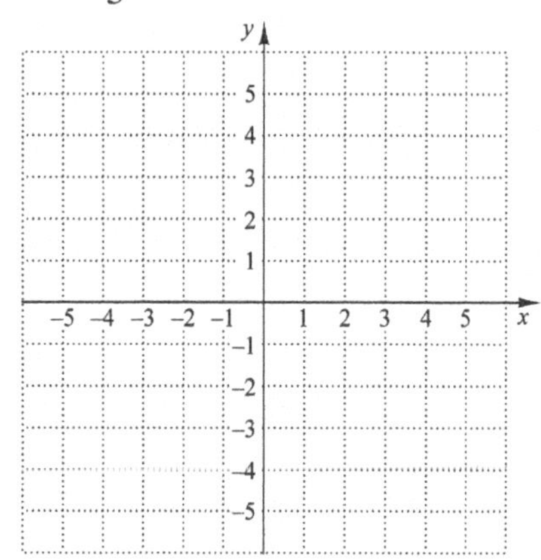

14. $y = x - 2$

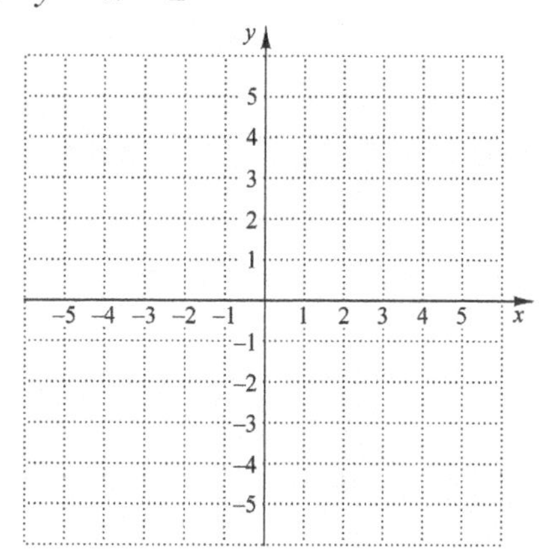

15. $x + y = 3$

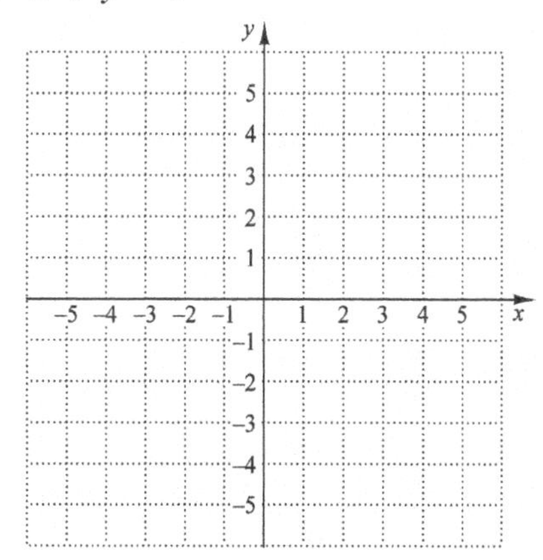

16. $5x - 2y = 10$

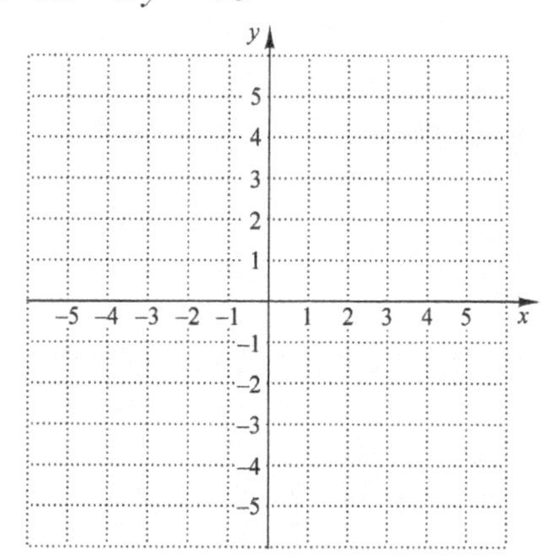

Name: Date:
Instructor: Section:

Objective E Solve applied problems involving graphs of linear equations.
For extra help, see Example 11 on pages 664–666 of your text and the Section 9.1 lecture video.

The average annual expenditure on home insurance can be approximated by $H = 21n + 414$, *where n is the number of years since 1995.*

17. Find the average annual home insurance cost in 1996 ($n = 1$), 1998, 2004, and 2007.

18. Graph the equation and use the graph to estimate what the average annual home insurance expenditure was in 2001.

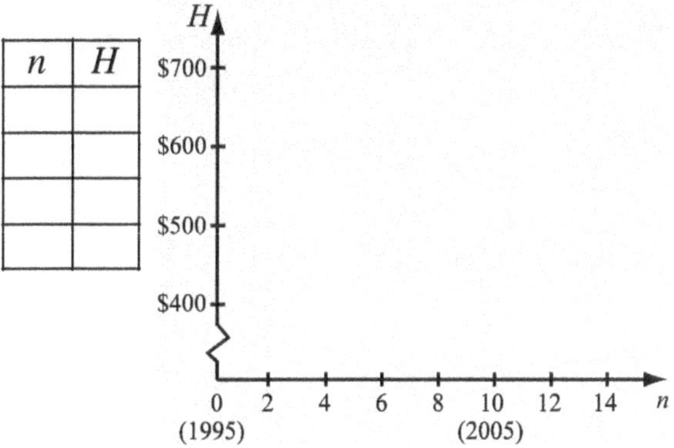

The value V, *in dollars, of a digital camera is given by* $V = -75t + 450$, *where* t *is the number of years since the camera was purchased.*

19. Find the value of the camera after 0 yr, 3 yr, and 6 yr.

20. Graph the equation and then use the graph to estimate the value of the camera after 4 yr.

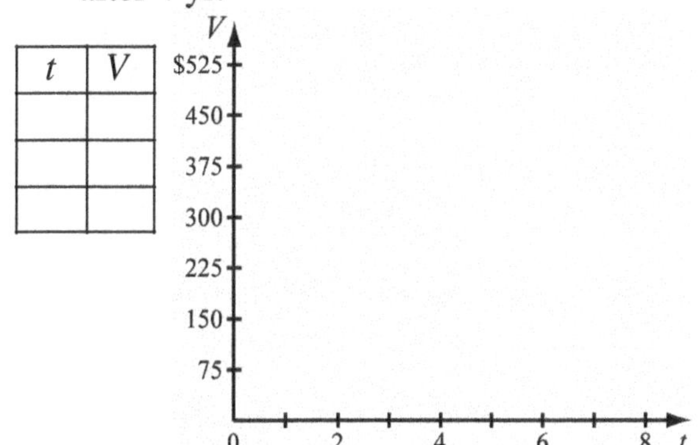

Name: Date:
Instructor: Section:

Chapter 9 GRAPHS OF LINEAR EQUATIONS

9.2 More with Graphing and Intercepts

Learning Objectives
A Find the intercepts of a linear equation, and graph using intercepts.
B Graph equations equivalent to those of the type $x = a$ and $y = b$.

Key Terms
Use the vocabulary terms listed below to complete each statement in Exercises 1–4.

horizontal line **vertical line** ***x*-intercept** ***y*-intercept**

1. The ____________________ occurs where a graph crosses the x-axis.

2. The ____________________ occurs where a graph crosses the y-axis.

3. The graph of $x = a$ is a(n) ____________________ .

4. The graph of $y = b$ is a(n) ____________________ .

GUIDED EXAMPLES AND PRACTICE

Objective A Find the intercepts of a linear equation, and graph using intercepts.

Review this example for Objective A:

1. Find the intercepts of $6x - 12 = 6y$. Then use the intercepts to graph the equation.

To find the x-intercept, let $y = 0$. Then solve for x.

$$6x - 12 = 6y$$
$$6x - 12 = 6 \cdot 0$$
$$6x - 12 = 0$$
$$6x = 12$$
$$x = 2$$

Thus, (2, 0) is the x-intercept.
To find the y-intercept, let $x = 0$. Then solve for y.

$$6x - 12 = 6y$$
$$6 \cdot 0 - 12 = 6y$$
$$-12 = 6y$$
$$-2 = y$$

Thus, (0, –2) is the y-intercept.

Practice this exercise:

1. Find the intercepts of $4x - 5y = 20$. Then use the intercepts to graph the equation.

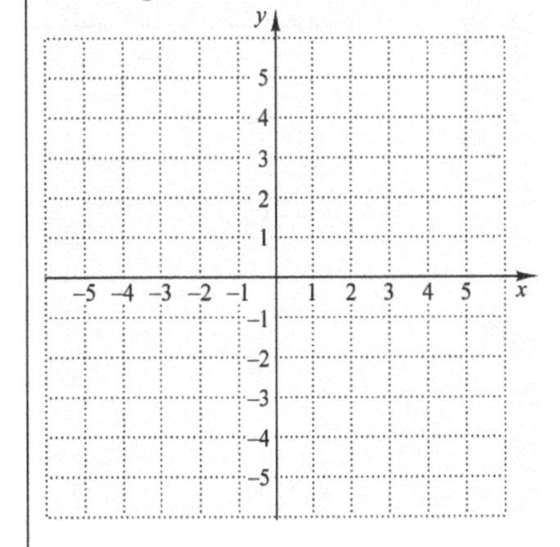

Plot these points and draw the line.

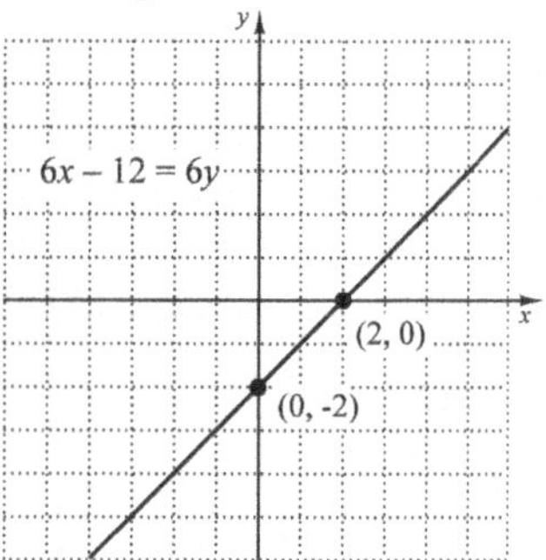

A third point should be used as a check. We substitute any value for x and solve for y.

Let $x = 4$. Then

$$6x - 12 = 6y$$
$$6 \cdot 4 - 12 = 6y$$
$$24 - 12 = 6y$$
$$12 = 6y$$
$$2 = y$$

The point (4, 2) is on the graph, so the graph is probably correct.

Objective B Graph equations equivalent to those of the type $x = a$ and $y = b$.

Review this example for Objective B:

2. Graph $y = 4$.

We can think of this equation as $0 \cdot x + y = 4$. No matter what number we choose for x, y must be 4. We make a table of values and plot and connect the corresponding points.

x	y
-2	4
0	4
3	4

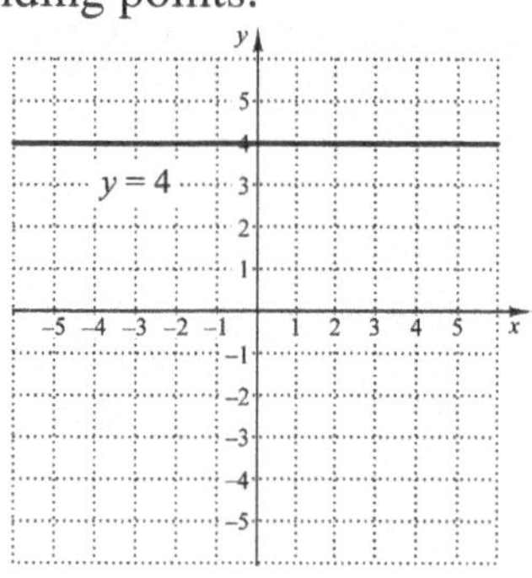

Practice this exercise:

2. Graph $x = -3$.

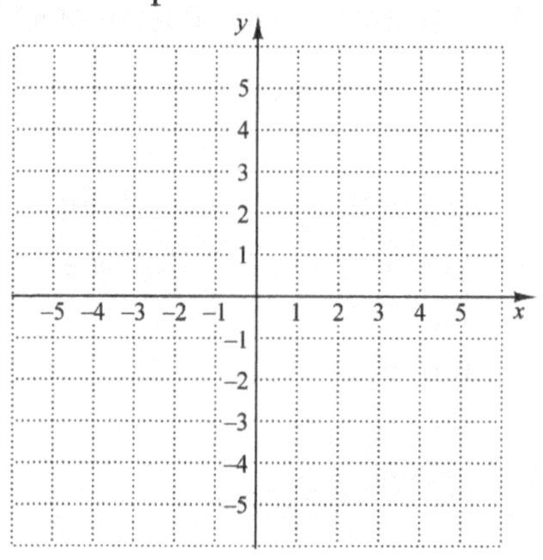

Name: Date:
Instructor: Section:

ADDITIONAL EXERCISES

Objective A Find the intercepts of a linear equation, and graph using intercepts.

For extra help, see Examples 1–2 on pages 673–675 of your text and the Section 9.2 lecture video.

Find the coordinates of the y-intercept and the coordinates of the x-intercept. Do not graph.

1. $6x+5y=30$

2. $2x-5y=40$

For each equation, find the intercepts. Then use the intercepts to graph the equation.

3. $x+2y=4$

x	y	
0		← y-intercept
	0	← x-intercept

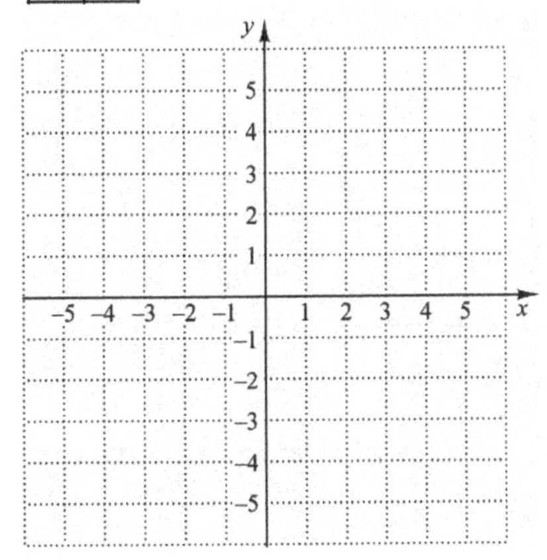

4. $x+5=y$

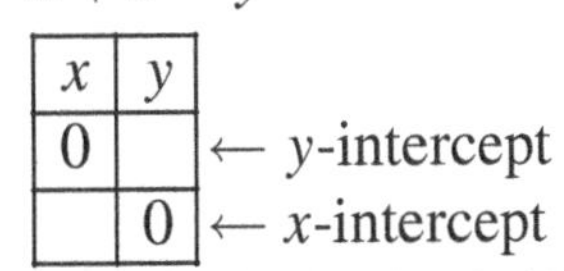

x	y	
0		← y-intercept
	0	← x-intercept

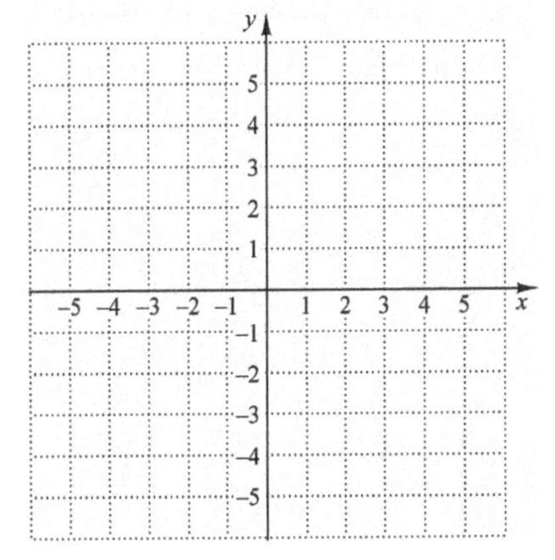

Objective B Graph equations equivalent to those of the type $x = a$ and $y = b$.
For extra help, see Examples 3–4 on pages 675–676 of your text and the Section 9.2 lecture video.

Graph.

5. $2y = -5$

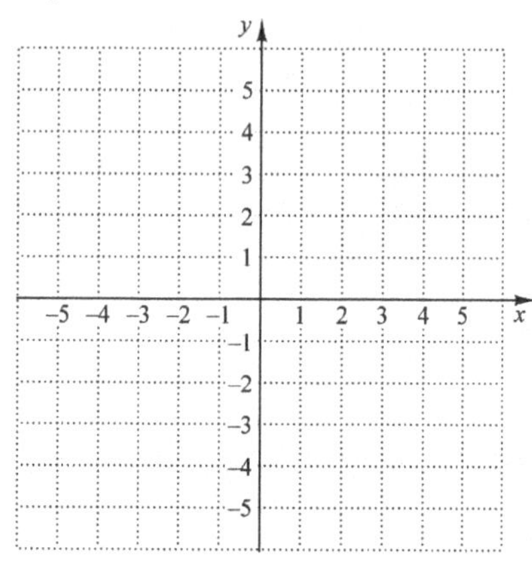

6. $3x - 15 = 0$

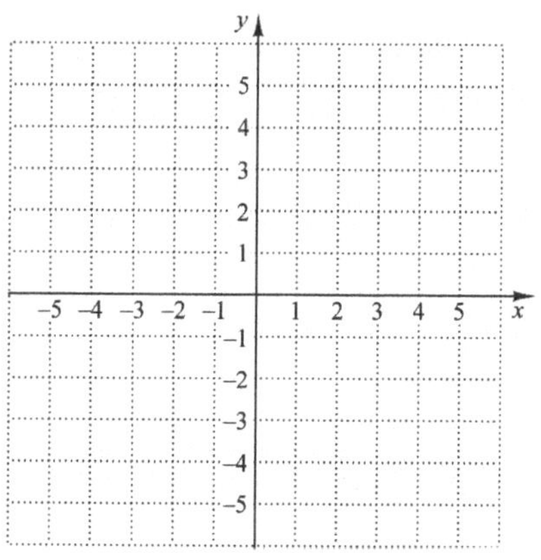

Write an equation for the graph shown.

7.

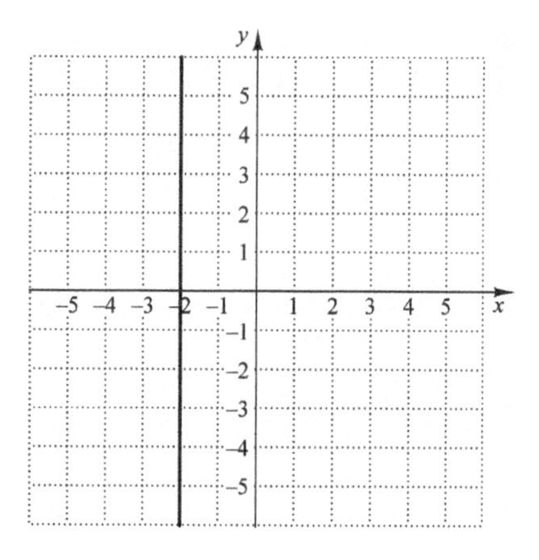

8.

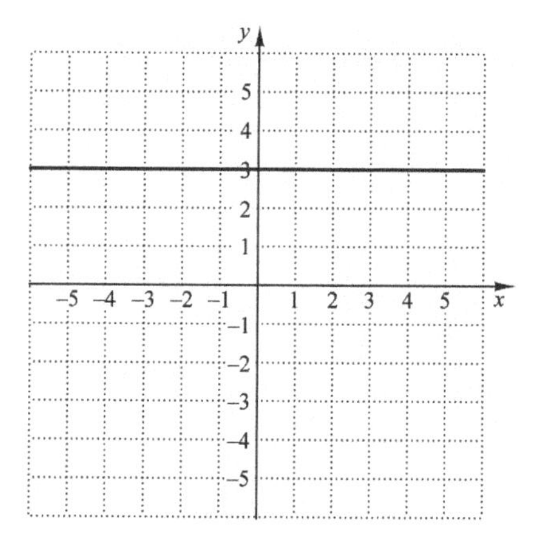

Name: Date:
Instructor: Section:

Chapter 9 GRAPHS OF LINEAR EQUATIONS

9.3 Slope and Applications

Learning Objectives

A Given the coordinates of two points on a line, find the slope of the line, if it exists.

B Find the slope of a line from an equation.

C Find the slope, or rate of change, in an applied problem involving slope.

Key Terms

Use the vocabulary terms listed below to complete each statement in Exercises 1–4.

grade **rise** **run** **slope**

1. The change in y as we move from one point to another is called the ________________ .

2. The change in x as we move from one point to another is called the ________________ .

3. The ________________ of a line is the ratio of the change in y to the change in x.

4. The ________________ of a road is a measure of how steep it is.

GUIDED EXAMPLES AND PRACTICE

Objective A Given the coordinates of two points on a line, find the slope of the line, if it exists.

Review this example for Objective A:

1. Find the slope, if it exists, of the line containing the points (–1, 5) and (2, –3).

Consider (x_1, y_1) to be (–1, 5) and (x_2, y_2) to be (2, –3).

$$\text{Slope} = \frac{\text{the change in } y}{\text{the change in } x} = \frac{y_2 - y_1}{x_2 - x_1}$$

$$= \frac{-3-5}{2-(-1)}$$

$$= \frac{-8}{3}, \text{ or } -\frac{8}{3}$$

Note that we would have gotten the same result if we had considered (x_1, y_1) to be (2, –3) and (x_2, y_2) to be (–1, 5). We can subtract in either order as long as the x-coordinates are subtracted in the same order in which the y-coordinates are subtracted.

Practice this exercise:

1. Find the slope, if it exists, of the line containing the points (6, –2) and (8, –1).

Objective B Find the slope of a line from an equation.

Review this example for Objective B:

2. Find the slope, if it exists, of each line.

a) $3x+4y=8$ b) $y=-1$ c) $x=2$

a) We solve for y to get the equation in the form $y=mx+b$.

$$3x+4y=8$$
$$4y=-3x+8$$
$$y=\frac{-3x+8}{4}$$
$$y=-\frac{3}{4}x+2$$

The slope is $-\frac{3}{4}$.

b) We can think of $y=-1$ as $y=0x-1$. Then we see that the slope is 0. Note that the graph of this equation is a horizontal line. The slope of any horizontal line is 0.

c) The graph of $x=2$ is a vertical line, so the slope is undefined.

Practice this exercise:

2. Find the slope, if it exists, of the line $2x-3y=12$.

Objective C Find the slope, or rate of change, in an applied problem involving slope.

Review this example for Objective C:

3. A road rises 40 m over a horizontal distance of 1250 m. Find the grade of the road.

$$\text{Slope}=\frac{\text{rise}}{\text{run}}=\frac{40}{1250}=0.032=3.2\%$$

Practice this exercise:

3. A set of stairs rises 12 ft over a horizontal distance of 150 ft. Find the grade of the stairs.

Name: Date:
Instructor: Section:

ADDITIONAL EXERCISES

Objective A Given the coordinates of two points on a line, find the slope of the line, if it exists.

For extra help, see Example 1 on page 685 of your text and the Section 9.3 lecture video.

Find the slope, if it exists, of each line.

1.

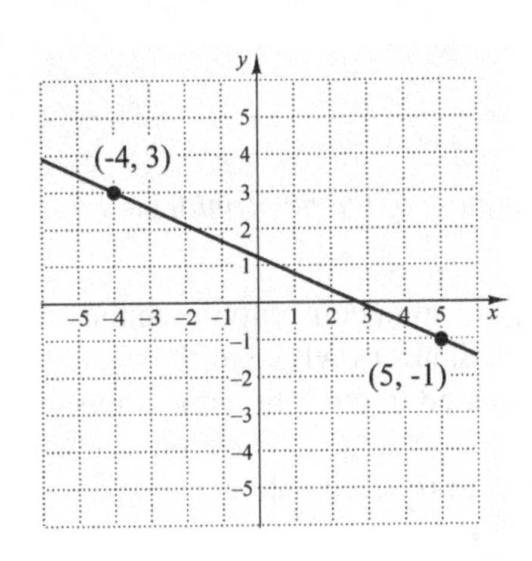

2.

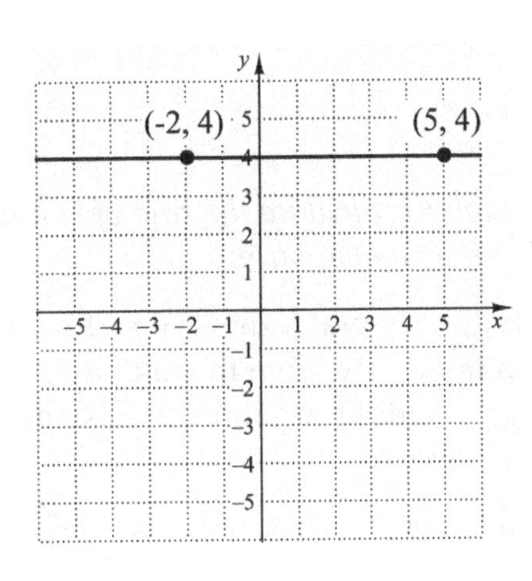

Find the slope, if it exists, of the line containing the given pair of points.

3. $\left(\frac{1}{2},-2\right), \left(\frac{1}{2},-5\right)$

4. $(0.6,-0.8), (0.1,-0.8)$

Objective B Find the slope of a line from an equation.

For extra help, see Examples 2–8 on pages 686–687 of your text and the Section 9.3 lecture video.

Find the slope, if it exists, of each line.

5. $y=-6x+8$

6. $x=\frac{5}{9}$

7. $4x+8y=16$

8. $y=18$

Objective C Find the slope, or rate of change, in an applied problem involving slope.
For extra help, see Examples 9–11 on pages 688–690 of your text and the Section 9.3 lecture video.

9. A road rises 7 ft over a horizontal distance of 280 ft. Find the slope (or grade) of the road.

10. A ramp rises 3 ft over a horizontal distance of 50 ft. Find the grade of the ramp.

Use the graph to calculate the rate of change in which the units of the horizontal axis are used in the denominator.

11. The following graph shows data for a car driven in the city. Find the rate of change in miles per gallon, that is, the gas mileage.

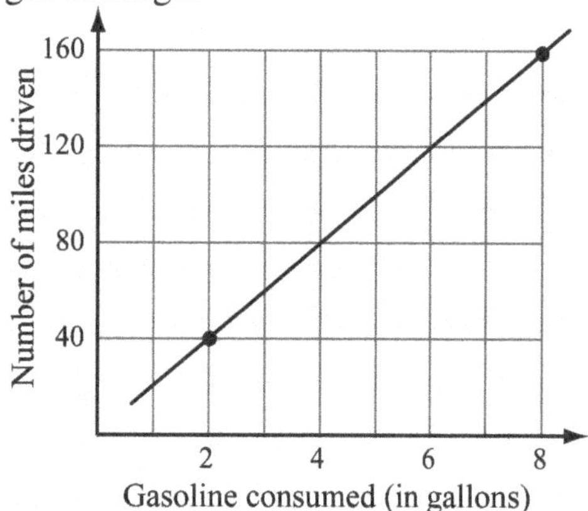

12. The per capita consumption in the United States of whole milk is represented in the following graph. Find the rate of change in consumption with respect to time, in gallons per year.

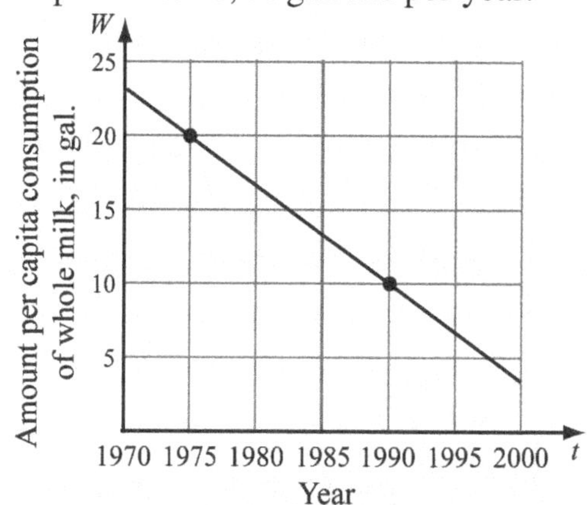

Source: U.S. Department of Agriculture

Name: Date:
Instructor: Section:

Chapter 9 GRAPHS OF LINEAR EQUATIONS

9.4 Equations of Lines

Learning Objectives

A Given an equation in the form $y = mx + b$, find the slope and the y-intercept; find an equation of a line when the slope and the y-intercept are given.

B Find an equation of a line when the slope and a point on the line are given.

C Find an equation of a line when two points on the line are given.

Key Terms

Use the vocabulary terms listed below to complete each statement in Exercises 1–3.

slope **slope-intercept equation** ***y*-intercept**

1. The equation $y = mx + b$ is called the ________________ .

2. The graph of an equation $y = mx + b$ has ________________ m.

3. The graph of an equation $y = mx + b$ has ________________ $(0, b)$.

GUIDED EXAMPLES AND PRACTICE

Objective A Given an equation in the form $y = mx + b$, find the slope and the y-intercept; find an equation of a line when the slope and the y-intercept are given.

Review these examples for Objective A:

1. Find the slope and y-intercept of $3x+5y=15$.

We solve the equation for y:

$$3x+5y=15.$$
$$5y=-3x+15$$
$$\frac{5y}{5}=\frac{-3x+15}{5}$$
$$y=\frac{-3x}{5}+\frac{15}{5}$$
$$y=-\frac{3}{5}x+3$$

Now that the equation is in the form $y = mx + b$, we see that the slope is $-\frac{3}{5}$ and the y-intercept is $(0, 3)$.

Practice these exercises:

1. Find the slope and y-intercept of $4x-3y=12$

2. A line has slope –3 and y-intercept (0, 2). Find an equation of the line.

We substitute –3 for m and 2 for b in the slope-intercept equation.

$$y = mx + b$$
$$y = -3x + 2$$

2. A line has slope 4 and y-intercept (0, –1). Find an equation of the line.

Objective B Find an equation of a line when the slope and a point on the line are given.

Review this example for Objective B:

3. Find an equation of the line with slope –2 that contains the point (3, –1).

We know that the slope is –2, so the equation is $y = -2x + b$. Using the point (3, –1), we substitute 3 for x and –1 for y in $y = -2x + b$.

$$y = -2x + b$$
$$-1 = -2 \cdot 3 + b$$
$$-1 = -6 + b$$
$$5 = b$$

Then the equation is $y = -2x + 5$.

Practice this exercise:

3. Find an equation of the line with slope 4 that contains the point (–2, –5).

Objective C Find an equation of a line when two points on the line are given.

Review this example for Objective C:

4. Find an equation of the line containing the points (4, 3) and (–2, 5).

First, we find the slope.

$$m = \frac{3-5}{4-(-2)} = \frac{-2}{6} = -\frac{1}{3}$$

Thus, $y = -\frac{1}{3}x + b$. Now use either of the given points to find b. We use (4, 3) and substitute 4 for x and 3 for y.

$$y = -\frac{1}{3}x + b$$
$$3 = -\frac{1}{3} \cdot 4 + b$$
$$3 = -\frac{4}{3} + b$$
$$\frac{13}{3} = b$$

Then the equation of the line is $y = -\frac{1}{3}x + \frac{13}{3}$.

Practice this exercise:

4. Find an equation of the line containing the points (–3, –2) and (3, 4).

Name: Date:
Instructor: Section:

ADDITIONAL EXERCISES

Objective A Given an equation in the form $y = mx + b$, find the slope and the y-intercept; find an equation of a line when the slope and the y-intercept are given.

For extra help, see Examples 1–4 on page 695–696 of your text and the Section 9.4 lecture video.

Find the slope and the y-intercept.

1. $y = -5x + 3$

2. $-7x - 3y = 12$

3. $y = 56$

Find an equation of the line with the given slope and y-intercept

4. Slope = –15, y-intercept = (0, 12)

5. Slope = 2.7, y-intercept = (0, –4.3)

Objective B Find an equation of a line when the slope and a point on the line are given.

For extra help, see Examples 5–6 on pages 696–697 of your text and the Section 9.4 lecture video.

Find an equation of the line containing the given point and having the given slope.

6. $(-4, 0)$, $m = -3$

7. $(1, 6)$, $m = \frac{2}{3}$

8. $(5, -1)$, $m = 2$

9. $(0, 7)$, $m = -5$

Objective C Find an equation of a line when two points on the line are given.
For extra help, see Example 7 on pages 697–698 of your text and the Section 9.4 lecture video.

Find an equation of the line containing the given pair of points.

10. $(3, 8)$ and $(1, 10)$

11. $(0, 3)$ and $(5, 2)$

12. $(4, 7)$ and $(-1, 3)$

13. $(-7, 5)$ and $(-1, 3)$

The line graph below describes the amount A, in billions of dollars, of online holiday spending in the United States in years x since 1999.

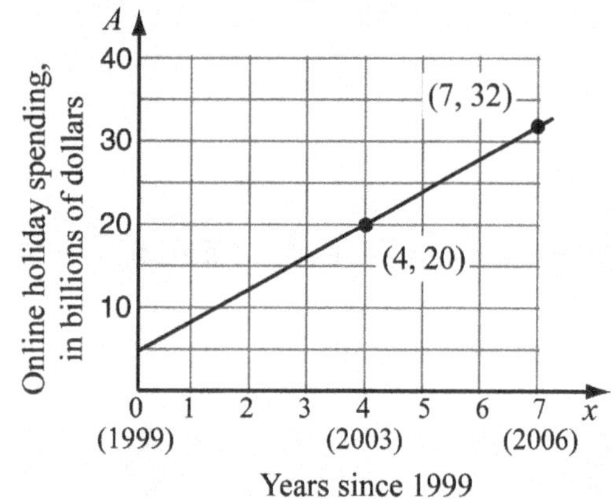

14. a) Find an equation of the line.
b) What is the rate of change of the amount of online holiday spending with respect to time?
c) Use the equation to predict the amount of online holiday spending in 2009.

Name: Date:
Instructor: Section:

Chapter 9 GRAPHS OF LINEAR EQUATIONS

9.5 Graphing Using Slope and the y-intercept

Learning Objectives
A Use the slope and the y-intercept to graph a line.

GUIDED EXAMPLES AND PRACTICE

Objective A Use the slope and the y-intercept to graph a line.

Review this example for Objective A:

1. Graph $y=-\frac{1}{2}x+3$ using the slope and the y-intercept.

   First, we plot the y-intercept (0, 3). Then we think of the slope as $\frac{-1}{2}$. Starting at (0, 3), we find another point by moving 1 unit down (since the numerator is negative and corresponds to the change in y) and 2 units to the right (since the denominator is positive and corresponds to the change in x). We get to the point (2, 2). Returning to the y-intercept, we consider the slope as $\frac{1}{-2}$, and move 1 unit up and 2 units to the left, to find the point (–2, 4).
   We draw a line through the three points we found.

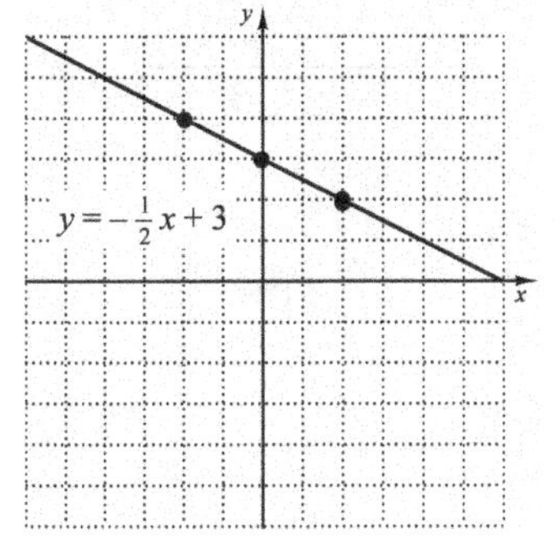

Practice this exercise:

1. Graph $y=\frac{3}{4}x+2$ using the slope and the y-intercept.

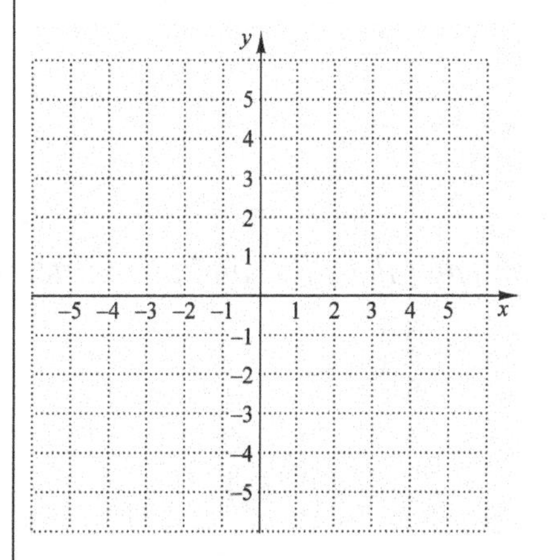

ADDITIONAL EXERCISES

Objective A Use the slope and the *y*-intercept to graph a line.

For extra help, see Examples 1–4 on pages 703–705 of your text and the Section 9.5 lecture video.

Draw a line that has the given slope and y-intercept.

1. Slope $= -2$; y-intercept $(0, 3)$

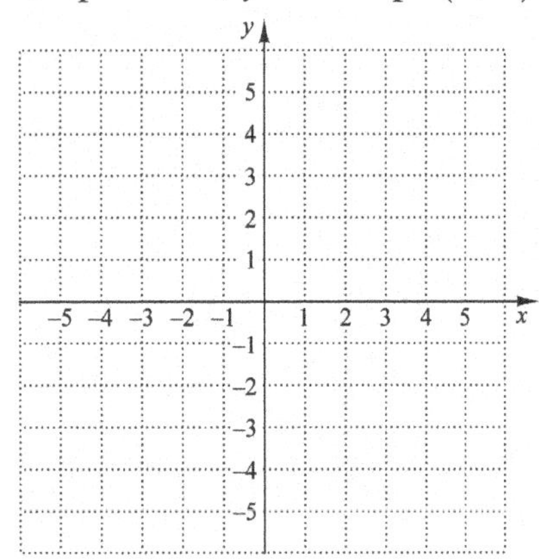

2. Slope $= 3$; y-intercept $(0, -5)$

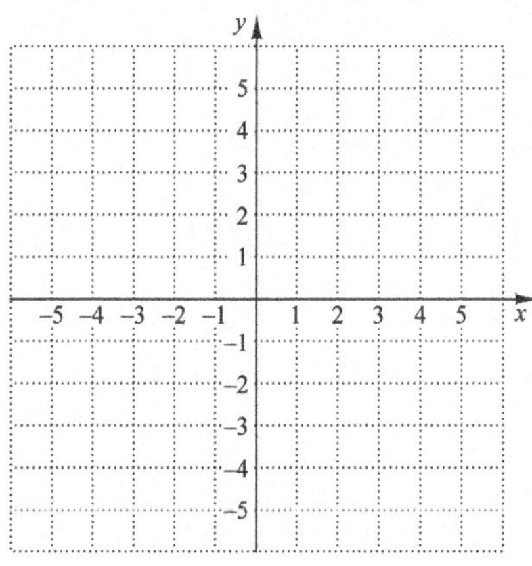

Graph using the slope and the y-intercept.

3. $y = -\frac{2}{5}x - 2$

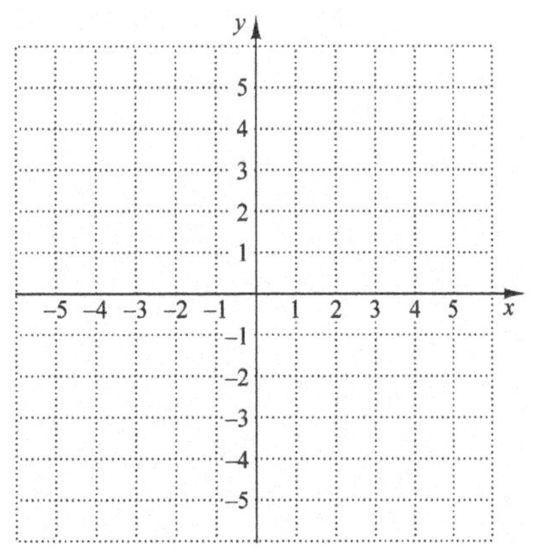

4. $2x - y = 1$

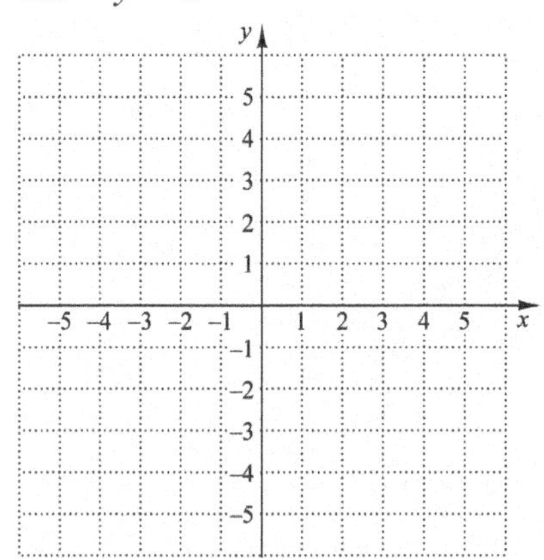

Name: Date:
Instructor: Section:

Chapter 9 GRAPHS OF LINEAR EQUATIONS

9.6 Parallel and Perpendicular Lines

Learning Objectives
A Determine whether the graphs of two linear equations are parallel.
B Determine whether the graphs of two linear equations are perpendicular.

Key Terms
Use the vocabulary terms listed below to complete each statement in Exercises 1–4.

horizontal **parallel** **perpendicular** **vertical**

1. Two nonvertical lines are ____________________ if they have the same slope and different y-intercepts.

2. Two nonvertical lines are ____________________ if the product of their slopes is –1.

3. Parallel ____________________ lines have equations $x = p$ and $x = q$, where $p \neq q$.

4. If one equation in a pair of perpendicular lines is vertical, then the other is ____________.

GUIDED EXAMPLES AND PRACTICE

Objective A Determine whether the graphs of two linear equations are parallel.

Review this example for Objective A:

1. Determine whether the graphs of the lines $y = -2x + 1$ and $4x + 2y = 5$ are parallel.

The first equation is in slope-intercept form $(y = mx + b)$, so we see that it has slope –2 and y-intercept (0, 1).
We solve the second equation for y.

$$4x + 2y = 5$$
$$2y = -4x + 5$$
$$y = \frac{1}{2}(-4x + 5)$$
$$y = -2x + \frac{5}{2}$$

Thus, the slope of the second line is –2 and its y-intercept is $\left(0, \frac{5}{2}\right)$. Since the two lines have the same slope, –2, and different y-intercepts, (0, 1) and $\left(0, \frac{5}{2}\right)$, they are parallel.

Practice this exercise:

1. Determine whether the graphs of the lines $x + y = 3$ and $x - y = 3$ are parallel.

Objective B Determine whether the graphs of two linear equations are perpendicular.

Review this example for Objective B:

2. Determine whether the graphs of the lines $2x + y = 4$ and $x + 2y = 3$ are perpendicular.

We solve each equation for y in order to determine the slope.

a) $2x + y = 4$

$y = -2x + 4$

b) $x + 2y = 3$

$2y = -x + 3$

$y = \frac{1}{2}(-x + 3)$

$y = -\frac{1}{2}x + \frac{3}{2}$

The slopes are -2 and $-\frac{1}{2}$. The product of the slopes is $-2\left(-\frac{1}{2}\right) = 1$. Since the product of the slopes is not -1, the lines are not perpendicular.

Practice this exercise:

2. Determine whether the graphs of the lines $3x - 2y = 4$ and $4x + 6y = 3$ are perpendicular.

ADDITIONAL EXERCISES

Objective A Determine whether the graphs of two linear equations are parallel.

For extra help, see Example 1 on page 710 of your text and the Section 9.6 lecture video.

Determine whether the graphs of the equations are parallel lines.

1. $x - y = 7,$
$y = x + 3$

2. $y + 2 = 5x,$
$5x + y = 3$

3. $4y + 18x = 12,$
$y + 2.5 = 5x$

4. $y = 3,$
$y = -5$

Objective B Determine whether the graphs of two linear equations are perpendicular.

For extra help, see Example 2 on page 711 of your text and the Section 9.6 lecture video.

Determine whether the graphs of the equations are perpendicular lines.

5. $y = 5 - 2x,$
$2y + x = 1$

6. $y = x + 1,$
$x + y = 4$

7. $y = 0.4x + 1.3$
$x + 2.5y = 10.7$

8. $x = 7,$
$y = 0$

Determine whether the graphs of the equations are parallel, perpendicular, or neither.

9. $5y = 4x - 7$
$4x + 5y = 8$

Name: Date:
Instructor: Section:

Chapter 9 GRAPHS OF LINEAR EQUATIONS

9.7 Graphing Inequalities in Two Variables

Learning Objectives

A Determine whether an ordered pair of numbers is a solution of an inequality in two variables.

B Graph linear inequalities.

Key Terms

Use the vocabulary terms listed below to complete each statement in Exercises 1–4.

linear inequality **half-plane** **solution** **test point**

1. Since $3 - 4 < 0$, (3, 4) is a(n) ________________ of $x - y < 0$.
2. The inequality $x - y < 0$ is an example of a(n) ________________ .
3. The graph of $x - y < 0$ is a(n) ________________ .
4. When graphing $x - y < 0$, we use a(n) ________________ to determine whether to shade above or below the line $x - y = 0$.

GUIDED EXAMPLES AND PRACTICE

Objective A Determine whether an ordered pair of numbers is a solution of an inequality in two variables.

Review this example for Objective A:

1. Determine whether (4, –1) is a solution of $x + 3y \geq 5$.

   Use alphabetical order to replace x with 4 and y with –1.

$$\begin{array}{r|l} \multicolumn{2}{c}{x + 3y \geq 5} \\ \hline 4 + 3(-1) & 5 \\ 4 - 3 & \\ 1 & \quad \text{FALSE} \end{array}$$

   Since $1 \geq 5$ is false, (4, –1) is not a solution.

Practice this exercise:

1. Determine whether (–2, 5) is a solution of $3x + y \leq -1$.

Objective B Graph linear inequalities.

Review this example for Objective B:

2. Graph: $y < x - 2$.

First graph the line $y = x - 2$. The intercepts are (0, –2) and (2, 0). We draw a dashed line since the inequality symbol is <. Next, choose a test point not on the line and determine if it is a solution of the inequality. We choose (0, 0), since it is usually an easy point to use.

y	$< x - 2$	
0	$0 - 2$	
0	-2	FALSE

Since (0, 0) is not a solution, we shade the half-plane that does not contain (0, 0).

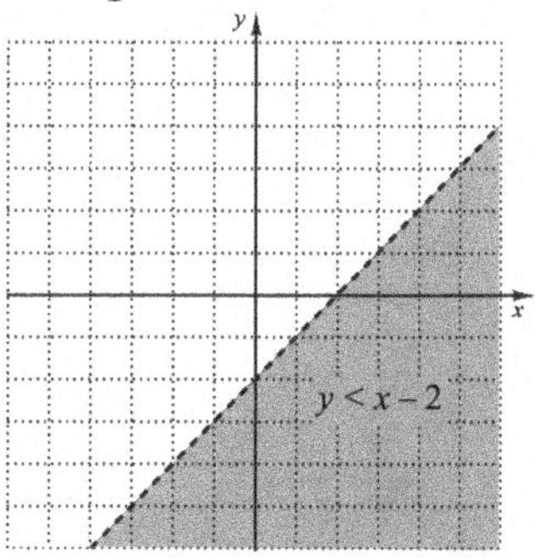

Practice this exercise:

2. Graph: $x + y \leq 5$.

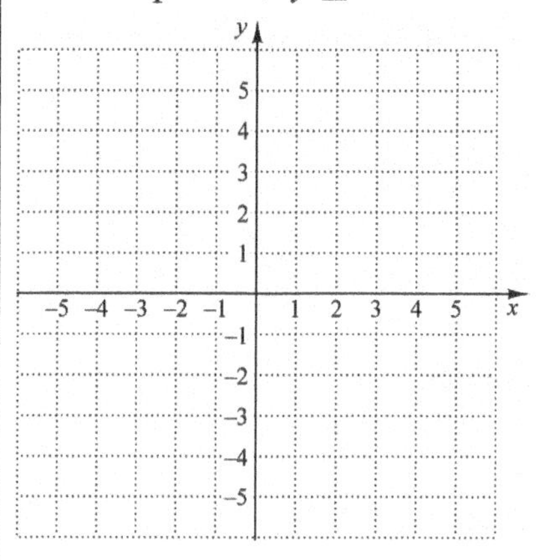

ADDITIONAL EXERCISES

Objective A Determine whether an ordered pair of numbers is a solution of an inequality in two variables.

For extra help, see Examples 1–2 on page 714 of your text and the Section 9.7 lecture video.

1. Determine whether (–4, 1) is a solution of $-x - y < 0$.

2. Determine whether (3, –5) is a solution of $2x - 3y \geq 10$.

3. Determine whether $\left(-\frac{1}{3}, -\frac{1}{2}\right)$ is a solution of $4x - 3y \leq 1$.

4. Determine whether (–10, 6) is a solution of $x + 0y > 8$.

Name: Date:
Instructor: Section:

Objective B Graph linear inequalities.

For extra help, see Examples 3–7 on pages 714–717 of your text and the Section 9.7 lecture video.

Graph on a plane.

5. $x - y > 3$

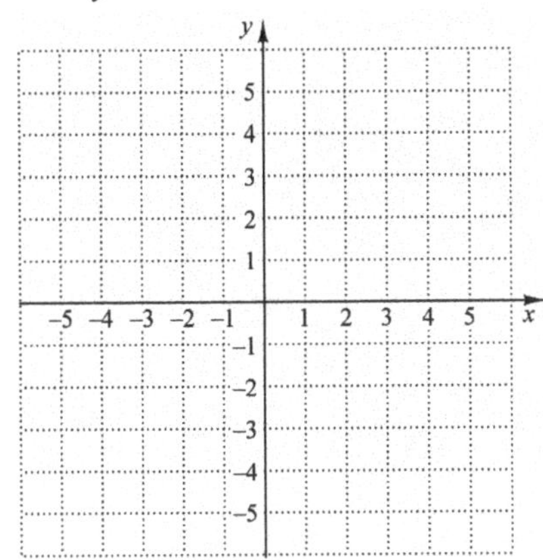

6. $y \geq 3 - 4x$

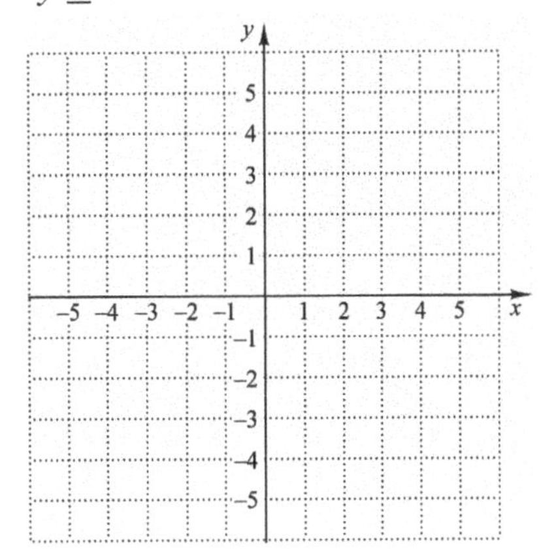

5. $2x - 5y > 10$

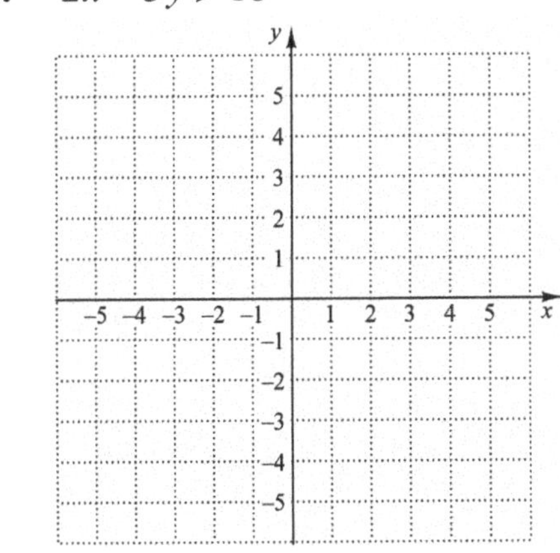

6. $y \leq -1$

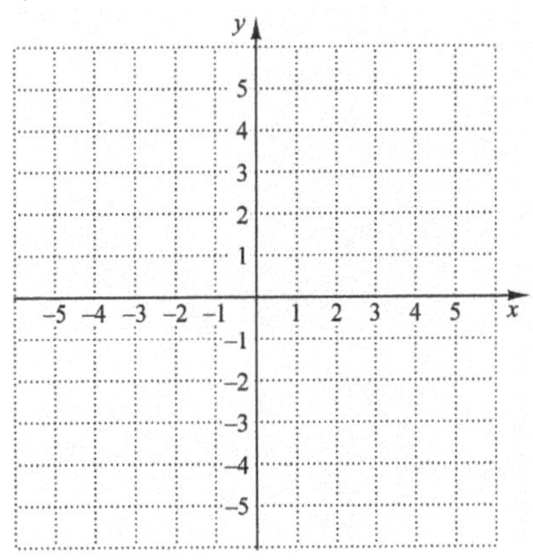

7. $x \geq 3$

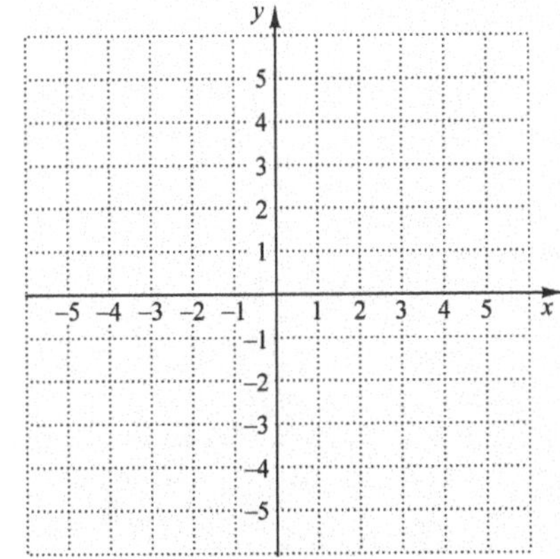

Name: Date:
Instructor: Section:

Chapter 10 POLYNOMIALS: OPERATIONS

10.1 Integers and Exponents

Learning Objectives
A Tell the meaning of exponential notation.
B Evaluate exponential expressions with exponents of 0 and 1.
C Evaluate algebraic expressions containing exponents.
D Use the product rule to multiply exponential expressions with like bases.
E Use the quotient rule to divide exponential expressions with like bases.
F Express an exponential expression involving negative exponents with positive exponents.

Key Terms
Use the vocabulary terms listed below to complete each statement in Exercises 1–4.

base **exponent** ***x*-cubed** ***x*-squared**

1. In the expression 3^7, 7 is the ________________.
2. In the expression 3^7, 3 is the ________________.
3. We often read x^2 as "________________."
4. We often read x^3 as "________________."

GUIDED EXAMPLES AND PRACTICE

Objective A Tell the meaning of exponential notation.

Review this example for Objective A:

1. What is the meaning of 2^4? Of $(5x)^3$?

   2^4 means $2 \cdot 2 \cdot 2 \cdot 2$.

   $(5x)^3$ means $5x \cdot 5x \cdot 5x$.

Practice this exercise:

1. What is the meaning of y^5?

Objective B Evaluate exponential expressions with exponents of 0 and 1.

Review this example for Objective B:

2. Evaluate 3^1 and $(-4)^0$.

   $3^1 = 3$

   $(-4)^0 = 1$

Practice this exercise:

2. Evaluate 2.8^0.

Objective C Evaluate algebraic expressions containing exponents.

Review this example for Objective C:

3. Evaluate n^5 for $n = -1$.

We substitute –1 for n and then evaluate the power.

$$\begin{aligned} n^5 &= (-1)^5 \\ &= (-1)\cdot(-1)\cdot(-1)\cdot(-1)\cdot(-1) \\ &= -1 \end{aligned}$$

Practice this exercise:

3. Evaluate $4t^3$ for $t = -2$.

Objective D Use the product rule to multiply exponential expressions with like bases.

Review this example for Objective D:

4. Multiply and simplify: $y^2 \cdot y^6$.

Since the bases are the same, we add the exponents.

$$y^2 \cdot y^6 = y^{2+6} = y^8$$

Practice this exercise:

4. Multiply and simplify: $x^3 \cdot x^4$.

Objective E Use the quotient rule to divide exponential expressions with like bases.

Review this example for Objective E:

5. Divide and simplify: $\dfrac{a^{10}b^4}{a^2b}$.

$$\begin{aligned} \frac{a^{10}b^4}{a^2b} &= \frac{a^{10}}{a^2}\cdot\frac{b^4}{b} \\ &= a^{10-2}b^{4-1} \\ &= a^8b^3 \end{aligned}$$

Practice this exercise:

5. Divide and simplify: $\dfrac{x^3y^7}{x^2y^4}$.

Objective F Express an exponential expression involving negative exponents with positive exponents.

Review this example for Objective F:

6. Express using positive exponents.

a) $3x^{-8}$ b) $\dfrac{1}{y^{-2}}$

a) $3x^{-8} = 3\cdot\dfrac{1}{x^8} = \dfrac{3}{x^8}$

b) $\dfrac{1}{y^{-2}} = y^{-(-2)} = y^2$

Practice this exercise:

6. Express $2n^{-5}$ using positive exponents.

Name: Date:
Instructor: Section:

ADDITIONAL EXERCISES

Objective A Tell the meaning of exponential notation.

For extra help, see Example 1 on page 734 of your text and the Section 10.1 lecture video.
What is the meaning of the following?

1. 7^3

2. $\left(\frac{3}{4}\right)^5$

3. $(8x)^4$

4. $-9y^2$

Objective B Evaluate exponential expressions with exponents of 0 and 1.

For extra help, see Example 2 on page 735 of your text and the Section 10.1 lecture video.
Evaluate.

5. $\left(\frac{3}{4}\right)^0$

6. 5.17^0

7. $(mn)^0$, $m,n \neq 0$

8. mn^0, $m,n \neq 0$

Objective C Evaluate algebraic expressions containing exponents.

For extra help, see Examples 3–6 on pages 735–736 of your text and the Section 10.1 lecture video.
Evaluate.

9. x^5, when $x = 2$

10. $-t^2$, when $t = -5$

11. $a^2 - 3$, when $a = -4$

12. $x^1 + 10$ and $x^0 + 10$, when $x = 12$

Objective D Use the product rule to multiply exponential expressions with like bases.
For extra help, see Examples 7–11 on page 736 of your text and the Section 10.1 lecture video.
Multiply and simplify.

13. $7^4 \cdot 7^6$

14. $(4p)^5(4p)^3$

15. $x^{11} \cdot x^{10}$

16. $n^0 \cdot n^9$

Objective E Use the quotient rule to divide exponential expressions with like bases.
For extra help, see Examples 12–15 on page 737 of your text and the Section 10.1 lecture video.
Divide and simplify.

17. $\frac{11^8}{11^3}$

18. $\frac{x^6}{x^2}$

19. $\frac{t^{10}}{t}$

20. $\frac{(3w)^5}{(3w)^5}$

Objective F Express an exponential expression involving negative exponents with positive exponents.
For extra help, see Examples 16–27 on page 739 of your text and the Section 10.1 lecture video.

21. Express $\frac{1}{x^{-5}}$ using positive exponents.

22. Express $\frac{1}{a^6}$ using negative exponents.

23. Multiply and simplify: $c^8 \cdot c^{-4} \cdot c^{-5}$.

24. Divide and simplify: $\frac{x}{x^{-3}}$.

Name: Date:
Instructor: Section:

Chapter 10 POLYNOMIALS: OPERATIONS

10.2 Exponents and Scientific Notation

Learning Objectives
A Use the power rule to raise powers to powers.
B Raise a product to a power and a quotient to a power.
C Convert between scientific notation and decimal notation.
D Multiply and divide using scientific notation.
E Solve applied problems using scientific notation.

Key Terms
Use the vocabulary terms listed below to complete each statement in Exercises 1–2.

exponential notation **scientific notation**

1. The expression 4^5 is written in ________________ .

2. The expression 1.3×10^8 is written in ________________ .

GUIDED EXAMPLES AND PRACTICE

Objective A Use the power rule to raise powers to powers.

Review this example for Objective A:

1. Simplify $\left(y^{-3}\right)^2$.

$$\left(y^{-3}\right)^2 = y^{-3\cdot 2} = y^{-6} = \frac{1}{y^6}$$

Practice this exercise:

1. Simplify $\left(b^{-4}\right)^{-3}$.

Objective B Raise a product to a power and a quotient to a power.

Review these examples for Objective B:

2. Simplify $\left(3x^{-4}y^2\right)^3$.

$$\begin{aligned}\left(3x^{-4}y^2\right)^3 &= 3^3\left(x^{-4}\right)^3\left(y^2\right)^3\\ &= 27x^{-12}y^6\\ &= \frac{27y^6}{x^{12}}\end{aligned}$$

3. Simplify $\left(\frac{4}{a^5}\right)^3$.

$$\left(\frac{4}{a^5}\right)^3 = \frac{4^3}{\left(a^5\right)^3} = \frac{64}{a^{15}}$$

Practice these exercises:

2. Simplify $\left(8a^3b^{-5}\right)^2$.

3. Simplify $\left(\frac{y^4}{7}\right)^2$.

Objective C Convert between scientific notation and decimal notation.

Review these examples for Objective C:

4. Convert 4.208×10^6 to decimal notation.

The exponent is positive, so the number is large. We move the decimal point right 6 places.

$$4.208 \times 10^6 = 4,208,000$$

5. Convert 0.00048 to scientific notation.

The number is small, so the exponent is negative. Move the decimal point right 4 places.

$$0.00048 = 4.8 \times 10^{-4}$$

Practice these exercises:

4. Convert 3.01×10^{-4} to decimal notation.

5. Convert 567,000 to scientific notation.

Objective D Multiply and divide using scientific notation.

Review these examples for Objective D:

6. Multiply: $(4.2 \times 10^8) \cdot (3.1 \times 10^{-3})$.

$(4.2 \times 10^8) \cdot (3.1 \times 10^{-3})$

$= (4.2 \cdot 3.1) \times (10^8 \cdot 10^{-3})$

$= 13.02 \times 10^5$ Not in scientific notation; 13.02 is greater than 10.

$= (1.302 \times 10^1) \times 10^5$ Substituting 1.302×10^1 for 13.02

$= 1.302 \times (10^1 \times 10^5)$ Using the associative law

$= 1.302 \times 10^6$ Adding exponents; the answer is now in scientific notation

7. Divide: $(5.7 \times 10^6) \div (9.5 \times 10^{-1})$.

$(5.7 \times 10^6) \div (9.5 \times 10^{-1})$

$= \dfrac{5.7 \times 10^6}{9.5 \times 10^{-1}}$

$= \dfrac{5.7}{9.5} \times \dfrac{10^6}{10^{-1}}$

$= 0.6 \times 10^7$ Not in scientific notation

$= (6.0 \times 10^{-1}) \times 10^7$ Substituting 6.0×10^{-1} for 0.6

$= 6.0 \times (10^{-1} \times 10^7)$ Using the associative law

$= 6.0 \times 10^6$ Adding exponents

Practice these exercises:

6. Multiply and express the result in scientific notation: $(2.1 \times 10^3) \cdot (4.5 \times 10^{-8})$.

7. Divide and express the result in scientific notation: $(3.3 \times 10^2) \div (4.4 \times 10^{-10})$.

Name: Date:
Instructor: Section:

Objective E Solve applied problems using scientific notation.

Review this example for Objective E:

8. In the summer, about 1.3088×10^8 L of water spill over the Canadian side of Niagra Falls in 1 min. How much water falls in 1 sec? Express the answer in scientific notation.

We divide 1.3088×10^8 by 60, expressing 60 in scientific notation as 6×10.

$$\frac{1.3088\times10^8}{6\times10}=\frac{1.3088}{6}\times\frac{10^8}{10}\approx0.218\times10^7$$
$$\approx(2.18\times10^{-1})\times10^7$$
$$\approx2.18\times(10^{-1}\times10^7)$$
$$\approx2.18\times10^6$$

About 2.18×10^6 L of water spill over the falls in 1 sec.

Practice this exercise:

8. Using the information given in the example at the left, find the amount of water that spills over the falls in 1 hr. Express the answer in scientific notation.

ADDITIONAL EXERCISES

Objective A Use the power rule to raise powers to powers.

For extra help, see Examples 1–5 on page 744 of your text and the Section 10.2 lecture video.

Simplify.

1. $(3^2)^4$

2. $(a^5)^{-3}$

3. $(x^{-2})^{-7}$

4. $(t^8)^{-5}$

Objective B Raise a product to a power and a quotient to a power.

For extra help, see Examples 6–17 on pages 745–746 of your text and the Section 10.2 lecture video.

Simplify.

5. $(m^2n^{-5})^{-4}$

6. $(-5x^4y^{-3})^2$

7. $\left(\frac{x^5}{3}\right)^2$

8. $\left(\frac{xy^2}{w^3z}\right)^{-3}$

Objective C Convert between scientific notation and decimal notation.
For extra help, see Examples 18–21 on page 747 of your text and the Section 10.2 lecture video.
Convert each number to scientific notation.

9. 640,000,000

10. 0.00000214

Convert each number to decimal notation.

11. 4.32×10^{-5}

12. 7.439×10^{9}

Objective D Multiply and divide using scientific notation.
For extra help, see Examples 22–25 on pages 748–749 of your text and the Section 10.2 lecture video.
Multiply or divide and write scientific notation for each result.

13. $(4.1 \times 10^{-4})(3.5 \times 10^{6})$

14. $(3.8 \times 10^{2})(5.4 \times 10^{-6})$

15. $\dfrac{7.2 \times 10^{5}}{1.8 \times 10^{-2}}$

16. $\dfrac{2.4 \times 10^{-3}}{4.8 \times 10^{-7}}$

Objective E Solve applied problems using scientific notation.
For extra help, see Examples 26–27 on pages 749–750 of your text and the Section 10.2 lecture video.
Solve.

17. A nanometer is 0.000000001 m. A nanowire with a diameter of 360 nanometers has been used in experiments on the transmission of light. Find the diameter of such a wire in meters.

18. The area of Hong Kong is 412 mi^2. It is estimated that the population of Hong Kong will be 9,600,000 in 2050. Find the number of people per square mile in 2050.

19. Every second the sun releases 2×10^{29} tiny particles called neutrinos. How many neutrinos are released in 1 hour?

Name: Date:
Instructor: Section:

Chapter 10 POLYNOMIALS: OPERATIONS

10.3 Introduction to Polynomials

Learning Objectives
A Evaluate a polynomial for a given value of the variable.
B Identify the terms of a polynomial.
C Identify the like terms of a polynomial.
D Identify the coefficients of a polynomial.
E Collect the like terms of a polynomial.
F Arrange a polynomial in descending order, or collect the like terms and then arrange in descending order.
G Identify the degree of each term of a polynomial and the degree of the polynomial.
H Identify the missing terms of a polynomial.
I Classify a polynomial as a monomial, a binomial, a trinomial, or none of these.

Key Terms
Use the vocabulary terms listed below to complete each statement in Exercises 1–8.

binomial	**coefficient**	**degree**	**descending order**
like terms	**monomial**	**trinomial**	**value**

1. A polynomial with just one terms is a(n) ___________________ .

2. When we replace the variable in a polynomial with a number, the polynomial represents a(n) ___________________ of the polynomial.

3. Terms that have the same variable raised to the same power are ___________________ .

4. The ___________________ of $4x^3$ is 4.

5. The ___________________ of $4x^3$ is 3.

6. When a polynomial is arranged in ___________________, the exponents decrease from left to right.

7. A polynomial with two terms is a(n) ___________________ .

8. A polynomial with three terms is a(n) ___________________ .

GUIDED EXAMPLES AND PRACTICE

Objective A Evaluate a polynomial for a given value of the variable.

Review this example for Objective A:

1. Evaluate $3x^2 - 5x + 7$ for $x = -2$.

Replace x with -2 and simplify.

$$\begin{aligned} 3(-2)^2 - 5(-2) + 7 &= 3 \cdot 4 - 5(-2) + 7 \\ &= 12 + 10 + 7 \\ &= 29 \end{aligned}$$

Practice this exercise:

1. Evaluate $-x^2 + 3x - 4$ for $x = -1$.

Objective B Identify the terms of a polynomial.

Review this example for Objective B:

2. Identify the terms of the polynomial $3y^3 - 2y^2 - 5y + 1$.

$$3y^3 - 2y^2 - 5y + 1 = 3y^3 + (-2y^2) + (-5y) + 1$$

Then the terms are $3y^3, -2y^2, -5y,$ and 1.

Practice this exercise:

2. Identify the terms of the polynomial $-5y^4 + 3y^2 - 2$.

Objective C Identify the like terms of a polynomial.

Review this example for Objective C:

3. Identify the like terms of the polynomial $3x^2 - 4x + 5 - 6x^2 - 2x + 7$.

$3x^2$ and $-6x^2$ have the same variable raised to the same power, so they are like terms.

$-4x$ and $-2x$ have the same variable raised to the same power, so they are like terms.

The constant terms 5 and 7 are also like terms, because they can be thought of as $5x^0$ and $7x^0$, respectively.

Practice this exercise:

3. Identify the like terms of the polynomial $4y^5 - 7 - 3y^5 + 4$.

Objective D Identify the coefficients of a polynomial.

Review this example for Objective D:

4. Identify the coefficients of each term of the polynomial $5y^6 - 10y^2 + 4$.

The coefficient of $5y^6$ is 5.
The coefficient of $-10y$ is -10.
The coefficient of 4 is 4.

Practice this exercise:

4. Identify the coefficients of each term of the polynomial $-8x^3 + 4x^2 - 7$.

Name: Date:
Instructor: Section:

Objective E Collect the like terms of a polynomial.

Review this example for Objective E:

5. Collect like terms: $5x^4 - 6x^2 - 3x^4 + 1$.

$$5x^4 - 6x^2 - 3x^4 + 1 = (5-3)x^4 - 6x^2 + 1$$
$$= 2x^4 - 6x^2 + 1$$

Practice this exercise:

5. Collect like terms:
$4x^3 - 2x^2 + 3x^2 - 5$.

Objective F Arrange a polynomial in descending order, or collect the like terms and then arrange in descending order.

Review this example for Objective F:

6. Collect like terms and then arrange in descending order: $8 + 3x^2 - 4x - x^2 - 4 + 5x$.

$$8 + 3x^2 - 4x - x^2 - 4 + 5x$$
$$= 4 + 2x^2 + x$$
$$= 2x^2 + x + 4$$

Practice this exercise:

6. Collect like terms and then arrange in descending order:
$x - x^2 + x + 7x - 9 - 2x^2$.

Objective G Identify the degree of each term of a polynomial and the degree of the polynomial.

Review this example for Objective G:

7. Identify the degree of each term and the degree of the polynomial: $2x^4 - 6x^3 + x - 4$.

The degree of $2x^4$ is the exponent of the variable, 4.

The degree of $-6x^3$ is the exponent of the variable, 3.

The degree of x is the exponent of the variable, 1, since $x = x^1$.

The degree of –4 is the exponent of the variable 0, since $-4 = -4x^0$.

The largest of the degrees of the terms is 4, so the degree of the polynomial is 4.

Practice this exercise:

7. Identify the degree of the polynomial:
$-5x - x^3 + 8x^2 + 7$.

Objective H Identify the missing terms of a polynomial.

Review this example for Objective H:

8. Identify the missing terms in the polynomial $4x^3 - x$.

There are no terms with degree 2 or 0. (A term with degree 0 is a constant term.) Thus the x^2- and x^0-terms are missing.

Practice this exercise:

8. Identify the missing terms in the polynomial $6x^4 - x^2 + 7$.

Objective I Classify a polynomial as a monomial, a binomial, a trinomial, or none of these.

Review this example for Objective I:

9. Classify each of the following as a monomial, binomial, trinomial, or none of these.

a) $x^2 - 7$ b) $2x^3 - x^2 + 5x + 6$

a) $x^2 - 7$ has two terms, so it is a binomial.

b) $2x^3 - x^2 + 5x + 6$ has more than three terms, so it is none of these.

Practice this exercise:

9. Classify $-6x^7$ as a monomial, binomial, trinomial, or none of these.

ADDITIONAL EXERCISES

Objective A Evaluate a polynomial for a given value of the variable.

For extra help, see Examples 1–5 on pages 756–758 of your text and the Section 10.3 lecture video.

Evaluate each polynomial for x = 3.

1. $-5x + 2$

2. $6 - 8x + x^2$

Evaluate each polynomial for x = –2.

3. $2x^2 + 6x - 1$

4. $-x^3 + 3x^2 + 7x + 5$

Objective B Identify the terms of a polynomial.

For extra help, see Examples 6–9 on page 759 of your text and the Section 10.3 lecture video.

Identify the terms of each polynomial.

5. $5 - 4x^2 + x$

6. $-5x^6 + \frac{2}{3}x^4 - 3x + 7$

Objective C Identify the like terms of a polynomial.

For extra help, see Examples 10–11 on page 759 of your text and the Section 10.3 lecture video.

Identify the like terms of each polynomial.

7. $4x^3 - 6x^2 + 2x^3 - x^2$

8. $2x^4 + 3x - 7 + x^4 - 5x + 11$

9. $y^4 + 3y - y^4 - 5y + 1$

Name: Date:
Instructor: Section:

Objective D Identify the coefficients of a polynomial.
For extra help, see Example 12 on page 760 of your text and the Section 10.3 lecture video.
Identify the coefficients of each term of the polynomial.

10. $3x^2+\frac{1}{4}x-7$

11. $-4x^3-5x^2+1.3x+8$

12. $-x^3+8x-9$

Objective E Collect the like terms of a polynomial.
For extra help, see Examples 13–18 on pages 760–761 of your text and the Section 10.3 lecture video.
Collect like terms.

13. $x-10x$

14. $2x^3+3x-x^3-10x$

15. $\frac{1}{2}x^4+7+\frac{3}{4}x^4-x^2-18$

16. $10x^2-8x^3+3x^2+11x^3-12x^2$

Objective F Arrange a polynomial in descending order, or collect the like terms and then arrange in descending order.
For extra help, see Examples 19–21 on page 761 of your text and the Section 10.3 lecture video.
Arrange the polynomial in descending order.

17. $2x^2-x+5x^3-6+x^5$

18. $12y^3+y^7-y+6+18y^2$

Collect like terms and then arrange in descending order.

19. $-x-5x^2-8x+14x^2+11$

20. $3x-\frac{2}{3}+5x^3-\frac{1}{3}+8x^5-15x$

Objective G Identify the degree of each term of a polynomial and the degree of the polynomial.

For extra help, see Examples 22–23 on pages 761–762 of your text and the Section 10.3 lecture video.

Identify the degree of each term of the polynomial and the degree of the polynomial.

21. $6x^4 - 5x + 7$

22. $3x + 1 + 8x^3 - 4x^6$

23. $3 - 7x^3 + 6x^5 + 2x$

Objective H Identify the missing terms of a polynomial.

For extra help, see Examples 24–26 on page 762 of your text and the Section 10.3 lecture video.

Identify the missing terms in each polynomial.

24. $x^6 - x^2 + x$

25. $3x^4 - 2x^3 - x - 4$

Write each polynomial in two ways: with its missing terms and by leaving space for them.

26. $x^3 - 8$

27. $4x^5 - x^3 + x^2 - 1$

Objective I Classify a polynomial as a monomial, a binomial, a trinomial, or none of these.

For extra help, see Example 27 on page 763 of your text and the Section 10.3 lecture video.

Classify each polynomial as a monomial, a binomial, a trinomial, or none of these.

28. $-3x^2$

29. $2x^2 - 8x + 5$

30. $4x^3 + 6x^2 + 6x + 4$

31. $x^2 - 1$

Name: Date:
Instructor: Section:

Chapter 10 POLYNOMIALS: OPERATIONS

10.4 Addition and Subtraction of Polynomials

Learning Objectives
A Add polynomials.
B Simplify the opposite of a polynomial.
C Subtract polynomials.
D Use polynomials to represent perimeter and area.

Key Terms
Use the vocabulary terms listed below to complete each statement in Exercises 1–2.

opposite **sign**

1. To find the additive inverse of a polynomial, change the ________________ of every term.

2. To subtract polynomials, add the ________________ of the polynomial being subtracted.

GUIDED EXAMPLES AND PRACTICE

Objective A Add polynomials.

Review this example for Objective A:

1. Add: $(5x^3+x-7)+(2x^3-4x+3)$.

Use the commutative property to pair like terms.
$(5x^3+x-7)+(2x^3-4x+3)$
$=(5x^3+2x^3)+(x-4x)+(-7+3)$
$=7x^3-3x-4$

Practice this exercise:

1. Add:
$(6x^3-5x^2-1)+(x^3-3x^2+4)$.

Objective B Simplify the opposite of a polynomial.

Review this example for Objective B:

2. Simplify: $-(10x^2-5x+2)$.

$-(10x^2-5x+2)=-10x^2+5x-2$

Practice this exercise:

2. Simplify: $-(-x^3+3x-4)$.

Objective C Subtract polynomials.

Review this example for Objective C:

3. Subtract: $(4x^2 - x + 3) - (6x^2 - 4x - 1)$.

Add the opposite, change the sign of every term in the second polynomial, and combine like terms.

$(4x^2 - x + 3) - (6x^2 - 4x - 1)$
$= 4x^2 - x + 3 - 6x^2 + 4x + 1$
$= (4x^2 - 6x^2) + (-x + 4x) + (3 + 1)$
$= -2x^2 + 3x + 4$

Practice this exercise:

3. Subtract:
$(x^3 - x + 2) - (5x^3 + x^2 - 8)$.

Objective D Use polynomials to represent perimeter and area.

Review this example for Objective D:

4. A square sandbox that is x ft on a side is placed on a lawn that is 12 ft by 18 ft. Find a polynomial for the area of the lawn not covered by the sandbox.

First we make a drawing.

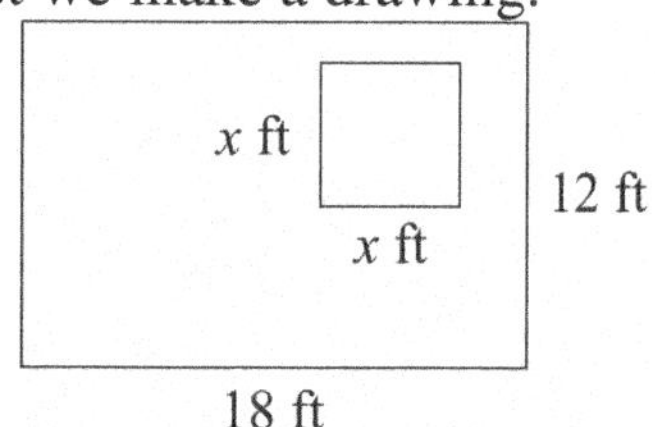

Then we reword the problem and find the polynomial.

$$\underbrace{\text{Area of lawn}}_{\downarrow} - \underbrace{\text{Area of sandbox}}_{\downarrow} = \underbrace{\text{Area left over}}_{\downarrow}$$

$$18 \cdot 12 - x \cdot x = \text{Area left over}$$

Then $216 - x^2 =$ Area left over.

Practice this exercise:

4. One rectangle has length $3y$ and width $2y$. Another has length $5y$ and width y. Find a polynomial for the sum of the perimeters of the rectangles.

ADDITIONAL EXERCISES

Objective A Add polynomials.

For extra help, see Examples 1–4 on page 769 of your text and the Section 10.4 lecture video.

Add.

1. $(x^3 - 9) + (2x^3 + 5)$

2. $(3y^3 - y + 2) + (4y^2 + 3y + 10)$

Name: Date:
Instructor: Section:

3. $(4t^4 - 8t^2 - t + 3) + (3t^4 + t^3 - 5t + 4)$

4. $(3.5a^3b^2c + 1.7abc^2) + (4.2abc^2 - 2.9abc)$

Objective B Simplify the opposite of a polynomial.

For extra help, see Examples 5–7 on page 770 of your text and the Section 10.4 lecture video.

Find two equivalent expressions for the opposite of each polynomial.

5. $2y^2 - 4y$

6. $4x^3 - 5x^2 - 8x + 7$

Simplify.

7. $-(-3x^2 - 2x + 14)$

8. $-(x^3 + 4x^2 - 5x - 1)$

Objective C Subtract polynomials.

For extra help, see Examples 8–11 on pages 770–771 of your text and the Section 10.4 lecture video.

Subtract.

9. $(-3t^2 + 2t - 4) - (8t^2 + 3t - 4)$

10. $(x^2 - 7x - 8) - (4x - 5)$

11. $(1.5y^4 - 3.2y^2 + 1) - (8.4y^4 - 1.8y^2 + 4)$

12. $(3a^2b^3 - 4ab^4 + 2ab) - (7a^2b^3 - 5ab^3 - ab)$

Objective D Use polynomials to represent perimeter and area.
For extra help, see Examples 12–13 on pages 771–772 of your text and the Section 10.4 lecture video.
Solve.

13. Find a polynomial for the perimeter of the figure.

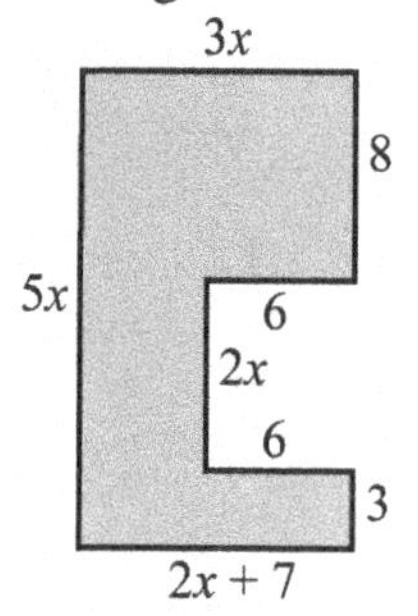

14. Find a polynomial for the sum of the areas of these rectangles.

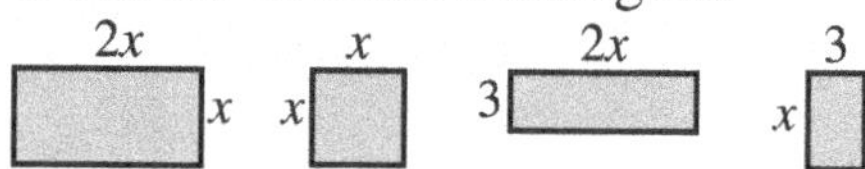

15. Find two algebraic expressions for the area of the figure. First, regard the figure as one large rectangle, and then regard the figure as a sum of four smaller rectangles.

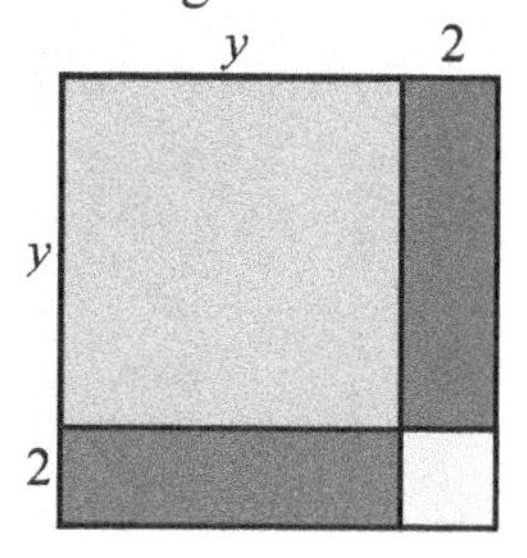

16. Find a polynomial for the shaded area of the figure.

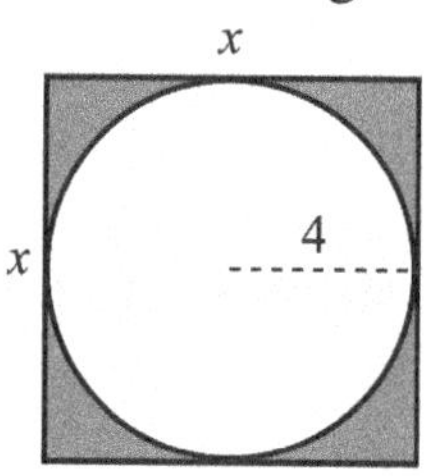

Name: Date:
Instructor: Section:

Chapter 10 POLYNOMIALS: OPERATIONS

10.5 Multiplication of Polynomials

Learning Objectives
A Multiply monomials.
B Multiply a monomial and any polynomial.
C Multiply two binomials.
D Multiply any two polynomials.

Key Terms
Use the vocabulary terms listed below to complete each statement in Exercises 1–2.

collect **term**

1. To multiply a monomial and a polynomial, multiply each ________________ of the polynomial by the monomial.

2. After multiplying, if possible, ________________ like terms.

GUIDED EXAMPLES AND PRACTICE

Objective A Multiply monomials.

Review this example for Objective A:

1. Multiply: $(-3y^2)(6y^5)$.

$(-3y^2)(6y^5) = (-3\cdot 6)(y^2 \cdot y^5)$ Multiplying coefficients

$= -18y^{2+5}$ Adding exponents

$= -18y^7$

Practice this exercise:

1. Multiply: $(5n^4)(-2n^2)$.

Objective B Multiply a monomial and any polynomial.

Review this example for Objective B:

2. Multiply: $3x(2x^3 - x)$.

$3x(2x^3 - x) = 3x\cdot 2x^3 - 3x\cdot x$ Using the distributive law

$= 6x^4 - 3x^2$ Multiplying each pair of monomials

Practice this exercise:

2. Multiply: $2x^2(x^2 - 3x - 5)$.

Objective C Multiply two binomials.

Review this example for Objective C:

3. Multiply: $(x-3)(x+2)$.

$$\begin{aligned}(x-3)(x+2) &= x(x+2)-3(x+2)\\ &= x^2+2x-3x-6\\ &= x^2-x-6\end{aligned}$$

Practice this exercise:

3. Multiply: $(2x-7)(x+1)$.

Objective D Multiply any two polynomials.

Review this example for Objective D:

4. Multiply: $(x^2-2x+3)(x-1)$.

We use columns. First we multiply the top row by −1 and then by *x*, placing like terms of the product in the same column. Finally we collect like terms.

$$\begin{array}{rrrr} & x^2 & -2x & +3 \\ & & x & -1 \\ \hline & -x^2 & +2x & -3 \\ x^3 & -2x^2 & +3x & \\ \hline x^3 & -3x^2 & +5x & -3 \end{array}$$

Practice this exercise:

4. Multiply:
$(x^3-3x+1)(x^2+4)$.

ADDITIONAL EXERCISES

Objective A Multiply monomials.

For extra help, see Examples 1–3 on page 779 of your text and the Section 10.5 lecture video.

Multiply.

1. $(-2x)(8x)$

2. $(5x^3)(7x^2)$

3. $(6x^3y^4)(3xy^2)$

4. $(-3y^2)(6y^3)(-2y)$

Objective B Multiply a monomial and any polynomial.

For extra help, see Examples 4–6 page 779–780 of your text and the Section 10.5 lecture video.

Multiply.

5. $4x(3x+2)$

6. $-2x(-2x+1)$

Name: Date:
Instructor: Section:

7. $6x(3x^2 - 5x + 4)$

8. $4xy^2(5x^2y + 7x^3y^2)$

Objective C Multiply two binomials.

For extra help, see Examples 7–8 on pages 780–781 of your text and the Section 10.5 lecture video.

Multiply.

9. $(x+1)(x+8)$

10. $(x+5)(x-7)$

11. $(2x-3)(2x-3)$

12. $\left(x-\frac{2}{3}\right)\left(x+\frac{3}{2}\right)$

Objective D Multiply any two polynomials.

For extra help, see Examples 9–12 pages 781–782 of your text and the Section 10.5 lecture video.

Multiply.

13. $(x^2+x+2)(x-2)$

14. $(x^2-4)(3x^2-2x+1)$

15. $(3n^2-n-5)(2n^2+5n-3)$

16. $(y+1)(y^3+4y^2-2y-7)$

Name: Date:
Instructor: Section:

Chapter 10 POLYNOMIALS: OPERATIONS

10.6 Special Products

Learning Objectives
A Multiply two binomials mentally using the FOIL method.
B Multiply the sum and the difference of two terms mentally.
C Square a binomial mentally.
D Find special products when polynomial products are mixed together.

Key Terms
Use the vocabulary terms listed below to complete each statement in Exercises 1–4.

binomials **difference** **FOIL** **square**

1. The expression $(x+5)(x^2+2)$ is a product of two ________________ .

2. The multiplication $(x+y)(a+b)=xa+xb+ya+yb$ illustrates the ________________ method.

3. The expression $(x+2)(x-2)$ is the product of the sum and the ________________ of the same two terms.

4. The expression $(x+3)^2$ is the ________________ of a binomial.

GUIDED EXAMPLES AND PRACTICE

Objective A Multiply two binomials mentally using the FOIL method.

Review this example for Objective A:

1. Multiply: $(2x+3)(x-4)$.

$$(2x+3)(x-4)$$
F O I L
$$=2x\cdot x+2x\cdot(-4)+3\cdot x+3\cdot(-4)$$
$$=2x^2-8x+3x-12$$
$$=2x^2-5x+12$$

Practice this exercise:

1. Multiply: $(y+2)(3y-5)$.

Objective B Multiply the sum and the difference of two terms mentally.

Review this example for Objective B:

2. Multiply: $(2x+1)(2x-1)$.

$$(2x+1)(2x-1)=(2x)^2-1^2$$
$$=4x^2-1$$

Practice this exercise:

2. Multiply: $(x+6)(x-6)$.

Objective C **Square a binomial mentally.**

Review this example for Objective C:

3. Multiply: $(3x-4)^2$.

$$(3x-4)^2 = (3x)^2 - 2\cdot 3x\cdot 4 + 4^2$$
$$= 9x^2 - 24x + 16$$

Practice this exercise:

3. Multiply: $(2x+1)^2$.

Objective D **Find special products when polynomial products are mixed together.**

Review this example for Objective D:

4. Multiply: $(n-4)(n+3)$.

This is the product of two binomials, but it is not the square of a binomial nor the product of the sum and difference of the same two terms. We use FOIL. $(n-4)(n+3) = n^2 + 3n - 4n - 12$
$= n^2 - n - 12$

Practice this exercise:

4. Multiply: $(-3y+1)(3y+1)$.

ADDITIONAL EXERCISES

Objective A **Multiply two binomials mentally using the FOIL method.**

For extra help, see Examples 1–10 on pages 786–787 of your text and the Section 10.6 lecture video.

Multiply. Try to write only the answer. If you need more steps, be sure to use them.

1. $(a+6)(a-7)$

2. $(-2n+3)(n+8)$

3. $(1+5x)(1-8x)$

4. $(9x^6+2)(x^6+9)$

Objective B **Multiply the sum and the difference of two terms mentally.**

For extra help, see Examples 11–15 on page 788 of your text and the Section 10.6 lecture video.

Multiply mentally, if possible. If you need extra steps, be sure to use them.

5. $(3x+7)(3x-7)$

6. $(4b^3-5)(4b^3+5)$

7. $(x^5+2x)(x^5-2x)$

8. $(3x^6+1.2)(3x^6-1.2)$

Name: Date:
Instructor: Section:

Objective C Square a binomial mentally.

For extra help, see Examples 16–20 on page 789 of your text and the Section 10.6 lecture video.

Multiply mentally, if possible. If you need extra steps, be sure to use them.

9. $(x+5)^2$

10. $(2x^2+3)^2$

11. $(6-2x^5)^2$

12. $\left(a-\frac{3}{4}\right)^2$

Objective D Find special products when polynomial products are mixed together.

For extra help, see Examples 21–28 on page 791 of your text and the Section 10.6 lecture video.

Multiply mentally, if possible.

13. $-2x^2(3x^2-4x-7)$

14. $\left(6q^2+\frac{1}{5}\right)\left(6q^2-\frac{1}{5}\right)$

15. $(y+2)(y^2-2y+4)$

16. $(7m^4+10)^2$

Name: Date:
Instructor: Section:

Chapter 10 POLYNOMIALS: OPERATIONS

10.7 Operations with Polynomials in Several Variables

Learning Objectives

A Evaluate a polynomial in several variables for given values of the variables.
B Identify the coefficients and the degrees of the terms of a polynomial and the degree of a polynomial.
C Collect like terms of a polynomial.
D Add polynomials.
E Subtract polynomials.
F Multiply polynomials.

Key Terms

Use the vocabulary terms listed below to complete each statement in Exercises 1–2.

degree **like term**

1. The ____________________ of a term is the sum of the exponents of the variables.

2. ____________________s have exactly the same variables with exactly the same exponents.

GUIDED EXAMPLES AND PRACTICE

Objective A Evaluate a polynomial in several variables for given values of the variables.

Review this example for Objective A:

1. Evaluate the polynomial $x^2y^2 - 3xy + 2xy^3$ for $x = 2$ and $y = -1$.

We replace x with 2 and y with –1.

$x^2y^2 - 3xy + 2xy^3$
$= 2^2(-1)^2 - 3(2)(-1) + 2(2)(-1)^3$
$= 4(1) - 3(2)(-1) + 2(2)(-1)$
$= 4 + 6 - 4$
$= 6$

Practice this exercise:

1. Evaluate the polynomial $2xy^2 - 4x^3y + 5$ for $x = -1$ and $y = 3$.

Objective B Identify the coefficients and the degrees of the terms of a polynomial and the degree of a polynomial.

Review this example for Objective B:

2. Identify the coefficient and the degree of each term and the degree of the polynomial $8xy^3-7x^2y^4+5xy-4$.

Term	Coefficient	Degree
$8xy^3$	8	4
$-7x^2y^4$	-7	6
$5xy$	5	2
-4	-4	0

The degree of the term of highest degree is 6, so the degree of the polynomial is 6.

Practice this exercise:

2. Identify the degree of the polynomial $2x^2y-8x^3y^4+9x+6x^2y^3-1$.

Objective C Collect like terms of a polynomial.

Review this example for Objective C:

3. Collect like terms: $7ab-ab^2-2ab+5a^2b+4ab^2$.

$$7ab-ab^2-2ab+5a^2b+4ab^2$$
$$=(7-2)ab+(-1+4)ab^2+5a^2b$$
$$=5ab+3ab^2+5a^2b$$

Practice this exercise:

3. Collect like terms: $5xy^2-4x^2y+3+2x^2y+7-xy^2$

Objective D Add polynomials.

Review this example for Objective D:

4. Add: $(3x^3-2xy+4)+(x^3-xy^2+5)$.

$$(3x^3-2xy+4)+(x^3-xy^2+5)$$
$$=(3+1)x^3-2xy-xy^2+(4+5)$$
$$=4x^3-2xy-xy^2+9$$

Practice this exercise:

4. Add: $(2x^3y^2-3x^2y+xy^2)+(5x^2y^2+4x^2y-8xy^2)$.

Objective E Subtract polynomials.

Review this example for Objective E:

5. Subtract: $(m^4n+2m^3n^2-m^2n^3)-(3m^4n+2m^3n^2-4m^2n^2)$.

$$(m^4n+2m^3n^2-m^2n^3)-(3m^4n+2m^3n^2-4m^2n^2)$$
$$=m^4n+2m^3n^2-m^2n^3+3m^4n-2m^3n^2+4m^2n^2$$
$$=-2m^4n-m^2n^3+4m^2n^2$$

Practice this exercise:

5. Subtract: $(a^3b^2-5a^2b+2ab)-(3a^3b^2-ab^2+4ab)$.

Name: Date:
Instructor: Section:

Objective F Multiply polynomials.

Review this example for Objective F:

6. Multiply: $(xy^2 - 3x)(xy + y^2)$.

We use FOIL.

$(xy^2 - 3x)(xy + y^2)$

F O I L

$= x^2y^3 + xy^4 - 3x^2y - 3xy^2$

Practice this exercise:

6. Multiply: $(2x+5y)^2$.

ADDITIONAL EXERCISES

Objective A Evaluate a polynomial in several variables for given values of the variables.

For extra help, see Examples 1–2 on page 797 of your text and the Section 10.7 lecture video.

Evaluate the polynomial when $x = -2$, $y = 3$, *and* $z = -1$.

1. $x^2 + 2xy^2 - y^3$

2. $3xyz - 2z^2$

3. The polynomial equation $C = 370 + 21.6 \times w \times (1 - p)$ can be used to estimate the number of daily calories needed to maintain weight for a person with mass w, in kilograms, and body fat percentage p (written in decimal form). How many calories are needed to maintain the weight of a 65-kg adult with a 25% body fat percentage?

Objective B Identify the coefficients and the degrees of the terms of a polynomial and the degree of a polynomial.

For extra help, see Example 3 on page 798 of your text and the Section 10.7 lecture video.

Identify the coefficient and the degree of each term of the polynomial. Then find the degree of the polynomial.

4. $2xy^2 - 3xy + 4x^4 - 7$

5. $26x^3y^4 - 10x^2y + 15$

6. $5xy + 6x^2 + 3y^2 + 1$

Objective C Collect like terms of a polynomial.
For extra help, see Examples 4–6 on page 798 of your text and the Section 10.7 lecture video.
Collect like terms.

7. $2c+d-c-4d$

8. $8x^3+3x^2y-5xy^2$

9. $4xy+3xz+18xy+10xz$

10. $4cd^2-8c^2d-2c^2d-3cd^2$

Objective D Add polynomials.
For extra help, see Examples 7–8 on page 799 of your text and the Section 10.7 lecture video.
Add.

11. $\left(-2x^2+5xy+y^2\right)+\left(x^2-7xy-6y^2\right)$

12. $(ab+2a+3b)+(a-5b-6ab)$
$+(7b-4a-ab)$

13. $\left(8u^2+3uv\right)+\left(-10u^2-8uv\right)+\left(u^2+uv\right)$

14. $\left(r^3s^2-6rs-7\right)+\left(2r^3s^2+5rs+10\right)$

Objective E Subtract polynomials.
For extra help, see Example 9 on page 799 of your text and the Section 10.7 lecture video.
Subtract.

15. $\left(x^3+2y^3\right)-\left(3x^3y-xy^3-x^3+2y^3\right)$

16. $(rs-tu-7)-(rs+3tu-10)$

17. $\left(6a^5+8a^4b-4b\right)-\left(a^5+8a^4b-6a-2b\right)$

18. $(-3x+7y+z)-(-5x+8y-10z)$

Objective F Multiply polynomials.
For extra help, see Examples 10–17 on page 800 of your text and the Section 10.7 lecture video.
Multiply.

19. $(4x-3y)(2x+y)$

20. $\left(6-x^2y^2\right)\left(8+x^2y^2\right)$

21. $(c+2d)^2$

22. $(a+b+5)(a+b-5)$

Name: Date:
Instructor: Section:

Chapter 10 POLYNOMIALS: OPERATIONS

10.8 Division of Polynomials

Learning Objectives
A Divide a polynomial by a monomial.
B Divide a polynomial by a divisor that is a binomial.

Key Terms
Use the vocabulary terms listed below to complete each statement in Exercises 1–6.

dividend **divisor** **quotient** **remainder**

1. ______________________.
2. ______________________.
3. ______________________.
4. ______________________

$$\begin{array}{r} \boxed{x+2} \leftarrow ① \\ ② \rightarrow \boxed{x+3}\,\overline{)\boxed{x^2+5x+10}} \leftarrow ③ \\ \underline{x^2+3x} \\ 2x+10 \\ \underline{2x+\ \ 6} \\ \boxed{4} \leftarrow ④ \end{array}$$

GUIDED EXAMPLES AND PRACTICE

Objective A Divide a polynomial by a monomial.

Review this example for Objective A:

1. Divide: $(6x^3-8x^2+15x)\div(3x)$.

$$\frac{6x^3-8x^2+15x}{3x}=\frac{6x^3}{3x}-\frac{8x^2}{3x}+\frac{15x}{3x}$$
$$=\frac{6}{3}x^{3-1}-\frac{8}{3}x^{2-1}+\frac{15}{3}$$
$$=2x^2-\frac{8}{3}x+5$$

Practice this exercise:

1. Divide: $(4y^2-5y+12)\div 4$.

Objective B Divide a polynomial by a divisor that is a binomial.

Review this example for Objective B:

2. Divide: x^2-3x+7 by $x+1$.

$$\begin{array}{r} x-4 \\ x+1\overline{)x^2-3x+7} \\ \underline{x^2+\ x} \\ -4x+7 \\ \underline{-4x-4} \\ 11 \end{array}$$

The answer is $x-4+\frac{11}{x+1}$.

Practice this exercise:

2. Divide:
$(x^2-8x+5)\div(x-2)$.

ADDITIONAL EXERCISES

Objective A Divide a polynomial by a monomial.

For extra help, see Examples 1–6 on pages 806–807 of your text and the Section 10.8 lecture video.

Divide.

1. $\dfrac{-18x^5}{3x^2}$

2. $\dfrac{48x^4 - 36x^3 - 16x^2 + 60}{4}$

3. $(100x^6 - 80x^5 - 45x^4) \div (-10x^4)$

4. $\dfrac{8a^2b^3 + 10ab^2 - 4ab}{2ab}$

ADDITIONAL EXERCISES

Objective B Divide a polynomial by a divisor that is a binomial.

For extra help, see Examples 7–10 on pages 808–809 of your text and the Section 10.8 lecture video.

Divide.

5. $(x^2 + 4x + 3) \div (x - 1)$

6. $\dfrac{3x^3 - 7x^2 - 21x - 10}{3x + 2}$

7. $(x^6 - x^3 - 30) \div (x^3 - 6)$

8. $(4x^4 - 5x^2 - 8x + 1) \div (2x - 3)$

Name: Date:
Instructor: Section:

Chapter 11 POLYNOMIALS: FACTORING

11.1 Introduction to Factoring

Learning Objectives

A Find the greatest common factor, the GCF, of monomials.

B Factor polynomials when the terms have a common factor, factoring out the greatest common factor.

C Factor certain expressions with four terms using factoring by grouping.

Key Terms

Use the vocabulary terms listed below to complete each statement in Exercises 1–4. Terms may be used more than once.

factor **factoring by grouping** **factorization**

1. To ________________ a polynomial is to express it as a product.

2. A(n) ________________ of a polynomial P is a polynomial that can be used to express P as a product.

3. A(n) ________________ of a polynomial is an expression that names that polynomial as a product.

4. Certain polynomials with four terms can be factored using ________________ .

GUIDED EXAMPLES AND PRACTICE

Objective A Find the greatest common factor, the GCF, of monomials.

Review this example for Objective A:

1. Find the GCF of $6pq^3$, $24p^2q^2$, and $18p^3q$.

$$6pq^2 = 2\cdot 3\cdot p\cdot q^3$$
$$24p^2q^2 = 2\cdot 2\cdot 2\cdot 3\cdot p^2\cdot q^2$$
$$18p^3q = 2\cdot 3\cdot 3\cdot p^3\cdot q$$

The GCF is $2\cdot 3\cdot p\cdot q$, or $6pq$.

Practice this exercise:

1. Find the GCF of $16x^3y^2$, $24xy^3$, and $12x^2y^2$.

Objective B Factor polynomials when the terms have a common factor, factoring out the greatest common factor.

Review this example for Objective B:

2. Factor $4x^2+4x-20$.

$$4x^2+4x-20 = 4\cdot 4x+4\cdot x-4\cdot 5$$
$$=4(4x+x-5)$$

Practice this exercise:

2. Factor $3x^8-30x^6+15x^5$.

Objective C Factor certain expressions with four terms using factoring by grouping.

Review this example for Objective C:

3. Factor $3x^3 - 6x^2 - x + 2$.

$$\begin{aligned} 3x^3 - 6x^2 - x + 2 &= (3x^3 - 6x^2) + (-x + 2) \\ &= 3x^2(x-2) - (x-2) \\ &= (x-2)(3x^2 - 1) \end{aligned}$$

Practice this exercise:

3. Factor $x^3 + 3x^2 - 7x - 21$.

ADDITIONAL EXERCISES

Objective A Find the greatest common factor, the GCF, of monomials.

For extra help, see Examples 1–6 on page 822–824 of your text and the Section 11.1 lecture video.

Find the GCF.

1. $x^3, -10x$

2. $4x^3,\ 20x^5,\ 12x^4$

3. $-11x^4y^3,\ 33x^3y^5,\ 55x^2y$

4. $-15x,\ 12x^2,\ 18x^7$

Objective B Factor polynomials when the terms have a common factor, factoring out the greatest common factor.

For extra help, see Examples 7–13 on pages 825–826 of your text and the Section 11.1 lecture video.

Factor. Check by multiplying.

5. $10x^5 - 5x^2$

6. $11x^4y^3 - 33x^3y^5 - 55x^2y$

7. $1.4x^4 - 3.5x^3 + 4.2x^2 + 7.0x$

8. $\frac{3}{2}x^7 + \frac{7}{2}x^5 - \frac{1}{2}x^3 + \frac{1}{2}x^2$

Objective C Factor certain expressions with four terms using factoring by grouping.

For extra help, see Examples 14–19 on pages 827–828 of your text and the Section 11.1 lecture video.

Factor.

9. $x^3(x+1) + 3(x+1)$

10. $7p^2(4p-5) - (4p-5)$

Factor by grouping.

11. $2z^3 + 2z^2 + 3z + 3$

12. $20x^3 - 15x^2 + 8x - 6$

Name: Date:
Instructor: Section:

Chapter 11 POLYNOMIALS: FACTORING

11.2 Factoring Trinomials of the Type $x^2 + bx + c$

Learning Objectives
A Factor trinomials of the type $x^2 + bx + c$ by examining the constant term c.

Key Terms
Use the vocabulary terms listed below to complete each statement in Exercises 1–4.

leading coefficient **negative** **positive** **prime**

1. In the expression $ax^2 + bx + c$, a is called the __________________ .
2. A(n) __________________ polynomial cannot be factored further.
3. When the constant term of a trinomial is __________________ , the constant terms of the binomial factors have the same sign.
4. When the constant term of a trinomial is __________________ , the constant terms of the binomial factors have opposite signs.

GUIDED EXAMPLES AND PRACTICE

Objective A Factor trinomials of the type $x^2 + bx + c$ by examining the constant term c.

Review this example for Objective A:

1. Factor $x^2 - 2x - 15$.

Since the constant term, –15, is negative, we look for a factorization of –15 in which one factor is positive and one factor is negative. The sum of the factors must be the coefficient of the middle term, –2, so the negative factor must have the larger absolute value. The possible pairs of factors that meet these criteria are 1, –15 and 3, –5. The numbers we need are 3 and –5.

$$x^2 - 2x - 15 = (x+3)(x-5)$$

Practice this exercise:

1. Factor $x^2 - 9x + 8$.

ADDITIONAL EXERCISES

Objective A Factor trinomials of the type $x^2 + bx + c$ by examining the constant term c.

For extra help, see Examples 1–9 on pages 832–836 of your text and the Section 11.2 lecture video.

Factor. Remember that you can check by multiplying.

1. $p^2 - 5p - 14$

2. $x^2 + 3x + 1$

3. $x^2 + 22x + 121$

4. $y^2 - 16y + 60$

5. $x^2 - 0.1x - 0.06$

Name: Date:
Instructor: Section:

Chapter 11 POLYNOMIALS: FACTORING

11.3 Factoring $ax^2 + bx + c$: The FOIL Method

Learning Objectives
A Factor trinomials of the type $ax^2 + bx + c$, $a \neq 1$ using the FOIL method.

Key Terms
Use the vocabulary terms listed below to complete each statement in Exercises 1–4.

First **Inside** **Last** **Outside**

1. In the product $(x-3)(2x+1)$, x and $2x$ are the ________________ terms.
2. In the product $(x-3)(2x+1)$, –3 and 1 are the ________________ terms.
3. In the product $(x-3)(2x+1)$, –3 and $2x$ are the ________________ terms.
4. In the product $(x-3)(2x+1)$, x and 1 are the ________________ terms.

GUIDED EXAMPLES AND PRACTICE

Objective A Factor trinomials of the type $ax^2 + bx + c$, $a \neq 1$ using the FOIL method.

Review this example for Objective A:

1. Factor $2y^3 + 5y^2 - 3y$.

1. Factor out the largest common factor, y:
$$2y^3 + 5y^2 - 3y = y(2y^2 + 5y - 3)$$
2. Because $2y^2$ factors as $2y \cdot y$, we have this possibility for a factorization:
$$(2y+\quad)(y+\quad).$$
3. There are two pairs of factors of –3 and each can be written in two ways:
3, –1 –3, 1
and
–1, 3 1, –3.
4. From steps (2) and (3) we see that there are four possibilities for factorizations. We look for Outer and Inner products for which the sum is the middle term, $5y$. We try some possibilities.
$$(2y+3)(y-1) = 2y^2 + y - 3$$
$$(2y-1)(y+3) = 2y^2 + 5y - 3$$

Practice this exercise:

1. Factor $6z^2 + 14z + 4$.

The factorization of $2y^2+5y-3$ is

$$(2y-1)(y+3).$$

We must include the common factor to get a factorization of the original trinomial.

$$2y^3+5y^2-3y=y(2y-1)(y+3)$$

ADDITIONAL EXERCISES

Objective A Factor trinomials of the type $ax^2 + bx + c$, $a \neq 1$ using the FOIL method.

For extra help, see Examples 1–5 on pages 842–845 of your text and the Section 11.3 lecture video.

Factor.

1. $3x^2-2x-1$

2. $15x^2+55x+50$

3. $14a^4+23a^2+3$

4. $16x^2-8xy-8y^2$

Name:　　　　　　　　　　　　　　　　Date:
Instructor:　　　　　　　　　　　　　　Section:

Chapter 11 POLYNOMIALS: FACTORING

11.4 Factoring $ax^2 + bx + c$: The *ac*-Method

Learning Objectives
A　Factor trinomials of the type $ax^2 + bx + c$, $a \neq 1$ using the *ac*-method.

Key Terms
Use the vocabulary terms listed below to complete the steps for factoring using the *ac*-method in Exercises 1–6.

common factor	**grouping**	**leading coefficient**
multiplying	**split**	**sum**

1. Factor out a(n) ____________________ , if any.

2. Multiply the ____________________ a and the constant c.

3. Try to factor the product ac so that the ____________________ of the factors is b.

4. ____________________ the middle term.

5. Factor by ____________________ .

6. Check by ____________________ .

GUIDED EXAMPLES AND PRACTICE

Objective A　Factor trinomials of the type $ax^2 + bx + c$, $a \neq 1$ using the *ac*-method.

Review this example for Objective A:

1. Factor $5x^2 + 7x - 6$ by grouping.

   1. There is no common factor (other than 1 or –1).
   2. Multiply the leading coefficient 5 and the constant, –6:
   $$5(-6) = -30.$$
   3. Look for a factorization of –30 in which the sum of the factors is the coefficient of the middle term, 7.
   The numbers we need are 10 and –3.
   4. Split the middle term, writing it as a sum or difference using the factors found in step (3).
   $$7x = 10x - 3x$$

Practice this exercise:

1. Factor $8x^2 - 2x - 1$ by grouping.

5. Factor by grouping.

$$\begin{aligned}5x^2+7x-6&=5x^2+10x-3x-6\\&=5x(x+2)-3(x+2)\\&=(x+2)(5x-3)\end{aligned}$$

6. Check: $(x+2)(5x-3)=5x^2+7x-6$

ADDITIONAL EXERCISES

Objective A Factor trinomials of the type ax^2+bx+c, $a \neq 1$ using the *ac*-method.

For extra help, see Examples 1–2 on pages 849–850 of your text and the Section 11.4 lecture video.

Factor. Note that the middle term has already been split.

1. $6x^2+4x+15x+10$

Factor by grouping.

2. $4x^2+8x-21$

3. $14x^2-65x+9$

4. $12p^4-4p^3-5p^2$

5. $15y^2+10y+18$

Name: Date:
Instructor: Section:

Chapter 11 POLYNOMIALS: FACTORING

11.5 Factoring Trinomials Squares and Differences of Squares

Learning Objectives

A Recognize trinomial squares.
B Factor trinomial squares.
C Recognize differences of squares.
D Factor differences of squares, being careful to factor completely.

Key Terms

Use the vocabulary terms listed below to complete each statement in Exercises 1–4.

difference of squares **factored completely**
sum of squares **trinomial square**

1. The expression $x^2+10x+25$ is a(n) ________________.

2. The expression $9x^2-16$ is a(n) ________________.

3. The expression y^2+81 is a(n) ________________.

4. When no factor can be factored further, we have ________________.

GUIDED EXAMPLES AND PRACTICE

Objective A Recognize trinomial squares.

Review this example for Objective A:

1. Determine whether each of the following is a trinomial square.

   a) x^2-4x-4 b) $9x^2+1+6x$

   a) x^2 and 4 are squares, but there is a minus sign before 4. This is not a trinomial square.

   b) Write the trinomial in descending order: $9x^2+6x+1$. $9x^2$ and 1 are squares. There is no minus sign before $9x^2$ or 1. The middle term, $6x$, is $2\cdot 3x\cdot 1$. Thus $9x^2+1+6x$ is a trinomial square.

Practice this exercise:

1. Determine whether $y^2+32y+16$ is a trinomial square.

Objective B Factor trinomial squares.

Review this example for Objective B:

2. Factor $4x^2-12x+9$.

$$\begin{aligned}4x^2-12x+9&=(2x)^2-2\cdot 2x\cdot 3+3^2\\&=(2x-3)^2\end{aligned}$$

Practice this exercise:

2. Factor $16x^2+8x+1$.

Objective C Recognize differences of squares.

Review this example for Objective C:

3. Determine whether each of the following is a difference of squares.

a) $16t^4-9$ b) $25x^2-3$

a) The expressions $16t^4$ and 9 are squares:

$16t^4=(4t^2)^2$ and $9=3^2$

The expressions have different signs. Thus, $16t^4-9$ is a difference of squares.

b) 3 is not a square, so $25x^2-3$ is not a difference of squares.

Practice this exercise:

3. Determine whether $1-36y^6$ is a difference of squares.

Objective D Factor differences of squares, being careful to factor completely.

Review this example for Objective D:

4. Factor t^5-t.

$$\begin{aligned}t^5-t&=t(t^4-1)\\&=t(t^2+1)(t^2-1)\\&=t(t^2+1)(t+1)(t-1)\end{aligned}$$

Practice this exercise:

4. Factor $12-27x^2$ completely.

ADDITIONAL EXERCISES

Objective A Recognize trinomial squares.

For extra help, see Examples 1–3 on pages 857–858 of your text and the Section 11.5 lecture video.

Determine whether each of the following is a trinomial square. Answer "yes" or "no."

1. $x^2+10x-25$

2. $4x^2-12x+9$

3. x^2+4x+1

4. x^2-3x+8

Name: Date:
Instructor: Section:

Objective B Factor trinomial squares.

For extra help, see Examples 4–9 pages 858–859 of your text and the Section 11.5 lecture video.

Factor completely. Remember to look first for a common factor and to check by multiplying.

5. $36+12x+x^2$

6. $64-80x+25x^2$

7. $12a^2-84a+147$

8. $x^2-8xy+16y^2$

Objective C Recognize differences of squares.

For extra help, see Examples 10–12 on page 860 of your text and the Section 11.5 lecture video.

Determine whether each of the following is a difference of squares. Answer "yes" or "no."

9. x^2-100

10. $4x^2-10y^2$

11. $-1+64x^2$

12. $25x^2-909$

Objective D Factor differences of squares, being careful to factor completely.
For extra help, see Examples 13–21 pages 860–862 of your text and the Section 11.5 lecture video.

Factor completely. Remember to look first for a common factor.

13. $64x^2 - y^2$

14. $4m^2 - 49n^2$

15. $3x^4 - 48$

16. $36p^4 - 1$

Name: Date:
Instructor: Section:

Chapter 11 POLYNOMIALS: FACTORING

11.6 Factoring: A General Strategy

Learning Objectives
A Factor polynomials completely using any of the methods considered in this chapter.

Key Terms
Use the vocabulary terms listed below to complete each statement in Exercises 1–4.

completely **difference** **grouping** **perfect square**

1. When factoring a polynomial with two terms, determine whether you have a(n) ____________ of squares.

2. When factoring a polynomial with three terms, determine whether the trinomial is a ____________ .

3. When factoring a polynomial with four terms, try factoring by ____________ .

4. Always factor ____________ .

GUIDED EXAMPLES AND PRACTICE

Objective A Factor polynomials completely using any of the methods considered in this chapter.

Review this example for Objective A:

1. Factor $2y^3 - 12y^2 + 18y$ completely.

a) We look for a common factor.
$2y^3 - 12y^2 + 18y = 2y(y^2 - 6y + 9)$

b) The factor $y^2 - 6y + 9$ has three terms and is a trinomial square. We factor it.

$$2y(y^2 - 6y + 9) = 2y(y - 2 \cdot y \cdot 3 + 3^2)$$
$$= 2y(y-3)^2$$

Practice this exercise:

1. Factor $15x^2 + 5x - 20$ completely.

ADDITIONAL EXERCISES

Objective A Factor polynomials completely using any of the methods considered in this chapter.

For extra help, see Examples 1–8 on pages 867–869 of your text and the Section 11.6 lecture video.

Factor completely.

1. $x^4 + 5x^2 - 4x^3 - 20x$

2. $9c^2 + 4d^2 - 12cd$

3. $4x(u^2 + 3v) - (u^2 + 3v)$

4. $25x^2z^2 + 40xyz + 16y^2$

5. $n^2 + 1$

Name: Date:
Instructor: Section:

Chapter 11 POLYNOMIALS: FACTORING

11.7 Solving Quadratic Equations by Factoring

Learning Objectives
A Solve equations (already factored) using the principle of zero products.
B Solve quadratic equations by factoring and then using the principle of zero products.

Key Terms
Use the vocabulary terms listed below to complete each statement in Exercises 1–4.

factor **products** **quadratic equation** **zero**

1. $3x^2 - 5x + 3 = 0$ is an example of a(n) ________________ .

2. The principle of zero ________________ states that a product of two terms is 0 if and only if one or both of the factors is 0.

3. To use the principle of zero products, you must have ________________ on one side of the equation.

4. To use the principle of zero products, we ________________ a quadratic polynomial.

GUIDED EXAMPLES AND PRACTICE

Objective A Solve equations (already factored) using the principle of zero products.

Review this example for Objective A:

1. Solve: $(4x-3)(x+2)=0$.

$$(4x-3)(x+2)=0$$
$$4x-3=0 \quad or \quad x+2=0$$
$$x=\frac{3}{4} \quad or \quad x=-2$$

The solutions are $\frac{3}{4}$ and -2.

Practice this exercise:

1. Solve: $x(6x+5)=0$.

Objective B Solve quadratic equations by factoring and then using the principle of zero products.

Review this example for Objective B:

2. Solve: $x^2+2x=24$.

$$x^2+2x=24$$
$$x^2+2x-24=0$$
$$(x+6)(x-4)=0$$
$$x+6=0 \quad or \quad x-4=0$$
$$x=-6 \quad or \quad x=4$$

The solutions are -6 and 4.

Practice this exercise:

2. Solve: $16x^2=49$.

ADDITIONAL EXERCISES

Objective A Solve equations (already factored) using the principle of zero products.

For extra help, see Examples 1–3 on pages 875–876 of your text and the Section 11.7 lecture video.

Solve using the principle of zero products.

1. $(x+3)(x-10)=0$

2. $(4x+3)(2x-10)=0$

3. $\left(\frac{1}{2}-3x\right)\left(\frac{1}{3}-4x\right)=0$

4. $(0.1x+0.2)(0.5x-15)=0$

Objective B Solve quadratic equations by factoring and then using the principle of zero products.

For extra help, see Examples 4–10 on pages 877–879 of your text and the Section 11.7 lecture video.

Solve by factoring and using the principle of zero products. Remember to check.

5. $x^2+4x-21=0$

6. $3x^2+5x=2$

7. $2y^2+16y+30=0$

8. Find the x-intercepts for the graph of the equation. (The grid is intentionally not included.)

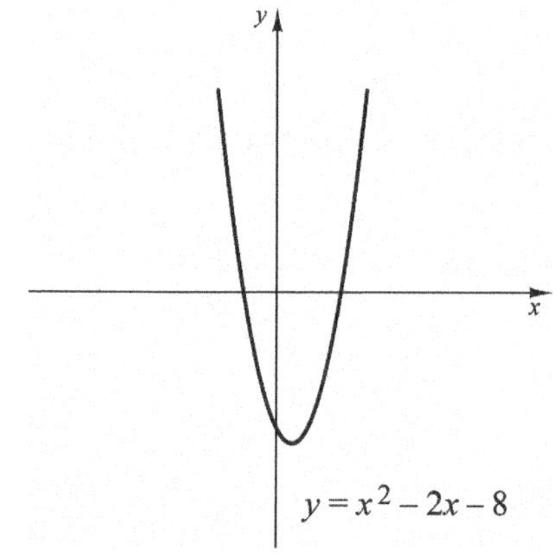

9. Use the following graph to solve $x^2-x-2=0$.

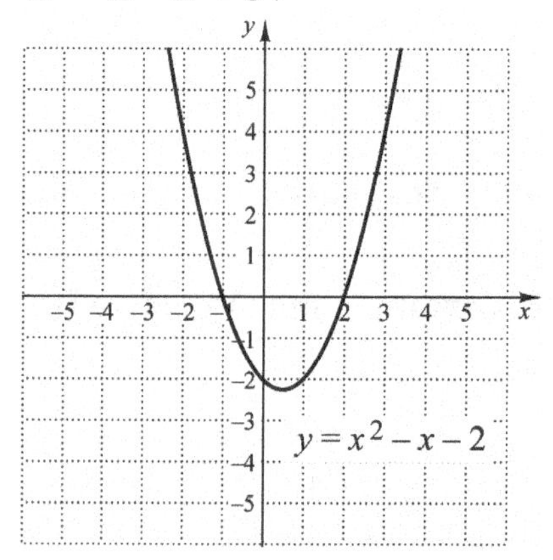

Name: Date:
Instructor: Section:

Chapter 11 POLYNOMIALS: FACTORING

11.8 Applications of Quadratic Equations

Learning Objectives
A Solve applied problems involving quadratic equations that can be solved by factoring.

Key Terms
Use the vocabulary terms listed below to complete each statement in Exercises 1–4.

hypotenuse **legs** **Pythagorean** **right**

1. A triangle that has a 90° angle is a ________________ triangle.
2. In a right triangle, the side opposite the 90° angle is the ________________ .
3. The sides that form the 90° angle in a right triangle are called ________________ .
4. The ________________ theorem states that, in a right triangle, $a^2 + b^2 = c^2$.

GUIDED EXAMPLES AND PRACTICE

Objective A Solve applied problems involving quadratic equations that can be solved by factoring.

Review this example for Objective A:

1. The length of a rectangular rug is 2 ft greater than the width. The area of the rug is 48 ft^2. Find the length and width.

   1. *Familiarize.* We make a drawing. Let w = the width of the rug. Then the length is $w + 2$.

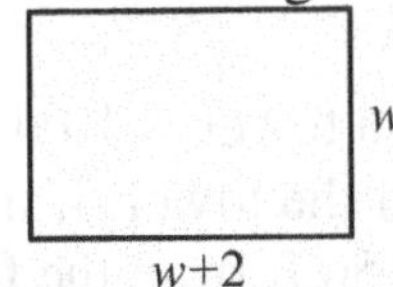

Recall that the area of a rectangle is length × width.

2. *Translate.* We reword the problem.

$$\begin{array}{ccccc} \text{Length} & \times & \text{width} & \text{is} & 48 \text{ ft}^2. \\ \downarrow & \downarrow & \downarrow & \downarrow & \downarrow \\ (w+2) & \times & w & = & 48 \end{array}$$

Practice this exercise:

1. The height of a triangle is 4 cm greater than the base. The area is 30 cm^2. Find the height and the base.

3. *Solve*. We solve the equation.

$$(w+2)\times w=48$$
$$w^2+2w=48$$
$$w^2+2w-48=0$$
$$(w+8)(w-6)=0$$
$$w+8=0 \quad or \quad w-6=0$$
$$w=-8 \quad or \quad w=6$$

4. *Check*. The width of a rectangle cannot be negative, so –8 cannot be a solution. Suppose the width is 6 ft. Then the length is 6 + 2, or 8 ft and the area is $6\cdot 8$, or 48 ft^2. These numbers check in the original problem.

5. *State*. The length is 8 ft and the width is 6 ft.

ADDITIONAL EXERCISES

Objective A Solve applied problems involving quadratic equations that can be solved by factoring.

For extra help, see Examples 1–6 on pages 884–889 of your text and the Section 11.8 lecture video.

Solve.

1. A rectangular serving tray is twice as long as it is wide. The area of the tray is 338 in^2. Find the dimensions of the tray.

2. A softball league plays a total of 210 games. How many teams are in the league if each team plays every other team twice? Use $x^2-x=N$, where x is the number of teams and N is the number of games played.

3. During a toast at a party, there were 66 "clinks" of glasses. How many people took part in the toast? Use $N=\frac{1}{2}(x^2-x)$, where N is the number of clinks and x is the number of people.

4. The guy wire on a tower is 10 ft longer than the height of the tower. If the guy wire is anchored 50 ft from the foot of the tower, how tall is the tower?

Name: Date:
Instructor: Section:

Chapter 12 RATIONAL EXPRESSIONS AND EQUATIONS

12.1 Multiplying and Simplifying Rational Expressions

Learning Objectives

A Find all numbers for which a rational expression is not defined.
B Multiply a rational expression by 1, using an expression such as *A*/*A*.
C Simplify rational expressions by factoring the numerator and the denominator and removing factors of 1.
D Multiply rational expressions and simplify.

Key Terms

Use the vocabulary terms listed below to complete each statement in Exercises 1–4.

equivalent **multiply** **rational** **simplify**

1. A quotient, or ratio, of polynomials is a(n) __________________ expression.

2. Expressions that have the same value for all allowable replacements are called

__________________ expressions.

3. To __________________ rational expressions, multiply numerators and multiply denominators.

4. To __________________ rational expressions, factor the numerator and the denominator and "remove" a factor of 1.

GUIDED EXAMPLES AND PRACTICE

Objective A Find all numbers for which a rational expression is not defined.

Review this example for Objective A:

1. Find all numbers for which the rational expression $\frac{5x+1}{x^2-x-6}$ is not defined.

We set the denominator equal to 0 and solve.

$x^2-x-6=0$

$(x+2)(x-3)=0$ Factoring

$x+2=0$ *or* $x-3=0$ Using the principle

$x=-2$ *or* $x=3$ of zero products

Practice this exercise:

1. Find all numbers for which the rational expression $\frac{n^2-n}{n^2-16}$ is not defined.

Objective B Multiply a rational expression by 1, using an expression such as *A/A*.

Review this example for Objective B:

2. Multiply: $\frac{x+5}{x+2} \cdot \frac{x-6}{x-6}$.

$$\frac{x+5}{x+2} \cdot \frac{x-6}{x-6} = \frac{(x+5)(x-6)}{(x+2)(x-6)}$$

Practice this exercise:

2. Multiply: $\frac{8x}{7y} \cdot \frac{x+1}{x+1}$.

Objective C Simplify rational expressions by factoring the numerator and the denominator and removing factors of 1.

Review this example for Objective C:

3. Simplify: $\frac{2x^2+14x+20}{6x^2+12x-90}$.

$\frac{2x^2+14x+20}{6x^2+12x-90}$

$= \frac{2(x^2+7x+10)}{6(x^2+2x-15)}$ Factoring the comon factor

$= \frac{2(x+2)(x+5)}{2 \cdot 3(x-3)(x+5)}$ Factoring the numerator and denominator

$= \frac{2(x+5)}{2(x+5)} \cdot \frac{x+2}{3(x-3)}$ Factoring the rational expression

$= 1 \cdot \frac{x+2}{3(x-3)}$ $\frac{2(x+5)}{2(x+5)} = 1$

$= \frac{x+2}{3(x-3)}$ Removing a factor of 1

Practice this exercise:

3. Simplify: $\frac{c^2-12c+36}{c^2-36}$.

Objective D Multiply rational expressions and simplify.

Review this example for Objective D:

4. Multiply and simplify: $\frac{6x^3}{5x^2+30x+45} \cdot \frac{5x+15}{3x}$.

$\frac{6x^3}{5x^2+30x+45} \cdot \frac{5x+15}{3x}$

$= \frac{(6x^3)(5x+15)}{(5x^2+30x+45)(3x)}$ Multiplying numerators and denominators

$= \frac{3x \cdot 2x^2 \cdot 5(x+3)}{3x \cdot 5(x+3)(x+3)}$ Factoring numerators and denominators

$= \frac{\cancel{3x} \cdot 2x^2 \cdot \cancel{5} \cancel{(x+3)}}{\cancel{3x} \cdot \cancel{5}(x+3)\cancel{(x+3)}}$ Removing factor of 1

$= \frac{2x^2}{x+3}$

Practice this exercise:

4. Multiply and simplify:

$\frac{t^2-25}{t^3} \cdot \frac{t^2-2t}{t^2+7t+10}$.

Name: Date:
Instructor: Section:

ADDITIONAL EXERCISES

Objective A Find all numbers for which a rational expression is not defined.

For extra help, see Example 1 on page 906 of your text and the Section 12.1 lecture video.

Find all numbers for which a rational expression is not defined.

1. $\frac{1}{5x}$

2. $\frac{3}{x+7}$

3. $\frac{2}{4a-5}$

4. $\frac{p+7}{20}$

Objective B Multiply a rational expression by 1, using an expression such as *A/A*.

For extra help, see Examples 2–4 on page 907 of your text and the Section 12.1 lecture video.

Multiply. Do not simplify. Note that in each case you are multiplying by1.

5. $\frac{5a^2}{5a^2}\cdot\frac{6c^2}{7d^4}$

6. $\frac{y-6}{y+1}\cdot\frac{y-7}{y-7}$

7. $\frac{1-x}{2-x}\cdot\frac{-1}{-1}$

Objective C Simplify rational expressions by factoring the numerator and the denominator and removing factors of 1.

For extra help, see Examples 5–10 on pages 908–910 of your text and the Section 12.1 lecture video.

Simplify.

8. $\dfrac{30a^8b^5}{12ab^4}$

9. $\dfrac{4x^2+8x}{12x^3+4x}$

10. $\dfrac{a^2+a-2}{a^2+2a-3}$

11. $\dfrac{x^2-100}{10-x}$

Objective D Multiply rational expressions and simplify.

For extra help, see Examples 11–13 on page 911 of your text and the Section 12.1 lecture video.

Multiply and simplify.

12. $\dfrac{4a}{b^3}\cdot\dfrac{3b}{10a^2}$

13. $\dfrac{x^2+6x+5}{x^2-4x+3}\cdot\dfrac{x-3}{x+5}$

14. $\dfrac{x^4-16}{x^4-81}\cdot\dfrac{x^2+9}{x^2+4}$

15. $\dfrac{3y^2-27}{2y^2-128}\cdot\dfrac{8y+64}{6y-6}$

Name: Date:
Instructor: Section:

Chapter 12 RATIONAL EXPRESSIONS AND EQUATIONS

12.2 Division and Reciprocals

Learning Objectives
A Find the reciprocal of a rational expression.
B Divide rational expressions and simplify.

Key Terms
Use the vocabulary terms listed below to complete each statement in Exercises 1–2.

interchange **multiply**

1. To find the reciprocal of a rational expression, ________________ the numerator and the denominator.

2. To divide by a rational expression, ________________ by its reciprocal.

GUIDED EXAMPLES AND PRACTICE

Objective A Find the reciprocal of a rational expression.

Review this example for Objective A:

1. Find the reciprocal of each expression.

a) $\frac{x}{5}$ b) $3y+1$ c) $\frac{1}{z-6}$

a) The reciprocal of $\frac{x}{5}$ is $\frac{5}{x}$.

b) The reciprocal of $3y+1$ is $\frac{1}{3y+1}$.

c) The reciprocal of $\frac{1}{z-6}$ is $\frac{z-6}{1}$, or $z-6$.

Practice this exercise:

1. Find the reciprocal of $\frac{x+y}{a+3}$.

Objective B Divide rational expressions and simplify.

Review this example for Objective B:

2. Divide and simplify: $\frac{a^2+3a+2}{a^2+5a+6} \div \frac{a^2-2a-3}{a^2+2a-3}$.

$$\frac{a^2+3a+2}{a^2+5a+6} \div \frac{a^2-2a-3}{a^2+2a-3}$$

$$= \frac{a^2+3a+2}{a^2+5a+6} \cdot \frac{a^2+2a-3}{a^2-2a-3}$$ Multiplying by the reciprocal

$$= \frac{(a^2+3a+2)(a^2+2a-3)}{(a^2+5a+6)(a^2-2a-3)}$$

$$= \frac{(a+2)(a+1)(a+3)(a-1)}{(a+2)(a+3)(a-3)(a+1)}$$ Factoring the numerator and denominator

$$= \frac{\cancel{(a+2)}\cancel{(a+1)}\cancel{(a+3)}(a-1)}{\cancel{(a+2)}\cancel{(a+3)}(a-3)\cancel{(a+1)}}$$ Removing a factor of 1

$$= \frac{a-1}{a-3}$$

Practice this exercise:

2. Divide and simplify:

$\frac{x^2+2x-15}{x^2+x-6} \div \frac{x^2+3x+2}{x^2+4x+3}$.

ADDITIONAL EXERCISES

Objective A Find the reciprocal of a rational expression.

For extra help, see Examples 1–3 on page 916 of your text and the Section 12.2 lecture video.

Find the reciprocal.

1. $\frac{3}{y}$

2. x^2-9

3. $\frac{1}{n+3}$

Objective B Divide rational expressions and simplify.

For extra help, see Examples 4–9 on pages 916–918 of your text and the Section 12.2 lecture video.

Divide and simplify.

4. $\frac{4x+16}{21} \div \frac{x+4}{14}$

5. $\frac{x^2-1}{x^2+1} \div \frac{x+1}{x-1}$

6. $\frac{r^2+9r}{r^2+7r+12} \div \frac{2r}{r+4}$

7. $\frac{y^2-4}{6y-12} \div \frac{3y+9}{2y^2+6y+4}$

Name: Date:
Instructor: Section:

Chapter 12 RATIONAL EXPRESSIONS AND EQUATIONS

12.3 Least Common Multiples and Denominators

Learning Objectives
A Find the LCM of several numbers by factoring.
B Add fractions, first finding the LCD.
C Find the LCM of algebraic expressions by factoring.

Key Terms
Use the vocabulary terms listed below to complete each statement in Exercises 1–2.

least common denominator **least common multiple**

1. The expression $12x^2y^3$ is the ________________ of $6xy^2$, $4x^2y$, and $2y^3$.

2. The expression $12x^2y^3$ is the ________________ of $\frac{5}{6xy^2}$, $\frac{1}{4x^2y}$, and $\frac{3x}{2y^3}$.

GUIDED EXAMPLES AND PRACTICE

Objective A Find the LCM of several numbers by factoring.

Review this example for Objective A:
1. Find the LCM of 15, 24, and 30.

$15 = 3\cdot5$
$24 = 2\cdot2\cdot2\cdot3$
$30 = 2\cdot3\cdot5$
The LCM is $2\cdot2\cdot2\cdot3\cdot5 = 120$

Practice this exercise:
1. Find the LCM of 12, 21, and 28.

Objective B Add fractions, first finding the LCD.

Review this example for Objective B:
2. Add: $\frac{5}{6}+\frac{5}{18}$.

$\frac{5}{6}+\frac{5}{18} = \frac{5}{2\cdot3}\cdot\frac{3}{3}+\frac{5}{2\cdot3\cdot3}$ Multiplying by 1
$= \frac{15}{2\cdot3\cdot3}+\frac{5}{2\cdot3\cdot3}$ Each denominator is now the LCD.
$= \frac{20}{2\cdot3\cdot3}$ Adding the numerators and keeping the LCD
$= \frac{\cancel{2}\cdot2\cdot5}{\cancel{2}\cdot3\cdot3}$ Factoring the numerator and removing a factor of 1
$= \frac{10}{9}$ Simplifying

Practice this exercise:
2. Add: $\frac{5}{12}+\frac{2}{15}$.

Objective C Find the LCM of algebraic expressions by factoring.

Review this example for Objective C:

3. Find the LCM of a^2-9 and a^2-2a-3.

$$a^2-9=(a+3)(a-3)$$
$$a^2-2a-3=(a-3)(a+1)$$

The LCM is $(a+1)(a+3)(a-3)$.

Practice this exercise:

3. Find the LCM of m^2-3m and m^3+6m^2+9m.

ADDITIONAL EXERCISES

Objective A Find the LCM of several numbers by factoring.

For extra help, see Example 1 on page 921 of your text and the Section 12.3 lecture video.

Find the LCM.

1. 3, 15

2. 6, 15, 20

3. 10, 50, 120

Objective B Add fractions, first finding the LCD.

For extra help, see Example 2 on page 922 of your text and the Section 12.3 lecture video.

Add, first finding the LCD. Simplify, if possible.

4. $\frac{1}{8}+\frac{3}{20}$

5. $\frac{5}{18}+\frac{11}{24}$

6. $\frac{1}{5}+\frac{7}{20}+\frac{2}{25}$

Objective C Find the LCM of algebraic expressions by factoring.

For extra help, see Examples 3–7 on page 922 of your text and the Section 12.3 lecture video.

Find the LCM.

7. $3y^3,\ 12x^2y,\ 15xy^4$

8. $x^2-x-30,\ x^2-7x+6$

9. $8a^2-16a,\ 2a^2+6a-20$

10. $4+5x,\ 16-25x^2,\ 4-5x$

Name: Date:
Instructor: Section:

Chapter 12 RATIONAL EXPRESSIONS AND EQUATIONS

12.4 Adding Rational Expressions

Learning Objectives
A Add rational expressions.

Key Terms
Use the vocabulary terms listed below to complete the steps for adding rational expressions with different denominators in Exercises 1–4.

equivalent expression **least common multiple**
numerators **simplify**

1. Find the __________________ of the denominators.

2. For each rational expression, find a(n) __________________ with the LCD.

3. Add the __________________ .

4. __________________ if possible.

GUIDED EXAMPLES AND PRACTICE

Objective A Add rational expressions.

Review this example for Objective A:

1. Add: $\dfrac{2x}{3x+6}+\dfrac{1}{x^2-4}$.

First we find the LCD:

$3x+6=3(x+2)$

$x^2-4=(x+2)(x-2)$

The LCD is $3(x+2)(x-2)$. Then we have:

$$\frac{2x}{3(x+2)}\cdot\frac{x-2}{x-2}+\frac{1}{(x+2)(x-2)}\cdot\frac{3}{3}$$

$$=\frac{2x(x-2)+3}{3(x+2)(x-2)}$$

$$=\frac{2x^2-4x+3}{3(x+2)(x-2)}$$

Practice this exercise:

1. Add:

$$\frac{3}{x^2-3x-4}+\frac{5}{x^2+3x+2}.$$

ADDITIONAL EXERCISES

Objective A Add rational expressions.

For extra help, see Examples 1–11 on pages 925–928 of your text and the Section 12.4 lecture video.

Add. Simplify if possible.

1. $\dfrac{4x}{x+5}+\dfrac{2x-3}{x+5}$

2. $\dfrac{8}{c^2d}+\dfrac{5}{cd^2}$

3. $\dfrac{2}{x^2-6x-7}+\dfrac{5}{x^2-2x-3}$

4. $\dfrac{3x}{x^2-4}+\dfrac{5x}{x^2+2x}$

Name: Date:
Instructor: Section:

Chapter 12 RATIONAL EXPRESSIONS AND EQUATIONS

12.5 Subtracting Rational Expressions

Learning Objectives
A Subtract rational expressions.
B Simplify combined additions and subtractions of rational expressions.

Key Terms
Use the vocabulary terms listed below to complete each statement in Exercises 1–4.

denominator **least common denominator** **multiply** **numerators**

1. To subtract rational expressions, they must be written with a common ______________ .

2. The least common multiple of the denominators is the ________________ .

3. To subtract rational expressions when the denominators are the same, subtract the

 ________________ .

4. When one denominator is the opposite of the other, we ________________ one expression by $-1/-1$ to obtain a common denominator.

GUIDED EXAMPLES AND PRACTICE

Objective A Subtract rational expressions.

Review this example for Objective A:

1. Subtract: $\frac{x-3}{x+5}-\frac{x-2}{x+1}$.

 The LCD is $(x+5)(x+1)$.

$$\frac{x-3}{x+5}-\frac{x-2}{x+1}=\frac{x-3}{x+5}\cdot\frac{x+1}{x+1}-\frac{x-2}{x+1}\cdot\frac{x+5}{x+5}$$
$$=\frac{(x-3)(x+1)}{(x+5)(x+1)}-\frac{(x-2)(x+5)}{(x+1)(x+5)}$$
$$=\frac{x^2-2x-3}{(x+5)(x+1)}-\frac{x^2+3x-10}{(x+5)(x+1)}$$
$$=\frac{x^2-2x-3-(x^2+3x-10)}{(x+5)(x+1)}$$
$$=\frac{x^2-2x-3-x^2-3x+10}{(x+5)(x+1)}$$
$$=\frac{-5x+7}{(x+5)(x+1)}$$

Practice this exercise:

1. Subtract: $\frac{4}{x^2-36}-\frac{1}{x-6}$.

Objective B   Simplify combined additions and subtractions of rational expressions.

Review this example for Objective B:

2. Perform the indicated operations and simplify, if possible:

$$\frac{4a}{a^2-4}-\frac{3}{a-2}+\frac{5}{a}.$$

The LCD is $a(a+2)(a-2)$.

$$\frac{4a}{a^2-4}\cdot\frac{a}{a}-\frac{3}{a-2}\cdot\frac{a(a+2)}{a(a+2)}+\frac{5}{a}\cdot\frac{(a+2)(a-2)}{(a+2)(a-2)}$$

$$=\frac{4a^2}{a(a+2)(a-2)}-\frac{3a(a+2)}{a(a+2)(a-2)}+\frac{5(a+2)(a-2)}{a(a+2)(a-2)}$$

$$=\frac{4a^2}{a(a+2)(a-2)}-\frac{3a^2+6a}{a(a+2)(a-2)}+\frac{5(a^2-4)}{a(a+2)(a-2)}$$

$$=\frac{4a^2-(3a^2+6a)+5a^2-20}{a(a+2)(a-2)}$$

$$=\frac{4a^2-3a^2-6a+5a^2-20}{a(a+2)(a-2)}$$

$$=\frac{6a^2-6a-20}{a(a+2)(a-2)}$$

Practice this exercise:

2. Perform the indicated operations and simplify, if possible:

$$\frac{3y}{y^2+y-20}+\frac{2}{y+5}-\frac{3}{y-4}.$$

ADDITIONAL EXERCISES

Objective A   Subtract rational expressions.

For extra help, see Examples 1–7 on pages 933–936 of your text and the Section 12.5 lecture video.

Subtract. Simplify if possible.

1. $\dfrac{2x+5}{x^2+6x-7}-\dfrac{x-2}{x^2+6x-7}$

2. $\dfrac{8}{x+1}-\dfrac{5}{x-1}$

3. $\dfrac{x-3}{x^2-16}-\dfrac{5-x}{16-x^2}$

4. $\dfrac{x}{x^2+4x+3}-\dfrac{1}{x^2-1}$

Objective B   Simplify combined additions and subtractions of rational expressions.

For extra help, see Examples 8–9 page 936 of your text and the Section 12.5 lecture video.

Subtract. Simplify if possible.

5. $\dfrac{2(t+1)}{t-2}-\dfrac{5(2t-3)}{2-t}+\dfrac{7t+6}{t-2}$

6. $\dfrac{8}{3x-1}-\dfrac{4}{1-3x}+\dfrac{2x}{3x-1}+\dfrac{x-5}{1-3x}$

7. $\dfrac{x+5}{x-3}-\dfrac{2-x}{x+3}-\dfrac{4x-18}{9-x^2}$

8. $\dfrac{1}{a-b}+\dfrac{2}{a-b}+\dfrac{2a}{a^2-b^2}$

Name: Date:
Instructor: Section:

Chapter 12 RATIONAL EXPRESSIONS AND EQUATIONS

12.6 Solving Rational Equations

Learning Objectives
A Solve rational equations.

Key Terms
Use the vocabulary terms listed below to complete each statement in Exercises 1–4.

check **clear** **LCM** **rational**

1. The equation $\frac{2}{x}=\frac{3}{x+1}$ is an example of a(n) ________________ equation.

2. To solve a rational equation, first ________________ the equation of fractions.

3. When solving a rational equation, multiply on both sides of the equation by the ________________ of all the denominators.

4. After solving a rational equation, always ________________ possible solutions in the original equation.

GUIDED EXAMPLES AND PRACTICE

Objective A Solve rational equations.

Review this example for Objective A:

1. Solve: $\frac{2x+1}{x-2}=\frac{x-1}{3x+2}$.

The LCM of the denominators is $(x-2)(3x+2)$. We multiply by the LCM on both sides.

$$\frac{2x+1}{x-2}=\frac{x-1}{3x+2}$$
$$(x-2)(3x+2)\frac{2x+1}{x-2}=(x-2)(3x+2)\frac{x-1}{3x+2}$$
$$(3x+2)(2x+1)=(x-2)(x-1)$$
$$6x^2+7x+2=x^2-3x+2$$
$$5x^2+10x=0$$
$$5x(x+2)=0$$

$5x=0$ *or* $x+2=0$
$x=0$ *or* $x=-2$

Both numbers check. The solutions are 0 and –2.

Practice this exercise:

1. Solve: $x-\frac{8}{x}=2$.

ADDITIONAL EXERCISES

Objective A Solve rational equations.

For extra help, see Examples 1–7 on pages 943–946 of your text and the Section 12.6 lecture video.

Solve. Don't forget to check!

1. $x+\frac{3}{x}=4$

2. $\frac{3}{x-1}=\frac{2}{x+2}$

3. $\frac{x-3}{x-8}=\frac{5}{x-8}$

4. $\frac{4}{x-3}=\frac{2x}{x^2-9}-\frac{7}{x+3}$

Name: Date:
Instructor: Section:

Chapter 12 RATIONAL EXPRESSIONS AND EQUATIONS

12.7 Applications Using Rational Equations and Proportions

Learning Objectives
A Solve applied problems using rational equations.
B Solve proportion problems.

Key Terms
Use the vocabulary terms listed below to complete each statement in Exercises 1–4.

proportion **ratio** **rate** **similar**

1. A(n) ________________ of two quantities is their quotient.
2. The ratio of two different kinds of measure is called a(n) ________________.
3. The equation $\frac{A}{B} = \frac{C}{D}$ is an example of a(n) ________________.
4. Two triangles are ________________ if their corresponding angles have the same measure.

GUIDED EXAMPLES AND PRACTICE

Objective A Solve applied problems using rational equations.

Review this example for Objective A:

1. One number is 3 more than another. The quotient of the larger number divided by the smaller is $\frac{3}{2}$. Find the numbers.

   1. *Familiarize.* Let x = the smaller number. Then $x+3=$ the larger number and the quotient of the larger number divided by the smaller number is $\frac{x+3}{x}$.

   2. *Translate.*

   The quotient is $\frac{3}{2}$

   $$\frac{x+3}{x} = \frac{3}{2}$$

Practice this exercise:

1. A passenger car travels 10 km/h faster than a delivery van. While the car travels 240 km, the van travels 200 km. Find their speeds.

3. *Solve.* We solve the equation.

$$\frac{x+3}{x}=\frac{3}{2}$$
$$2x\cdot\frac{x+3}{x}=2x\cdot\frac{3}{2}$$
$$2(x+3)=3x$$
$$2x+6=3x$$
$$6=x$$

If $x = 6$, then $x + 3 = 6 + 3$, or 9.

4. *Check.* The larger number, 9, is 3 more than the smaller number, 6. Also $\frac{9}{6}=\frac{3}{2}$, so the numbers check.

5. *State.* The numbers are 6 and 9.

Objective B Solve proportion problems.

Review this example for Objective B:

2. Melinda biked 78 mi in 5 days. At this rate, how far would she bike in 7 days?

1. *Familiarize.* We can set up ratios, letting m = the number of miles Melinda would bike in 7 days.

2. *Translate.* We assume Melinda bikes at the same rate during the entire 7 days. Thus, the ratios are the same and we can write a proportion.

$$\begin{array}{r} \text{Miles}\rightarrow \\ \text{Days}\rightarrow \end{array} \frac{78}{5}=\frac{m}{7} \begin{array}{l} \leftarrow\text{Miles} \\ \leftarrow\text{Days} \end{array}$$

3. *Solve.* We multiply by the LCM, $5\cdot 7$, or 35, on both sides.

$$35\cdot\frac{78}{5}=35\cdot\frac{m}{7}$$
$$7\cdot 78=5\cdot m$$
$$\frac{7\cdot 78}{5}=m$$
$$109.2=m$$

4. *Check.* $\frac{78}{5}=15.6$ and $\frac{109.2}{7}=15.6$; since the ratios are the same, the answer checks.

5. *State.* Melinda would bike 109.2 mi in 7 days.

Practice this exercise:

2. Jeremy can read 6 pages of his history textbook in 20 min. At this rate, how many pages can he read in 50 min?

Name: Date:
Instructor: Section:

ADDITIONAL EXERCISES

Objective A Solve applied problems using rational equations.

For extra help, see Examples 1–2 on pages 952–955 of your text and the Section 12.7 lecture video.

Solve.

1. It takes Alexis 3 hr to clean her family's garage. Alyssa takes 4 hr to do the same job. How long would it take them, working together, to clean the garage?

2. Abigail can file a week's worth of invoices in 75 min. Ava can do the same job in 90 min. How long would it take if they worked together?

3. Maya drives 20 km/h faster than Tara. While Tara travels 180 km, Maya travels 260 km. Find their speeds.

4. The speed of Tom's scooter is 16 mph less than the speed of Mary Lynn's motorcycle. The motorcycle can travel 290 mi in the same time that the scooter can travel 210 mi. Find the speed of each vehicle.

Objective B Solve proportion problems.

For extra help, see Examples 3–7 on pages 956–959 of your text and the Section 12.7 lecture video.

Solve.

5. Find the ratio of 360 mi, 15 gal. Simplify, if possible.

6. A sample of 48 memory cards contained 3 defective cards. How many defective cards would you expect in a sample of 192 cards?

7. To determine the number of trout in his pond, Phil catches 25 trout, tags them and lets them loose. Later he catches 18 trout; 10 of them have tags. Estimate the number of trout in the pond.

8. Linda walked 610 steps in 5 min on an elliptical trainer. At this rate, how many steps would she walk in 12 min?

9. For the pair of triangles, find the length of the indicated side.

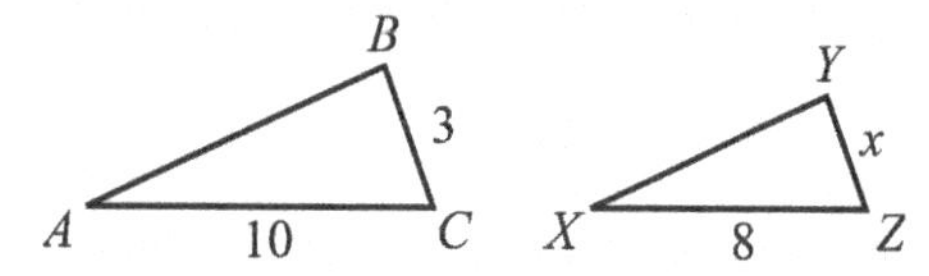

Name: Date:
Instructor: Section:

Chapter 12 RATIONAL EXPRESSIONS AND EQUATIONS

12.8 Complex Rational Expressions

Learning Objectives
A Simplify complex rational expressions.

GUIDED EXAMPLES AND PRACTICE

Objective A Simplify complex rational expressions.

Review this example for Objective A:

1. Simplify: $\dfrac{\frac{x}{10}-\frac{10}{x}}{\frac{1}{x}+\frac{1}{10}}$.

Method I. We find the LCM of all denominators: $10x$. We multiply by 1 using $10x/10x$.

$$\frac{\frac{x}{10}-\frac{10}{x}}{\frac{1}{x}+\frac{1}{10}}\cdot\frac{10x}{10x}=\frac{\left(\frac{x}{10}-\frac{10}{x}\right)10x}{\left(\frac{1}{x}+\frac{1}{10}\right)10x}$$

$$=\frac{\frac{x}{10}\cdot 10x-\frac{10}{x}\cdot 10x}{\frac{1}{x}\cdot 10x+\frac{1}{10}\cdot 10x}$$

$$=\frac{x^2-100}{10+x}=\frac{(x+10)(x-10)}{x+10}$$

$$=\frac{\cancel{(x+10)}(x-10)}{\cancel{x+10}}$$

$$=x-10$$

Method II. We write a single rational expression in the numerator and denominator.

$$\frac{\frac{x}{10}-\frac{10}{x}}{\frac{1}{x}+\frac{1}{10}}=\frac{\frac{x}{10}\cdot\frac{x}{x}-\frac{10}{x}\cdot\frac{10}{10}}{\frac{1}{x}\cdot\frac{10}{10}+\frac{1}{10}\cdot\frac{x}{x}}=\frac{\frac{x^2}{10x}-\frac{100}{10x}}{\frac{10}{10x}+\frac{x}{10x}}$$

$$=\frac{\frac{x^2-100}{10x}}{\frac{10+x}{10x}}=\frac{x^2-100}{10x}\cdot\frac{10x}{10+x}$$

$$=\frac{(x+10)(x-10)}{x+10}=x-10$$

Practice this exercise:

1. Simplify: $\dfrac{\frac{3}{y}+\frac{1}{y}}{y-\frac{y}{3}}$.

ADDITIONAL EXERCISES

Objective A Simplify complex rational expressions.

For extra help, see Examples 1–6 on pages 968–971 of your text and the Section 12.8 lecture video.

Simplify.

1. $$\frac{\frac{1}{2}-\frac{1}{x}}{\frac{2-x}{2}}$$

2. $$\frac{6+\frac{6}{n}}{2+\frac{2}{n}}$$

3. $$\frac{x-4-\frac{5}{x}}{x-2-\frac{3}{x}}$$

4. $$\frac{\frac{5}{n+1}+\frac{2}{n}}{\frac{3}{n+1}+\frac{5}{n}}$$

Name: Date:
Instructor: Section:

Chapter 12 RATIONAL EXPRESSIONS AND EQUATIONS

12.9 Direct Variation and Inverse Variation

Learning Objectives

A Find an equation of direct variation given a pair of values of the variables.
B Solve applied problems involving direct variation.
C Find an equation of inverse variation given a pair of values of the variables.
D Solve applied problems involving inverse variation.

Key Terms

Use the vocabulary terms listed below to complete each statement in Exercises 1–4.

direct **inverse** **proportionality** **variation**

1. The equation $y = kx$ is called an equation of ________________ variation.
2. In the equation $y = kx$, k is called the constant of ________________.
3. The equation $y = k / x$ is called an equation of ________________ variation.
4. In the equation $y = k / x$, k is called the ________________ constant.

GUIDED EXAMPLES AND PRACTICE

Objective A Find an equation of direct variation given a pair of values of the variables.

Review this example for Objective A:

1. Find an equation of variation in which y varies directly as x and $y = 20$ when $x = 4$.

   We substitute to find k:

   $$y = kx$$
   $$20 = k \cdot 4$$
   $$5 = k$$

   Then the equation of variation is $y = 5x$.

Practice this exercise:

1. Find an equation of variation in which y varies directly as x and $y = 3$ when $x = 2$.

Objective B Solve applied problems involving direct variation.

Review this example for Objective B:

2. Interest I earned in 1 yr on a fixed principal varies directly as the interest rate r. An investment earns \$56.25 at an interest rate of 3.75%. How much will the investment earn at a rate of 4.5%?

1., 2. *Familiarize* and *Translate*. The problem states that we have direct variation between the variables I and r. Thus, an equation $I = kr$, $k > 0$, applies. As the interest rate increases, the amount of interest earned increases.

3. *Solve*. First find an equation of variation.

$$I = kr$$
$$56.25 = k \cdot 0.0375$$
$$\frac{56.25}{0.0375} = k$$
$$1500 = k$$

The equation of variation is $I = 1500r$.

Now use the equation to find the interest earned when the interest rate is 4.5%.

$$I = 1500r$$
$$I = 1500(0.045)$$
$$I = 67.50$$

4. *Check*. This check might be done by repeating the computations. We might also do some reasoning about the answer. The interest rate increased from 3.75% to 4.5%. Similarly, the interest earned increased from \$56.25 to \$67.50.

5. *State*. When the interest rate is 4.5%, the investment earns \$67.50.

Practice this exercise:

2. The amount of Melissa's paycheck P varies directly as the number H of hours worked. For working 16 hr, her pay is \$132. Find her pay for 28 hr of work.

Objective C Find an equation of inverse variation given a pair of values of the variables.

Review this example for Objective C:

3. Find an equation of variation in which y varies inversely as x and $y = 10$ when $x = 0.5$.

We substitute to find k:

$$y = \frac{k}{x}$$
$$10 = \frac{k}{0.5}$$
$$5 = k$$

Then the equation of variation is $y = \frac{5}{x}$.

Practice this exercise:

3. Find an equation of variation in which y varies inversely as x and $y = 12$ when $x = 3$.

Name: Date:
Instructor: Section:

Objective D Solve applied problems involving inverse variation.

Review this example for Objective D:

4. The time t required to drive a fixed distance varies inversely as the speed r. It takes 4 hr at 60 mph to drive a fixed distance. How long would it take at 50 mph?

1. *Familiarize*. The problem states that we have inverse variation between the variables t and r. As the speed decreases, the time required to travel the fixed distance increases.
2. *Translate*. We write an equation of variation. Travel time varies inversely as speed. This translates to $t=\frac{k}{r}$.
3. *Solve*. First find an equation of variation.

$$t=\frac{k}{r}$$
$$4=\frac{k}{60}$$
$$240=k$$

The equation is $t=\frac{240}{r}$.

now use the equation to find the time required to travel the fixed distance at 50 mph.

$$t=\frac{240}{r}$$
$$t=\frac{240}{50}$$
$$t=4.8$$

4. *Check*. In addition to repeating the computations, we can analyze the results. The speed decreased from 60 mph to 50 mph, and the travel time increased from 4 hr to 4.8 hr. This is what we would expect with inverse variation.
5. *State*. It would take 4.8 hr to travel the fixed distance at a speed of 50 mph.

Practice this exercise:

4. It takes 4 days for 2 people to paint a house. How long will it take 3 people to do the job?

ADDITIONAL EXERCISES

Objective A Find an equation of direct variation given a pair of values of the variables.

For extra help, see Examples 1–2 on page 975 of your text and the Section 12.9 lecture video.

Find an equation of variation in which y varies directly as x and the following are true. Then find the value of y when $x = 30$.

1. $y = 42$ when $x = 6$

2. $y = 0.3$ when $x = 0.4$

3. $y = 30$ when $x = 25$

4. $y = 30$ when $x = 200$

Objective B Solve applied problems involving direct variation.

For extra help, see Example 3 on pages 976–977 of your text and the Section 12.9 lecture video.

Solve.

5. The number of teaspoons t of tea leaves varies directly as the number of cups C of tea made. Lisa uses 15 teaspoons of tea leaves to make 10 cups of tea.
a) Find an equation of variation.
b) How many teaspoons of tea leaves are needed to make 8 cups of tea?

6. The number of servings P of dried pineapple varies directly as the size C of the container. A 14-oz bag of pineapple contains 10 servings.
a) Find an equation of variation.
b) How many servings of dried pineapple are contained in a 20-oz carton?

7. The number of calories c burned by a person in a Zumba aerobic class is directly proportional to the time t spent exercising. It takes 10 min to burn 80 calories (*Source*: Family Fun and Fitness). How long would it take to burn 200 calories in the class?

8. The electrical current I, in amperes, in a circuit varies directly as the voltage V. When 12 volts are applied, the current is 3 amperes. What is the current when 16 volts are applied?

Name: Date:
Instructor: Section:

Objective C Find an equation of inverse variation given a pair of values of the variables.

For extra help, see Example 4 on page 978 of your text and the Section 12.9 lecture video.

Find an equation of variation in which y varies inversely as x and the following are true. Then find the value of x when $x = 20$.

9. $y = 4$ when $x = 15$

10. $y = 3.5$ when $x = 0.4$

11. $y = 30$ when $x = 12$

12. $y = 0.5$ when $x = 0.8$

Objective D Solve applied problems involving inverse variation.

For extra help, see Example 5 on pages 978–979 of your text and the Section 12.9 lecture video.

Solve.

13. The number of gallons N, Ash uses to drive to work is inversely proportional to the miles-per-gallon rating P of the vehicle he drives. When he drives his Chevrolet Suburban, rated at 16 mpg, he uses 2.5 gal of gas.
a) Find an equation of variation.
b) How much gas will he use if he rides his Yamaha V-Max, rated at 40 mpg?

14. The time T required to do a job varies inversely as the number of people P working. It takes 4 hr for 9 people to weed the community garden. How long would it take 10 people to complete the job?

15. The current I in an electrical conductor varies inversely as the resistance R in the conductor. If the current is $\frac{2}{3}$ ampere when the resistance is 120 ohms, what is the current when the resistance is 150 ohms?

16. The wavelength W of a radio wave varies inversely as its frequency F. A wave with a frequency of 1600 kilohertz has a length of 225 meters. What is the length of a wave with a frequency of 3000 kilohertz?

Name: Date:
Instructor: Section:

Chapter 13 SYSTEMS OF EQUATIONS

13.1 Systems of Equations in Two Variables

Learning Objectives
A Determine whether an ordered pair is a solution of a system of equations.
B Solve systems of equations in two variables by graphing.

Key Terms
Use the vocabulary terms listed below to complete each statement in Exercises 1–6.

graph	**intersection**	**solution**	**system**
are parallel	**intersect**	**have the same graph**	

1. The following is an example of a(n) ________________ of equations:
$x - y = 2,$
$2x + y = 1.$

2. A(n) ________________ of a system of two equations is an ordered pair that makes both equations true.

3. The ________________ of an equation is a drawing that represents its solution set.

4. In a system of equations with one solution, the graphs ________________.

5. In a system of equations with no solution, the graphs ________________.

6. In a system of equations with infinitely many solutions, the graphs ________________.

GUIDED EXAMPLES AND PRACTICE

Objective A Determine whether an ordered pair is a solution of a system of equations.

Review this example for Objective A:

1. Determine whether (2, 5) is a solution of the system
$3a - b = 1,$
$4a - 5b = -17.$

We check by substituting alphabetically 2 for *a* and 5 for *b*.

$3a - b = 1$		
$3 \cdot 2 - 5$	1	
$6 - 5$		
1		TRUE

$4a - 5b = -17$		
$4 \cdot 2 - 5 \cdot 5$	-17	
$8 - 25$		
-17		TRUE

This checks, so (2, 5) is a solution of the system of equations.

Practice this exercise:

1. Determine whether (12, 10) is a solution of the system
$5x = 6y,$
$3y = 2x + 6.$

Objective B Solve systems of equations in two variables by graphing.

Review this example for Objective B:

2. Solve the this system of equations by graphing:

$x - y = 4,$
$y = 2x - 5.$

We graph the equations.

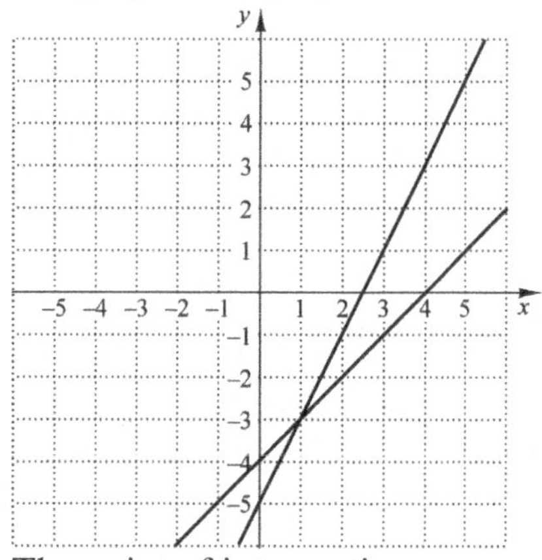

The point of intersection appears to be (1, –3). This checks in both equations, so it is the solution.

Practice this exercise:

2. Solve this system of equations by graphing:

$3x - 2y = 6,$
$y = x - 1.$

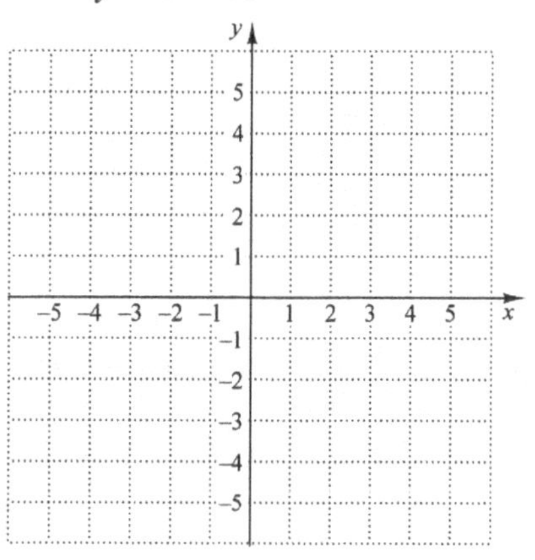

ADDITIONAL EXERCISES

Objective A Determine whether an ordered pair is a solution of a system of equations.

For extra help, see Examples 1–2 on pages 994–995 of your text and the Section 13.1 lecture video.

Determine whether the given ordered pair is a solution of the system of equations. Use alphabetical order of the variables.

1. $(-1, 4)$; $b - 3a = 1,$
$b + a = 3.$

2. $(8, -3)$; $s = r - 5,$
$r - 2s = 2.$

3. $(2, -3)$; $2x - y = 7,$
$x = y + 5.$

4. $(2, 0)$; $y = x - 2,$
$x + y = 2.$

Name: Date:
Instructor: Section:

Objective B Solve systems of equations in two variables by graphing.
For extra help, see Examples 3–6 on pages 995–996 of your text and the Section 13.1 lecture video.
Solve each system of equations by graphing.

5. $x = 2y,$
$x - 2y = 2$

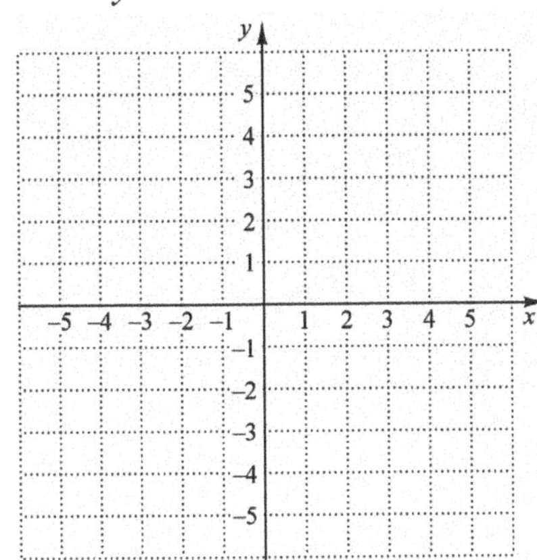

6. $x + y = 3,$
$x = 3 - y$

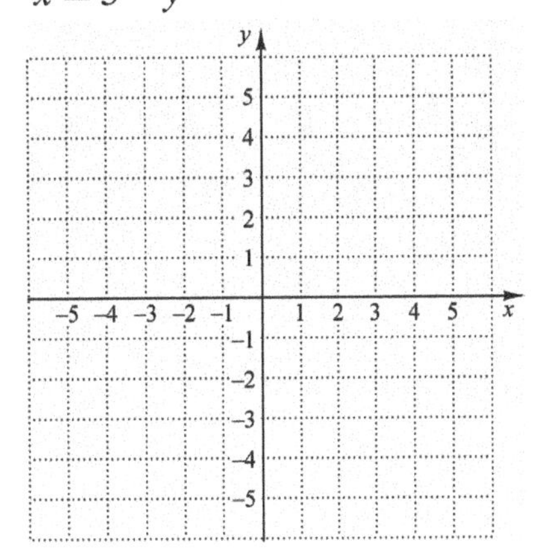

7. $x = -3,$
$y = 2$

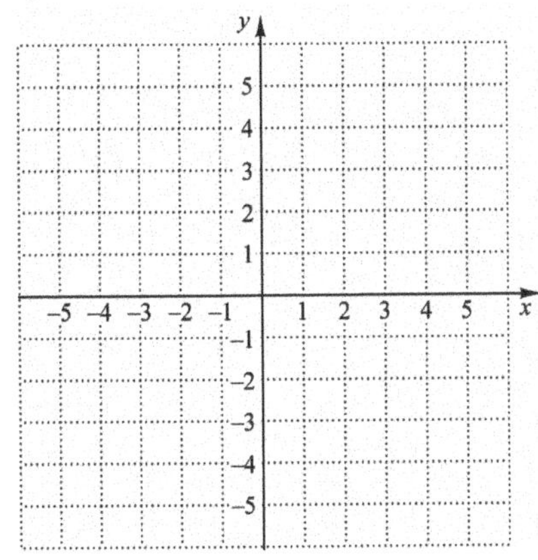

8. $6x - y = 1,$
$3x + y = 8$

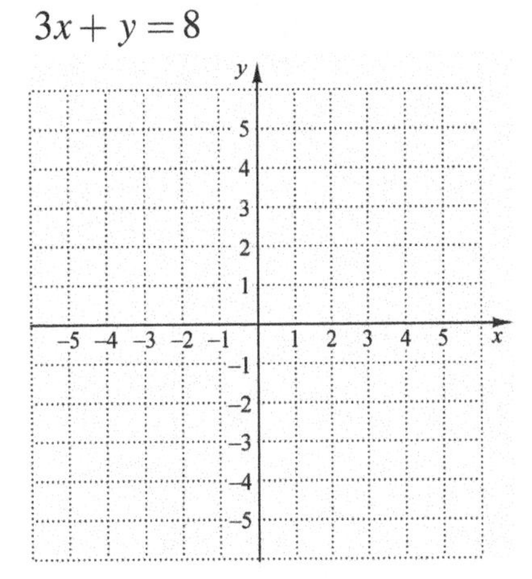

Name: Date:
Instructor: Section:

Chapter 13 SYSTEMS OF EQUATIONS

13.2 The Substitution Method

Learning Objectives

A Solve a system of two equations in two variables by the substitution method when one of the equations has a variable alone on one side.

B Solve a system of two equations in two variables by the substitution method when neither equation has a variable alone on one side.

C Solve applied problems by translating to a system of two equations and then solve using the substitution method.

Key Terms

Use the vocabulary terms listed below to complete each statement in Exercises 1–4.

algebraic **solve** **substitute** **translate**

1. Nongraphical methods for solving systems of equations are called ________________ methods.

2. One way to solve a system of two equations is to ________________ an equivalent expression for a variable into an equation.

3. If neither equation of a system has a variable alone on one side, we ________________ one equation for one of the variables.

4. Sometimes it is easier to ________________ a problem into two equations in two variables.

GUIDED EXAMPLES AND PRACTICE

Objective A Solve a system of two equations in two variables by the substitution method when one of the equations has a variable alone on one side.

Review this example for Objective A:

1. Solve the system

$$x + y = -2 \quad (1)$$
$$x = 2y + 7 \quad (2)$$

First substitute $2y + 7$ for x in Equation (1) and solve for y.

$$x + y = -2$$
$$(2y+7) + y = -2$$
$$3y + 7 = -2$$
$$3y = -9$$
$$y = -3$$

Practice this exercise:

1. Solve the system using substitution.

$$y = x - 2,$$
$$x - 2y = 6.$$

Now substitute –3 for y in either of the original equations and find x. We can choose Equation (2) because it has x alone on one side.

$$x = 2y + 7 = 2(-3) + 7 = -6 + 7 = 1$$

The ordered pair (1, –3) checks in both equations, so it is the solution of the system of equations.

Objective B Solve a system of two equations in two variables by the substitution method when neither equation has a variable alone on one side.

Review this example for Objective B:

2. Solve the system

$$x - 2y = 1, \quad (1)$$
$$2x - 3y = 3. \quad (2)$$

We solve Equation (1) for x, since the coefficient of x is 1 in that equation.

$$x - 2y = 1$$
$$x = 2y + 1 \quad (3)$$

Now substitute for x in Equation (2) and solve for y.

$$2x - 3y = 3$$
$$2(2y + 1) - 3y = 3$$
$$4y + 2 - 3y = 3$$
$$y + 2 = 3$$
$$y = 1$$

Now substitute 1 for y in Equation (1), (2), or (3) and find x. We choose Equation (3) since it is already solved for x.

$$x = 2y + 1 = 2 \cdot 1 + 1 = 2 + 1 = 3$$

The ordered pair (3, 1) checks in both equations, so it is the solution of the system of equations.

Practice this exercise:

2. Solve the system using substitution.

$$x + y = 3,$$
$$5x + 2y = 3.$$

Objective C Solve applied problems by translating to a system of two equations and then solve using the substitution method.

Review this example for Objective C:

3. The sum of two numbers is 1. One number is 11 more than the other. Find the numbers.

1. *Familiarize.* We let x = the smaller number and y = the larger number.
2. *Translate.* The first statement gives us one equation.

Practice this exercise:

3. The perimeter of a rectangular poster is 12 ft. The length is twice the width. Find the length and width.

Name: Date:
Instructor: Section:

The sum of two numbers is 1

$$x + y = 1$$

Now we translate the second statement.

One number is 11 more than the other

$$y = 11 + x$$

We now have a system of equations:

$$x + y = 1, \quad (1)$$
$$y = 11 + x. \quad (2)$$

3. *Solve*. First we substitute $11 + x$ for y in Equation (1).

$$x + (11 + x) = 1$$
$$2x + 11 = 1$$
$$2x = -10$$
$$x = -5$$

Now substitute –5 for x in Equation (2).

$$y = 11 + x = 11 + (-5) = 6$$

4. *Check*. The sum of –5 and 6 is 1. Also, 6 is 11 more than –5, so the numbers check.
5. *State*. The numbers are –5 and 6.

ADDITIONAL EXERCISES

Objective A Solve a system of two equations in two variables by the substitution method when one of the equations has a variable alone on one side.

For extra help, see Examples 1–2 on pages 1001–1002 of your text and the Section 13.2 lecture video.

Solve using the substitution method.

1. $x + y = 12,$
$y = x + 8$

2. $x = y - 8,$
$x + y = 2$

3. $x = 6y,$
$x - 7y = 1$

4. $x + y = 1,$
$y = 5 - 2x$

Objective B Solve a system of two equations in two variables by the substitution method when neither equation has a variable alone on one side.

For extra help, see Example 3 on page 1003 of your text and the Section 13.2 lecture video.

Solve using the substitution method. First, solve one equation for one variable.

5. $x + y = 3,$
$x - y = 5$

6. $y - x = 4,$
$2x + 5y = -8$

7. $2b - 3a = 0,$
$a - b = 1$

8. $3r + 4s = 0,$
$r + 2s = -10$

Objective C Solve applied problems by translating to a system of two equations and then solve using the substitution method.

For extra help, see Example 4 on page 1004 of your text and the Section 13.2 lecture video.

Solve.

9. The perimeter of the largest possible regulation-size rugby field is 428 m. The length is 4 m longer than twice the width. Find the dimensions.

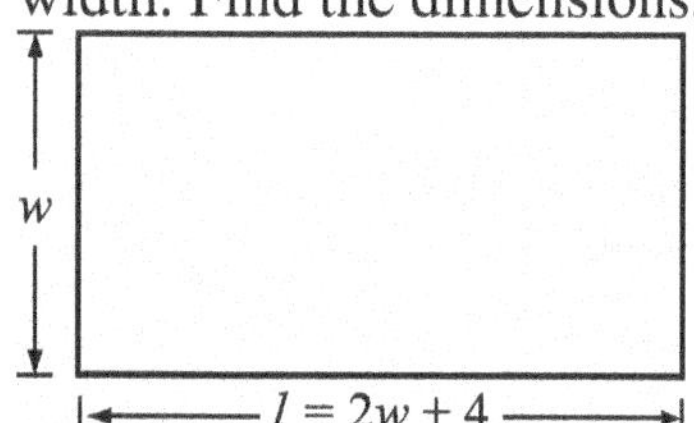

10. A rectangle has a perimeter of 124 ft. The width is 10 ft less than the length. Find the length and width.

11. The perimeter of a doubles-play badminton court is 128 ft. The width is 2 ft less than half the length. Find the length and the width.

12. Find two numbers whose sum is 85 and whose difference is 13.

Name: Date:
Instructor: Section:

Chapter 13 SYSTEMS OF EQUATIONS

13.3 The Elimination Method

Learning Objectives

A Solve a system of two equations in two variables using the elimination method when no multiplication is necessary.

B Solve a system of two equations in two variables using the elimination method when multiplication is necessary.

Key Terms

Use the vocabulary terms listed below to complete each statement in Exercises 1–4.

elimination **false** **opposites** **true**

1. The ____________________ method for solving systems of equations makes use of the addition principle.

2. To eliminate a variable when adding, the terms containing that variable must be ____________________ .

3. When solving using the elimination method, obtaining a(n) ____________________ equation means there is no solution.

4. When solving using the elimination method, obtaining a(n) ____________________ equation means there is an infinite number of solutions.

GUIDED EXAMPLES AND PRACTICE

Objective A Solve a system of two equations in two variables using the elimination method when no multiplication is necessary.

Review this example for Objective A:

1. Solve the system

$$\begin{aligned} 2x - y &= 5, \quad (1) \\ x + y &= 7. \quad (2) \end{aligned}$$

First we add.

$$\begin{array}{r} 2x - y = 5 \\ x + y = 7 \\ \hline 3x \quad\; = 12 \\ x = 4 \end{array}$$

Now substitute 4 for x in either of the original equations and solve for y.

Practice this exercise:

1. Solve the system

$$\begin{aligned} 3x + 2y &= 1, \quad (1) \\ x - 2y &= -13. \quad (2) \end{aligned}$$

We use Equation (2).

$$x + y = 7$$
$$4 + y = 7$$
$$y = 3$$

The ordered pair (4, 3) checks in both equations, so it is a solution of the system of equations.

Objective B Solve a system of two equations in two variables using the elimination method when multiplication is necessary.

Review this example for Objective B:

2. Solve the system

$$2a - 3b = 7, \quad (1)$$
$$3a - 2b = 8. \quad (2)$$

We could eliminate either *a* or *b*. Here we decide to eliminate the *a*-terms. Multiply Equation (1) by 3 and Equation (2) by –2. Then add and solve for *b*.

$$\begin{aligned} 6a - 9b &= 21 \\ -6a + 4b &= -16 \\ \hline -5b &= 5 \\ b &= -1 \end{aligned}$$

Next substitute –1 for *b* in either of the original equations.

$$\begin{aligned} 2a - 3b &= 7 \quad (1) \\ 2a - 3(-1) &= 7 \\ 2a + 3 &= 7 \\ 2a &= 4 \\ a &= 2 \end{aligned}$$

The ordered pair (2, –1) checks in both equations, so it is a solution of the system of equations.

Practice this exercise:

2. Solve the system

$$3x + 2y = 5, \quad (1)$$
$$x - y = 5. \quad (2)$$

ADDITIONAL EXERCISES

Objective A Solve a system of two equations in two variables using the elimination method when no multiplication is necessary.

For extra help, see Example 1 on pages 1008–1009 of your text and the Section 13.3 lecture video.

Solve using the elimination method.

1. $x - y = 3,$
$x + y = 11$

2. $3x - y = 3,$
$-5x + y = 3$

Name: Date:
Instructor: Section:

3. $3a + 2b = 5,$
$-2a - b = 1$

4. $5x - 3y = 4,$
$-5x + 3y = 4$

Objective B Solve a system of two equations in two variables using the elimination method when multiplication is necessary.

For extra help, see Examples 2–7 on pages 1009–1012 of your text and the Section 13.3 lecture video.

Solve using the multiplication principle first. Then add.

5. $x + y = -10,$
$2x - 5y = 8$

6. $c - d = 4,$
$3c + 2d = 17$

7. $3x - 2y = 5,$
$6x - 10 = 4y$

8. $8a + 12b = 10,$
$6a + 9b = 5$

Name: Date:
Instructor: Section:

Chapter 13 SYSTEMS OF EQUATIONS

13.4 Applications and Problem Solving

Learning Objectives
A Solve applied problems by translating to a system of two equations in two variables.

GUIDED EXAMPLES AND PRACTICE

Objective A Solve applied problems by translating to a system of two equations in two variables.

Review this example for Objective A:

1. Solution A is 40% acid and solution B is 55% acid. How much of each should be used in order to make 100 L of a solution that is 46% acid?

1. *Familiarize*. Let x and y represent the number of liters of 40% and 55% solution to be used, respectively. We organize the given information in a table.

Type of solution	A	B	Mixture
Amount of solution	x	y	100 L
Percent of acid	40%	55%	46%
Amount of acid in solution	$40\%x$	$55\%y$	$46\% \times 100$ or 46 L

2. *Translate*. The first row of the table gives us one equation.

$$x + y = 100$$

The last row gives us a second equation.

$$40\%x + 55\%y = 46, \text{ or}$$
$$0.4x + 0.55y = 46$$

After multiplying by 100 on both sides of the second equation to clear decimals, we have the following system of equations.

$$x + y = 100, \quad (1)$$
$$40x + 55y = 4600 \quad (2)$$

3. *Solve*. We use the elimination method. First multiply Equation (1) by –40 and then add to eliminate the x-terms.

$$-40x - 40y = -4000$$
$$40x + 55y = 4600$$
$$15y = 600$$
$$y = 40$$

Practice this exercise:

1. There were 220 tickets sold for a school play. The price for students was $3 and it was $7 for non-students. A total of $1080 was collected. How many of each type of ticket were sold?

Now substitute in Equation (2) and solve for x.

$$x + y = 100$$
$$x + 40 = 100$$
$$x = 60$$

4. *Check.* The sum of 60 and 40 is 100. Also, 40% of 60 L is 24 L and 55% of 40 L is 22 L. These add up to 46 L, so the answer checks.
5. *State.* 60 L of solution A and 40 L of solution B should be used.

ADDITIONAL EXERCISES

Objective A Solve applied problems by translating to a system of two equations in two variables.

For extra help, see Examples 1–5 on pages 1018–1023 of your text and the Section 13.4 lecture video.

Solve.

1. Admission to Mammoth Cave is \$12 for adults and \$8 for youth (*Source*: National Park Service). One day, 575 people entered the cave, paying a total of \$5600. How many adults and how many youth entered the cave?

2. Tropical Punch is 18% fruit juice and Caribbean Spring is 24% fruit juice. How many liters of each should be mixed to get 18 L of a mixture that is 20% juice?

3. Amazing Awards can pack 180 medals in a small box and 240 medals in a large box. A shipment of 2220 medals exactly filled 10 boxes. How many boxes of each size were used?

4. Supplementary angles are angles whose sum is 180°. Two angles are supplementary. One angle measures 20° more than four times the measure of the other. Find the measure of each angle.

Name: Date:
Instructor: Section:

Chapter 13 SYSTEMS OF EQUATIONS

13.5 Applications with Motion

Learning Objectives
A Solve motion problems using the formula $d = rt$.

Key Terms
Use the vocabulary terms listed below to complete each numbered blank in the statement.

distance **motion** **rate** **time**

The ________________ formula states that ________________ =
(1) (2)
________________ (or speed) · ________________ .
(3) (4)

GUIDED EXAMPLES AND PRACTICE

Objective A Solve motion problems using the formula $d = rt$.

Review this example for Objective A:

1. A canoeist paddles for 1 hr with a 3 mph current. The return trip against the current took 2 hr. Find the speed of the canoe in still water.

1. *Familiarize.* We first make a drawing. Let d = the distance traveled in one direction and let r = the speed of the canoe in still water. When the canoe travels with the current, its speed is $r + 3$ and traveling against the current, its speed is $r - 3$.

With the current $r+3$ →
1 hour d miles

← Against the current $r-3$
2 hours d miles

We can also organize the given information in a table.

	Distance	Speed	Time
With current	d	$r+3$	1
Against current	d	$r-3$	2

2. *Translate.* Using $d = rt$, we get an equation from each row of the table.

$d = (r+3)1$ (1)
$d = (r-3)2$ (2)

Practice this exercise:

1. A train leaves a station and travels west at 80 mph. One hour later a second train leaves the same station and travels west on a parallel track at 100 mph. When will it overtake the first train?

3. *Solve*. We use the substitution method, substituting

$$(r-3)2=(r+3)1$$
$$2r-6=r+3$$
$$r-6=3$$
$$r=9$$

4. *Check*. When $r = 9$, then $r + 3 = 12$, and $12 \cdot 1 = 12$, the distance traveled with the current. When $r = 9$, then $r - 3 = 6$ and $6 \cdot 2 = 12$, the distance traveled against the current. Since the distances are the same, the answer checks.

5. *State*. The speed of the canoe in still water is 9 mph.

ADDITIONAL EXERCISES

Objective A Solve motion problems using the formula $d = rt$.

For extra help, see Examples 1–3 on pages 1031–1034 of your text and the Section 13.5 lecture video.

Solve.

1. Two private airplanes leave an airport at the same time, both flying due east. One travels at 150 mph and the other travels at 165 mph. In how many hours will they be 45 mi apart?

2. Gavin's boat took 2 hr to make a trip downstream with a 4-mph current. The return trip against the same current took 3 hr. Find the speed of the boat in still water.

3. Two cars leave at the same time and travel directly toward each other from points 100 mi apart at rates of 50 mph and 70 mph. When will they meet?

4. Paul walks at a speed of 4 mph to the train station, rides the train averaging 40 mph, then walks from the terminal station at 4 mph to work. The total distance from home to work is 31 mi, and the total time for the trip is 1 hr. For how many miles does Paul ride the train?

Name: Date:
Instructor: Section:

Chapter 14 RADICAL EXPRESSIONS AND EQUATIONS

14.1 Introduction to Radical Expressions

Learning Objectives

A Find the principal square roots and their opposites of the whole numbers from 0^2 to 25^2.
B Approximate square roots of real numbers using a calculator.
C Solve applied problems involving square roots.
D Identify radicands of radical expressions.
E Determine whether a radical expression represents a real number.
F Simplify a radical expression with a perfect-square radicand.

Key Terms

Use the vocabulary terms listed below to complete each statement in Exercises 1–4.

principal **radical** **radicand** **square**

1. The number c is a(n) ________________ root of a if $c^2 = a$.
2. The ________________ square root is the positive square root.
3. The symbol $\sqrt{\ }$ is called a(n) ________________ symbol.
4. In the expression $\sqrt{17}$, 17 is called the ________________ .

GUIDED EXAMPLES AND PRACTICE

Objective A Find the principal square roots and their opposites of the whole numbers from 0^2 to 25^2.

Review these examples for Objective A:

1. Find the square roots of 49.

   The square roots of 49 are –7 and 7 because $7^2 = 49$ and $(-7)^2 = 49$.

2. Simplify: $-\sqrt{144}$.

   The symbol $\sqrt{144}$ represents the positive square root. Then $-\sqrt{144}$ represents the opposite of the square root.
   $-\sqrt{144} = -12$

Practice these exercises:

1. Find the square roots of 64.

2. Simplify: $-\sqrt{100}$.

Objective B Approximate square roots of real numbers using a calculator.

Review this example for Objective B:

3. Use a calculator to approximate $\sqrt{2.58}$. Round to three decimal places.

   $\sqrt{2.58} \approx 1.60623784$, using a calculator.
   Rounding to three decimal places, we have 1.606.

Practice this exercise:

3. Use a calculator to approximate $\sqrt{10.49}$. Round to three decimal places.

Objective C Solve applied problems involving square roots.

Review this example for Objective C:

4. The attendants at a parking lot park cars in temporary spaces before the care are taken to long-term parking spaces. The number N of such spaces needed is approximated by the formula $N = 2.5\sqrt{A}$, where A is the average number of arrivals during peak hours. Find the number of spaces needed when the average number of arrivals is 45.

   We substitute 45 for A and find an approximation.
   $N = 2.5\sqrt{A} = 2.5\sqrt{45} \approx 16.77$
   There will be 17 spaces needed.

Practice this exercise:

4. Use the formula described at left to find the number of spaces when the average number of arrivals is 57.

Objective D Identify radicands of radical expressions.

Review this example for Objective D:

5. Identify the radicand: $39\sqrt{2x+7}$.

   The radicand is $2x + 7$.

Practice this exercise:

5. Identify the radicand: $-12\sqrt{3a}$.

Objective E Determine whether a radical expression represents a real number.

Review this example for Objective E:

6. Determine whether the expression $\sqrt{-64}$ is a real number.

   No, $\sqrt{-64}$ does not represent a real number because there is no real number that when squared yields –64.

Practice this exercise:

6. Determine whether the expression $-\sqrt{25}$ is a real number.

Objective F Simplify a radical expression with a perfect-square radicand.

Review these examples for Objective F:

7. Simplify $\sqrt{x^2+14x+49}$, assume that radicands do not represent the square of a negative number.

   $\sqrt{x^2+14x+49} = \sqrt{(x+7)^2} = x+7$

Practice these exercises:

7. Simplify $\sqrt{x^2-24x+144}$. Assume that all expressions under the radical represent real numbers.

Name: Date:
Instructor: Section:

ADDITIONAL EXERCISES

Objective A Find the principal square roots and their opposites of the whole numbers from 0^2 to 25^2.

For extra help, see Examples 1–3 on page 1046 of your text and the Section 14.1 lecture video.

Find the square roots.

1. 400

2. 36

Simplify.

3. $-\sqrt{81}$

4. $\sqrt{324}$

Objective B Approximate square roots of real numbers using a calculator.

For extra help, see Examples 4–6 on page 1047 of your text and the Section 14.1 lecture video.

Use a calculator to approximate each square root. Round to three decimal places.

5. $\sqrt{485}$

6. $-\sqrt{85.6}$

7. $\sqrt{\dfrac{891}{362}}$

8. $-\sqrt{\dfrac{3\cdot 45}{8}}$

Objective C Solve applied problems involving square roots.

For extra help, see Example 7 on page 1047 of your text and the Section 14.1 lecture video.

The formula $D = 1.2\sqrt{h}$ can be used to approximate the distance, D, in miles, that a person can see to the horizon from a height, h, in feet. Find the distance, rounding to the nearest tenth of a mile, a person can see from the horizon, given the height.

9. 25 ft (from a building)

10. 10 ft (on a ladder)

11. 5 ft (on the ground)

12. 30,625 ft (from an airplane)

Objective D Identify radicands of radical expressions.
For extra help, see Examples 8–12 on page 1048 of your text and the Section 14.1 lecture video.
Identify the radicand.

13. $\sqrt{17w}$

14. $-\sqrt{5b+18}$

15. $-6\sqrt{x^2y-11}$

16. $x^2\sqrt{\frac{x+y}{3}}$

Objective E Determine whether a radical expression represents a real number.
For extra help, see page 1048 of your text and the Section 14.1 lecture video.
Determine whether each expression represents a real number. Write "yes" *or* "no."

17. $-\sqrt{9}$

18. $\sqrt{-9}$

19. $\sqrt{-81}$

20. $\sqrt{-(-25)}$

Objective F Simplify a radical expression with a perfect-square radicand.
For extra help, see Examples 13–20 on pages 1049–1050 of your text and the Section 14.1 lecture video.
Simplify. Remember that we have assumed that radicands do not represent the square of a negative number.

21. $\sqrt{36a^2}$

22. $\sqrt{(3p)^2}$

23. $\sqrt{(h-10)^2}$

24. $\sqrt{x^2+12x+36}$

Name: Date:
Instructor: Section:

Chapter 14 RADICAL EXPRESSIONS AND EQUATIONS

14.2 Multiplying and Simplifying with Radical Expressions

Learning Objectives
A Simplify radical expressions.
B Simplify radical expressions where radicands are powers.
C Multiply radical expressions and, if possible, simplify.

Key Terms
Use the vocabulary terms listed below to complete each statement in Exercises 1–4.

half **nonnegative** **perfect squares** **radicands**

1. For any __________________ radicands A and B, $\sqrt{A}\cdot\sqrt{B}=\sqrt{A\cdot B}$.

2. The product of square roots is the square root of the product of the

__________________.

3. The square-root radical expression is simplified when its radicand has no factors that are

__________________.

4. To take the square root of an even power, take __________________ the exponent.

GUIDED EXAMPLES AND PRACTICE

Objective A Simplify radical expressions.

Review this example for Objective A:

1. Simplify $\sqrt{32x^2}$ by factoring.

$\sqrt{32x^2}=\sqrt{16\cdot 2\cdot x^2}$ Identifying perfect-square factors and factoring the radicand.

$=\sqrt{16}\sqrt{2}\sqrt{x^2}$ Factoring into a product of several radicals

$=4x\sqrt{2}$ Taking square roots

No absolute-value signs are necessary since we have assumed that expression under radicals do not represent the square of a negative number.

Practice this exercise:

1. Simplify $\sqrt{48w^2}$ by factoring.

Objective B Simplify radical expressions where radicands are powers.

Review this example for Objective B:

2. Simplify $\sqrt{300b^5}$ by factoring.

We factor the radicand, looking for perfect-square factors. The largest even power of b is 4.

$\sqrt{300b^5} = \sqrt{100 \cdot 3 \cdot b^4 \cdot b}$

$= \sqrt{100}\sqrt{b^4}\sqrt{3b}$ Factoring into a product of radicals

$= 10b^2\sqrt{3b}$ Simplifying

Practice this exercise:

2. Simplify $\sqrt{512a^7}$ by factoring.

Objective C Multiply radical expressions and, if possible, simplify.

Review this example for Objective C:

3. Multiply and simplify $\sqrt{6xy^2}\sqrt{3x^2y}$ by factoring.

We multiply and look for perfect-square factors or largest even powers.

$\sqrt{6xy^2}\sqrt{3x^2y} = \sqrt{6xy^2 \cdot 3x^2y}$

$= \sqrt{2 \cdot 3 \cdot x \cdot y^2 \cdot 3 \cdot x^2 \cdot y}$

$= \sqrt{3 \cdot 3 \cdot x^2 \cdot y^2 \cdot 2 \cdot x \cdot y}$

$= 3xy\sqrt{2xy}$

Practice this exercise:

3. Multiply and simplify $\sqrt{20ab^4}\sqrt{12a^2b}$ by factoring.

ADDITIONAL EXERCISES

Objective A Simplify radical expressions.

For extra help, see Examples 1–10 on pages 1054–1055 of your text and the Section 14.2 lecture video.

Simplify by factoring.

1. $\sqrt{25t}$

2. $\sqrt{17a^2}$

3. $\sqrt{180d^2}$

4. $\sqrt{m^2 - 10m + 25}$

Name: Date:
Instructor: Section:

Objective B Simplify radical expressions where radicands are powers.

For extra help, see Examples 11–15 on page 1056 of your text and the Section 14.2 lecture video.

Simplify by factoring.

5. $\sqrt{z^{13}}$

6. $\sqrt{(m+5)^{6}}$

7. $\sqrt{25(a+9)^{16}}$

8. $\sqrt{320x^{10}y^{5}}$

Objective C Multiply radical expressions and, if possible, simplify.

For extra help, see Examples 16–18 on pages 1056–1057 of your text and the Section 14.2 lecture video.

Multiply and then, if possible, simplify by factoring.

9. $\sqrt{21}\sqrt{14}$

10. $\sqrt{13c}\sqrt{26c}$

11. $\sqrt{54s^{10}t^{12}}\sqrt{6st}$

12. $\sqrt{2x-7}\sqrt{2x-7}$

Name: Date:
Instructor: Section:

Chapter 14 RADICAL EXPRESSIONS AND EQUATIONS

14.3 Quotients Involving Radical Expressions

Learning Objectives
A Divide radical expressions.
B Simplify square roots of quotients.
C Rationalize the denominator of a radical expression.

Key Terms
Use the vocabulary terms listed below to complete each statement in Exercises 1–4.

perfect square **radicands** **rationalizing** **separately**

1. The statement $\frac{\sqrt{A}}{\sqrt{B}}=\sqrt{\frac{A}{B}}$ states that the quotient of two square roots is the square root of the quotients of the __________________ .

2. The statement $\sqrt{\frac{A}{B}}=\frac{\sqrt{A}}{\sqrt{B}}$ states that we can take the square roots of the numerator and the denominator __________________ .

3. When we find an equivalent expression without a radical in the denominator, we are __________________ the denominator.

4. To rationalize a denominator, we can multiply by 1 under the radical to make the denominator of the radicand a(n) __________________ .

GUIDED EXAMPLES AND PRACTICE

Objective A Divide radical expressions.

Review this example for Objective A:

1. Divide and simplify: $\frac{\sqrt{192x^5}}{\sqrt{48x}}$.

$$\frac{\sqrt{192x^5}}{\sqrt{48x}}=\sqrt{\frac{192x^5}{48x}}=\sqrt{4x^4}=2x^2$$

Practice this exercise:

1. Divide and simplify: $\frac{\sqrt{975w}}{\sqrt{39w^3}}$.

Objective B Simplify square roots of quotients.

Review this example for Objective B:

2. Simplify $\sqrt{\dfrac{250x^{12}}{640}}$.

We check for common factors then take the square root of the numerator and the square root of the denominator.

$$\sqrt{\frac{250x^{12}}{640}} = \sqrt{\frac{25x^{12}\cdot 10}{64\cdot 10}} = \sqrt{\frac{25x^{12}}{64}\cdot\frac{10}{10}}$$
$$= \sqrt{\frac{25x^{12}}{64}\cdot 1} = \frac{\sqrt{25x^{12}}}{\sqrt{64}}$$
$$= \frac{5x^6}{8}$$

Practice this exercise:

2. Simplify $\sqrt{\dfrac{80}{5y^4}}$.

Objective C Rationalize the denominator of a radical expression.

Review this example for Objective C:

3. Rationalize the denominator: $\dfrac{\sqrt{16}}{\sqrt{125}}$.

Factoring 125, we get $5\cdot 5\cdot 5$, so we need another factor of 5 in order for the radicand in the denominator to be a perfect square. We multiply by $\dfrac{\sqrt{5}}{\sqrt{5}}$.

$$\frac{\sqrt{16}}{\sqrt{125}} = \frac{\sqrt{16}}{\sqrt{5\cdot 5\cdot 5}} = \frac{\sqrt{16}}{\sqrt{5\cdot 5\cdot 5}}\cdot\frac{\sqrt{5}}{\sqrt{5}}$$
$$= \frac{\sqrt{16}\sqrt{5}}{\sqrt{625}}$$
$$= \frac{4\sqrt{5}}{25}$$

Practice this exercise:

3. Rationalize the denominator: $\sqrt{\dfrac{9}{20}}$.

ADDITIONAL EXERCISES

Objective A Divide radical expressions.

For extra help, see Examples 1–2 on page 1062 of your text and the Section 14.3 lecture video.

Divide and simplify.

1. $\dfrac{\sqrt{80}}{\sqrt{5}}$

2. $\dfrac{\sqrt{5}}{\sqrt{20}}$

Name: Date:
Instructor: Section:

3. $\dfrac{\sqrt{32x}}{\sqrt{8x}}$

4. $\dfrac{\sqrt{99c^3}}{\sqrt{11c}}$

Objective B Simplify square roots of quotients.

For extra help, see Examples 3–8 on page 1063 of your text and the Section 14.3 lecture video.

Simplify.

5. $\sqrt{\dfrac{49}{121}}$

6. $-\sqrt{\dfrac{81}{169}}$

7. $\sqrt{\dfrac{4x^2}{225}}$

8. $\dfrac{\sqrt{72v^{11}}}{\sqrt{2v^{25}}}$

Objective C Rationalize the denominator of a radical expression.

For extra help, see Examples 9–14 on pages 1064–1065 of your text and the Section 14.3 lecture video.

Rationalize the denominator.

9. $\dfrac{5}{\sqrt{5}}$

10. $\dfrac{\sqrt{32}}{\sqrt{24}}$

11. $\dfrac{\sqrt{y^7}}{\sqrt{xy}}$

12. $\dfrac{\sqrt{5x^3}}{\sqrt{75x}}$

Name: Date:
Instructor: Section:

Chapter 14 RADICAL EXPRESSIONS AND EQUATIONS

14.4 Addition, Subtraction, and More Multiplication

Learning Objectives

A Add or subtract with radical notation, using the distributive laws to simplify.
B Multiply expressions involving radicals, where some of the expressions contain more than one term.
C Rationalize denominators having two terms.

Key Terms

Use the vocabulary terms listed below to complete each statement in Exercises 1–2.

conjugates **like radicals**

1. ____________________ have the same radicands.

2. Expressions such as $12+\sqrt{2}$ and $12-\sqrt{2}$ are known as ____________________ .

GUIDED EXAMPLES AND PRACTICE

Objective A Add or subtract with radical notation, using the distributive laws to simplify.

Review this example for Objective A:

1. Add: $\sqrt{54}+\sqrt{24}$.

$$\sqrt{54}+\sqrt{24}=\sqrt{9\cdot 6}+\sqrt{4\cdot 6} \quad \text{Factoring 54 and 24}$$
$$=\sqrt{9}\sqrt{6}+\sqrt{4}\sqrt{6}$$
$$=3\sqrt{6}+2\sqrt{6}$$
$$=(3+2)\sqrt{6} \quad \text{Using a distributive law to factor out } \sqrt{6}$$
$$=5\sqrt{6}$$

Practice this exercise:

1. Subtract: $\sqrt{80}-\sqrt{45}$.

Objective B Multiply expressions involving radicals, where some of the expressions contain more than one term.

Review this example for Objective B:

2. Multiply: $(7+\sqrt{2})(3-\sqrt{2})$.

$$(7+\sqrt{2})(3-\sqrt{2})$$
$$=7\cdot 3-7\sqrt{2}+3\sqrt{2}-\sqrt{2}\sqrt{2} \quad \text{Using FOIL}$$
$$=21-7\sqrt{2}+3\sqrt{2}-2$$
$$=19-4\sqrt{2} \quad \text{Combining like terms}$$

Practice this exercise:

2. Multiply:
$(\sqrt{17}+\sqrt{3})(\sqrt{17}-\sqrt{3})$.

Objective C Rationalize denominators having two terms.

Review this example for Objective C:

3. Rationalize the denominator: $\dfrac{\sqrt{6}}{\sqrt{5}-2}$.

$$\frac{\sqrt{6}}{\sqrt{5}-2}=\frac{\sqrt{6}}{\sqrt{5}-2}\cdot\frac{\sqrt{5}+2}{\sqrt{5}+2}=\frac{\sqrt{6}\left(\sqrt{5}+2\right)}{\left(\sqrt{5}-2\right)\left(\sqrt{5}+2\right)}$$
$$=\frac{\sqrt{30}+2\sqrt{6}}{\left(\sqrt{5}\right)^2-2^2}=\frac{\sqrt{30}+2\sqrt{6}}{5-4}$$
$$=\sqrt{30}+2\sqrt{6}$$

Practice this exercise:

3. Rationalize the denominator: $\dfrac{7}{4-\sqrt{3}}$.

ADDITIONAL EXERCISES

Objective A Add or subtract with radical notation, using the distributive laws to simplify.

For extra help, see Examples 1–5 on pages 1071–1072 of your text and the Section 14.4 lecture video.

Add or subtract. Simplify by collecting like radical terms, if possible.

1. $3\sqrt{7}+8\sqrt{7}$

2. $2\sqrt{45}+6\sqrt{5}$

3. $3\sqrt{20}-4\sqrt{125}+2\sqrt{245}$

4. $5n\sqrt{mn^2}+n\sqrt{m^3n^4}-6m\sqrt{m^5}$

Objective B Multiply expressions involving radicals, where some of the expressions contain more than one term.

For extra help, see Examples 6–10 on pages 1072–1073 of your text and the Section 14.4 lecture video.

Multiply.

5. $\sqrt{6}\left(\sqrt{7}-3\right)$

6. $\left(3+\sqrt{2}\right)\left(5-\sqrt{5}\right)$

7. $\left(\sqrt{13}+2\right)\left(\sqrt{13}-2\right)$

8. $\left(\sqrt{w}-\sqrt{z}\right)^2$

Objective C Rationalize denominators having two terms.

For extra help, see Examples 11–13 on pages 1073—1074 of your text and the Section 14.4 lecture video.

Rationalize the denominator.

9. $\dfrac{\sqrt{6}}{\sqrt{5}+\sqrt{2}}$

10. $\dfrac{10}{\sqrt{15}+2}$

11. $\dfrac{7}{2-\sqrt{x}}$

12. $\dfrac{4+\sqrt{5}}{6+\sqrt{y}}$

Name: Date:
Instructor: Section:

Chapter 14 RADICAL EXPRESSIONS AND EQUATIONS

14.5 Radical Equations

Learning Objectives

A Solve radical equations with one or two radical terms isolated, using the principle of squaring once.
B Solve radical equations with two radical terms, using the principle of squaring twice.
C Solve applied problems using radical equations.

Key Terms

Use the vocabulary terms listed below to complete each statement in Exercises 1–2.

principle of squaring **radical equation**

1. A ____________________ has variables in one or more radicands.

2. The ____________________ states that if $a = b$ is true, then $a^2 = b^2$ is true.

GUIDED EXAMPLES AND PRACTICE

Objective A Solve radical equations with one or two radical terms isolated, using the principle of squaring once.

Review this example for Objective A:

1. Solve $x = 3 + \sqrt{x-1}$.

Subtract 3 on both sides to isolate the radical.

$$x = 3 + \sqrt{x-1}$$
$$x - 3 = \sqrt{x-1}$$
$$(x-3)^2 = \left(\sqrt{x-1}\right)^2$$
$$x^2 - 6x + 9 = x - 1$$
$$x^2 - 7x + 10 = 0$$
$$x - 2 = 0 \quad or \quad x - 5 = 0$$
$$x = 2 \quad or \quad x = 5$$

We check each possible solution.

For 2: $x = 3 + \sqrt{x-1}$

2	$3 + \sqrt{2-1}$
	$3 + \sqrt{1}$
	$3 + 1$
	4 FALSE

For 5: $x = 3 + \sqrt{x-1}$

5	$3 + \sqrt{5-1}$
	$3 + \sqrt{4}$
	$3 + 2$
	5 TRUE

The number 5 checks, but 2 does not.
Thus, the solution is 5.

Practice this exercise:

1. Solve $\sqrt{3x+7} = \sqrt{4x+1}$.

Objective B Solve radical equations with two radical terms, using the principle of squaring twice.

Review this example for Objective B:

2. Solve $1=\sqrt{x+9}-\sqrt{x}$.

$$
\begin{aligned}
1&=\sqrt{x+9}-\sqrt{x}\\
\sqrt{x}+1&=\sqrt{x+9}\\
\left(\sqrt{x}+1\right)^2&=\left(\sqrt{x+9}\right)^2\\
x+2\sqrt{x}+1&=x+9\\
2\sqrt{x}&=8\\
\sqrt{x}&=4\\
\left(\sqrt{x}\right)^2&=4^2\\
x&=16
\end{aligned}
$$

The number 16 checks. It is the solution.

Practice this exercise:

2. Solve $\sqrt{x+3}-\sqrt{x-2}=1$.

Objective C Solve applied problems using radical equations.

Review this example for Objective C:

3. At a height of h meters, a person can see V kilometers to the horizon, where $V=3.5\sqrt{h}$. Martin can see 53.2 km to the horizon from the top of a cliff. What is the altitude of Martin's eyes?

We substitute 53.2 for V in the equation $V=3.5\sqrt{h}$ and solve for h.

$$
\begin{aligned}
53.2&=3.5\sqrt{h}\\
\frac{53.2}{3.5}&=\sqrt{h}\\
15.2&=\sqrt{h}\\
(15.2)^2&=\left(\sqrt{h}\right)^2\\
231.04&=h
\end{aligned}
$$

The altitude of Martin's eyes is about 231 m.

Practice this exercise:

3. Solve using the formula given at the left. A passenger can see 301 km to the horizon through an airplane window. What is the altitude of the passenger's eyes?

Name: Date:
Instructor: Section:

ADDITIONAL EXERCISES

Objective A Solve radical equations with one or two radical terms isolated, using the principle of squaring once.

For extra help, see Examples 1–4 on pages 1079–1081 of your text and the Section 14.5 lecture video.

Solve.

1. $7+\sqrt{y-3}=11$

2. $7+2\sqrt{x+1}=x$

3. $\sqrt{5x-1}+3=x$

4. $\sqrt{x^2+5}-x+2=0$

Objective B Solve radical equations with two radical terms, using the principle of squaring twice.

For extra help, see Example 5 page 1081 of your text and the Section 14.5 lecture video.

Solve.

5. $\sqrt{5x+1}=1+\sqrt{4x-3}$

6. $7+\sqrt{10-z}=8+\sqrt{1-z}$

7. $2\sqrt{t-7}=1+\sqrt{2t+3}$

8. $\sqrt{2x+1}=3-\sqrt{x+4}$

Objective C   Solve applied problems using radical equations.
For extra help, see Examples 6–7 on pages 1082–1083 of your text and the Section 14.5 lecture video.
Solve

9. How far to the horizon can you see through an airplane window at a height, or altitude, of 29,000 ft?
Use the formula $D = \sqrt{2h}$.

10. A person can see 250 mi to the horizon through an airplane window. How high above sea level is the airplane?
Use the formula $D = \sqrt{2h}$

11. The formula $r = 2\sqrt{5L}$ can be used to approximate the speed, r, in miles per hour, of a car that has left a skid mark of length L, in feet. How far will a car skid at 60 mph? At 80 mph?

Name: Date:
Instructor: Section:

Chapter 14 RADICAL EXPRESSIONS AND EQUATIONS

14.6 Applications with Right Triangles

Learning Objectives
A Given lengths of any two sides of a right triangle, find the length of the third side.
B Solve applied problems involving right triangles.

Key Terms
Use the vocabulary terms listed below to complete each statement in Exercises 1–4.

hypotenuse **legs** **Pythagorean equation** **right triangle**

1. A ________________ is a triangle with a 90° angle.
2. In a right triangle, the longest side is called the ________________ .
3. In a right triangle, the sides forming the right angle are called the________________ .
4. The equation $a^2 + b^2 = c^2$ is called the ________________ .

GUIDED EXAMPLES AND PRACTICE

Objective A Given lengths of any two sides of a right triangle, find the length of the third side.

Review this example for Objective A:

1. In a right triangle, find the length of the side not given. Give an exact answer and an approximation to three decimal places.
   $a = 5,\ c = 9$

   We substitute in the Pythagorean equation.

$$a^2 + b^2 = c^2$$
$$5^2 + b^2 = 9^2$$
$$25 + b^2 = 81$$
$$b^2 = 56$$
$$b = \sqrt{56} \approx 7.483$$

Practice this exercise:

1. In a right triangle, find the length of the side not given.
   $b = 12,\ c = 13$

Objective B Solve applied problems involving right triangles.

Review this example for Objective B:

2. Find the length of a diagonal of a square whose sides are 8 yd long.

   We first make a drawing. We label the diagonal d.

Practice this exercise:

2. How long is a guy wire reaching from the top of a 10-ft pole to a point on the ground 7 ft from the pole?

We know that $8^2 + 8^2 = d^2$. We solve this equation.

$$\begin{aligned} 8^2 + 8^2 &= d^2 \\ 64 + 64 &= d^2 \\ 128 &= d^2 \\ \sqrt{128} &= d \\ 11.314 &\approx d \end{aligned}$$

The length of the diagonal is $\sqrt{128} \approx 11.314$ yd..

ADDITIONAL EXERCISES

Objective A Given lengths of any two sides of a right triangle, find the length of the third side.

For extra help, see Examples 1–4 on page 1088 of your text and the Section 14.6 lecture video.

Find the length of the third side of the right triangle. Where appropriate, give both an exact answer and an approximation to three decimal places.

1.

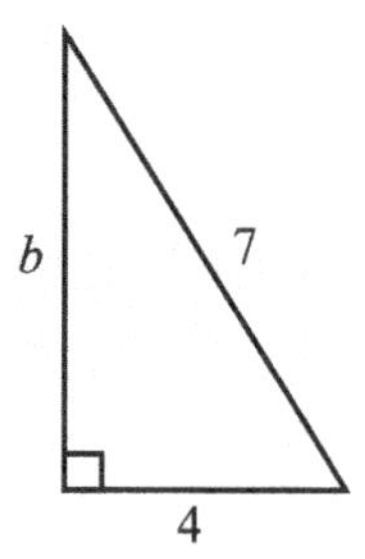

2.

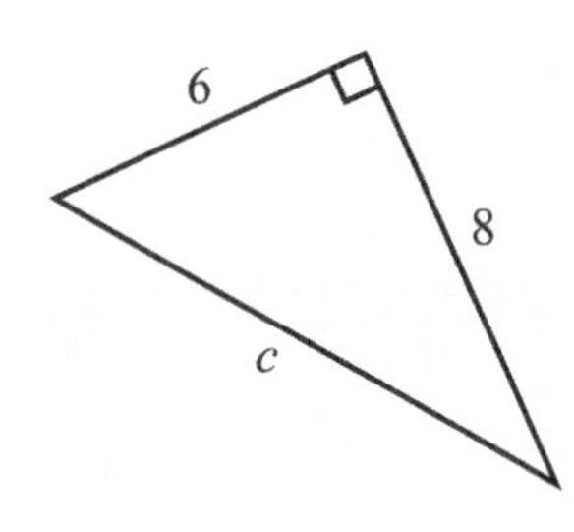

In a right triangle, find the length of the side not given.

3. $b = 3, c = \sqrt{10}$

4. $a = \sqrt{11}, b = \sqrt{5}$

Objective B Solve applied problems involving right triangles.

For extra help, see Example 5 on page 1089 of your text and the Section 14.6 lecture video.

Solve. Don't forget to use a drawing. Give an exact answer and an approximation to three decimal places.

5. Students have worn a path diagonally across a large grassy rectangle between classroom buildings. The rectangular area is 200 ft long and 100 ft wide. How long is the diagonal path?

6. An airplane is flying at an altitude of 3300 ft. The slanted distance to the airport is 14,600 ft. How far is the airplane horizontally from the airport?

7. A 15-ft guy wire reaches from the top of a pole to a point on the ground 6 ft from the pole. How tall is the pole?

8. Find the length of a diagonal of a square whose sides are 10 cm long.

Name: Date:
Instructor: Section:

Chapter 15 QUADRATIC EQUATIONS

15.1 Introduction to Quadratic Equations

Learning Objectives

A Write a quadratic equation in standard form $ax^2 + bx + c = 0$, $a > 0$, and determine the coefficients a, b, and c.

B Solve quadratic equations of the type $ax^2 + bx = 0$, where $b \neq 0$, by factoring.

C Solve quadratic equations of the type $ax^2 + bx + c = 0$ where $b \neq 0$ and $c \neq 0$ by factoring.

D Solve applied problems involving quadratic equations.

Key Terms

Use the vocabulary terms listed below to complete each statement in Exercises 1–4.

first **quadratic** **standard** **zero**

1. A(n) ________________ equation is an equation equivalent to an equation of the type $ax^2 + bx + c = 0$, $a > 0$.

2. The equation $ax^2 + bx + c = 0$ is written in ________________ form.

3. The solutions of $ax^2 + bx + c = 0$, if any exist, are the ________________ coordinates of the x-intercepts of the graph of $y = ax^2 + bx + c = 0$.

4. An equation of the type $ax^2 + bx = 0$, $a \neq 0$, $b \neq 0$ will always have ________________ as one solution.

GUIDED EXAMPLES AND PRACTICE

Objective A Write a quadratic equation in standard form $ax^2 + bx + c = 0, a > 0,$ and determine the coefficients a, b, and c.

Review this example for Objective A:

1. Write $-2x^2 + 3x = 5$ in standard form and determine a, b, and c.

   First, we subtract 5 on both sides of the equation. Then we multiply by –1 on both sides.

$$-2x^2 + 3x = 5$$
$$-2x^2 + 3x - 5 = 0$$
$$2x^2 - 3x + 5 = 0$$

   With the equation in standard form, we see that $a = 2$, $b = -3$, and $c = 5$.

Practice this exercise:

1. Write $x^2 + 6 = 4x$ in standard form and determine a, b, and c.

Objective B Solve quadratic equations of the type $ax^2 + bx = 0$ where $b \neq 0$, by factoring.

Review this example for Objective B:

2. Solve $5x^2 - 4x = 0$.

$$5x^2 - 4x = 0$$
$$x(5x-4) = 0$$
$$x = 0 \quad or \quad 5x - 4 = 0$$
$$x = 0 \quad or \quad 5x = 4$$
$$x = 0 \quad or \quad x = \frac{4}{5}$$

The solutions are 0 and $\frac{4}{5}$.

Practice this exercise:

2. Solve $2x^2 + 3x = 0$.

Objective C Solve quadratic equations of the type $ax^2 + bx + c = 0$ where $b \neq 0$ and $c \neq 0$ by factoring.

Review this example for Objective C:

3. Solve $2x^2 + 5x = 3$.

First, we write the equation in standard form. Then, we factor and use the principle of zero products.

$$2x^2 + 5x = 3$$
$$2x^2 + 5x - 3 = 0$$
$$(x+3)(2x-1) = 0$$
$$x + 3 = 0 \quad or \quad 2x - 1 = 0$$
$$x = -3 \quad or \quad 2x = 1$$
$$x = -3 \quad or \quad x = \frac{1}{2}$$

The solutions are -3 and $\frac{1}{2}$.

Practice this exercise:

3. Solve $x^2 + 20 = 9x$.

Objective D Solve applied problems involving quadratic equations.

Review this example for Objective D:

4. The number of diagonals d of a polygon of n sides is given by the formula $d = \frac{n^2 - 3n}{2}$.

If a polygon has 5 diagonals, how many sides does it have?

1. *Familiarize*. We will use the formula given above.

Practice this exercise:

4. If a polygon has 9 diagonals, how many sides does it have?

Name: Date:
Instructor: Section:

2. *Translate*. We substitute 5 for d in the formula.

$$5=\frac{n^2-3n}{2}$$

3. *Solve*. We solve the equation for n, first reversing the equation for convenience.

$$\frac{n^2-3n}{2}=5$$
$$n^2-3n=10$$
$$n^2-3n-10=0$$
$$(n-5)(n+2)=0$$
$$n-5=0 \quad or \quad n+2=0$$
$$n=5 \quad or \quad n=-2$$

4. *Check*. Since the number of sides cannot be negative, –2 cannot be a solution. We substitute 5 for n in the formula to verify that it is a solution.

$$d=\frac{n^2-3n}{2}=\frac{5^2-3\cdot 5}{2}=\frac{25-15}{2}=\frac{10}{2}=5$$

Since $d = 5$ when $n = 5$, the number 5 checks.

5. *State*. The polygon has 5 sides.

ADDITIONAL EXERCISES

Objective A Write a quadratic equation in standard form $ax^2+bx+c=0, a>0$, and determine the coefficients a, b, and c.

For extra help, see Examples 1–3 on page 1104 of your text and the Section 15.1 lecture video.

Write in standard form and determine a, b, and c.

1. $4x^2=x-3$

2. $8=3x-5x^2$

3. $2-6x=4+3x^2$

Objective B Solve quadratic equations of the type $ax^2+bx=0$ where $b\neq 0$, by factoring.

For extra help, see Examples 4–5 on pages 1104–1105 of your text and the Section 15.1 lecture video.

Solve.

4. $x^2+11x=0$

5. $5x^2+50x=0$

6. $3x^2 = 4x$

7. $0 = 13x^2 + 26x$

Objective C Solve quadratic equations of the type $ax^2 + bx + c = 0$ where $b \neq 0$ and $c \neq 0$ by factoring.

For extra help, see Examples 6–8 on pages 1105–1106 of your text and the Section 15.1 lecture video.

Solve.

8. $x^2 + 10x + 16 = 0$

9. $w(w+3) = 54$

10. $4(p-16) = p(p-12)$

11. $\frac{1}{x} + \frac{1}{x+8} = \frac{1}{3}$

12. $\frac{3-v}{v-3} = -\frac{v+7}{v-7}$

Objective D Solve applied problems involving quadratic equations.

For extra help, see Example 9 on pages 1106–1107 of your text and the Section 15.1 lecture video.

Solve. Use the formula $d = \frac{n^2 - 3n}{2}$.

13. A dodecagon is a figure with 12 sides. How many diagonals does a dodecagon have?

14. A polygon has 20 diagonals. How many sides does it have?

Name: Date:
Instructor: Section:

Chapter 15 QUADRATIC EQUATIONS

15.2 Solving Quadratic Equations by Completing the Square

Learning Objectives

A Solve quadratic equations of the type $ax^2 = p$.

B Solve quadratic equations of the type $(x+c)^2 = d$.

C Solve quadratic equations by completing the square.

D Solve certain applied problems involving quadratic equations of the type $ax^2 = p$.

Key Terms

Use the vocabulary terms listed below to complete each statement in Exercises 1–4.

complete **no real-number** **only one** **two real**

1. The equation $x^2 = d$ has ________________ solution(s) when $d > 0$.

2. The equation $x^2 = d$ has ________________ solution(s) when $d < 0$.

3. The equation $x^2 = d$ has ________________ solution(s) when $d = 0$

4. To ________________ the square for $x^2 + bx$, we add $\left(\frac{b}{2}\right)^2$.

GUIDED EXAMPLES AND PRACTICE

Objective A Solve quadratic equations of the type $ax^2 = p$.

Review this example for Objective A:

1. Solve $3x^2 = 15$.

$3x^2 = 15$

$x^2 = 5$

$x = \sqrt{5}$ or $x = -\sqrt{5}$

The solutions are $\sqrt{5}$ and $-\sqrt{5}$.

Practice this exercise:

1. Solve $3x^2 - 2 = 0$.

Objective B Solve quadratic equations of the type $(x+c)^2 = d$.

Review this example for Objective B:

2. Solve $x^2 - 2x + 1 = 25$.

$$x^2 - 2x + 1 = 25$$
$$(x-1)^2 = 25$$
$$x - 1 = 5 \quad or \quad x - 1 = -5$$
$$x = 6 \quad or \quad x = -4$$

The solutions are 6 and –4.

Practice this exercise:

2. Solve $(x+3)^2 = 7$.

Objective C Solve quadratic equations by completing the square.

Review this example for Objective C:

3. Solve $2x^2 + 2x - 3 = 0$ by completing the square.

First, we multiply by $\frac{1}{2}$ on both sides of the equation to make the x^2 coefficient 1.

$$2x^2 + 2x - 3 = 0$$
$$\frac{1}{2}(2x^2 + 2x - 3) = \frac{1}{2}\cdot 0$$
$$x^2 + x - \frac{3}{2} = 0$$
$$x^2 + x = \frac{3}{2}$$

Now we add $\left(\frac{b}{2}\right)^2$, or $\left(\frac{1}{2}\right)^2$, or $\frac{1}{4}$ on both sides.

$$x^2 + x + \frac{1}{4} = \frac{3}{2} + \frac{1}{4}$$
$$\left(x + \frac{1}{2}\right)^2 = \frac{7}{4}$$
$$x + \frac{1}{2} = \frac{\sqrt{7}}{2} \quad or \quad x + \frac{1}{2} = -\frac{\sqrt{7}}{2}$$
$$x = -\frac{1}{2} + \frac{\sqrt{7}}{2} \quad or \quad x = -\frac{1}{2} - \frac{\sqrt{7}}{2}$$

The solutions are $\frac{-1 \pm \sqrt{7}}{2}$.

Practice this exercise:

3. Solve $x^2 + 2x - 5 = 0$ by completing the square.

Name: Date:
Instructor: Section:

Objective D Solve certain applied problems involving quadratic equations of the type $ax^2 = p$.

Review this example for Objective D:

4. An object is dropped from the top of a 1214-m high building. How long will it take the object to reach the ground?

1. *Familiarize*. A formula that fits this situation is $s = 16t^2$, where s is the distance, in feet, traveled by a body falling freely from the rest in t seconds. Here we know that s is 1214 m and we want to find t.

2. *Translate*. We substitute 1214 for s in the formula.

$$1214 = 16t^2$$

3. *Solve*.

$$1214 = 16t^2$$
$$\frac{1214}{16} = t^2$$
$$75.875 = t^2$$
$$\sqrt{75.875} = t \quad or \quad \sqrt{75.875} = t$$
$$8.7 \approx t \quad or \quad -8.7 \approx t$$

4. *Check*. Time cannot be negative in this situation, so –8.7 cannot be a solution. We substitute 8.7 for t in the formula:

$$s = 16(8.7)^2 = 16(75.69) = 1211.04$$

Note that $1211.04 \approx 1214$; since we approximated the solution, we have a check.

5. *State*. It would take the object about 8.7 sec to reach the ground.

Practice this exercise:

4. The Chrysler Building in New York is 1046 ft tall. How long would it take an object to fall to the ground from the top?

ADDITIONAL EXERCISES

Objective A Solve quadratic equations of the type $ax^2 = p$.

For extra help, see Examples 1–3 on pages 1101–1111 of your text and the Section 15.2 lecture video.

Solve.

1. $x^2 = 144$

2. $2y^2 = 7$

3. $9w^2 - 16 = 0$

4. $4x^2 - 125 = 75$

Objective B   Solve quadratic equations of the type $(x+c)^2 = d$.
For extra help, see Examples 4–6 on page 1111 of your text and the Section 15.2 lecture video.
Solve.

5. $(x+7)^2 = 100$

6. $(v+1)^2 = 18$

7. $(x-6)^2 = 35$

8. $x^2 - 40x + 400 = 81$

Objective C   Solve quadratic equations by completing the square.
For extra help, see Examples 7–10 on pages 1113–1114 of your text and the Section 15.2 lecture video.
Solve by completing the square. Show your work.

9. $x^2 - 16x + 55 = 0$

10. $x^2 - 2x - 1 = 0$

11. $x^2 - 18x + 24 = 0$

12. $x^2 + 13x + 41 = 0$

Objective D   Solve certain applied problems involving quadratic equations of the type $ax^2 = p$.
For extra help, see Example 11 on page 1115 of your text and the Section 15.2 lecture video.
Solve.

13. The Taipei 101 in Taiwan is 1667 ft tall. How long would it take an object to fall from the top?

14. As part of a movie stunt, Corrie Janson made a 182-ft freefall from a cliff. Approximately how long did the fall take?

Name: Date:
Instructor: Section:

Chapter 15 QUADRATIC EQUATIONS

15.3 The Quadratic Formula

Learning Objectives

A Solve quadratic equations using the quadratic formula.
B Find approximate solutions of quadratic equations using a calculator.

Key Terms

Use the vocabulary terms listed below to complete each step in the procedure for solving quadratic equations in Exercises 1–4.

factoring **principle of square roots** **quadratic formula** **standard form**

1. If the equation is in the form $ax^2 = p$ or $(x+c)^2 = d$, use the __________________.
2. Write the equation in __________________ .
3. Try __________________ .
4. If it is not possible or it is difficult to factor, use the __________________ .

GUIDED EXAMPLES AND PRACTICE

Objective A Solve quadratic equations using the quadratic formula.

Review this example for Objective A:

1. Solve $x^2 + 4x = 3$ using the quadratic formula.

   First, we find standard form and determine *a*, *b*, and *c*.

   $x^2 + 4x - 3 = 0$
   $a = 1,\ b = 4,\ c = -3$

   Then we use the quadratic formula.

   $$x = \frac{-b \pm \sqrt{b^2 - 4ac}}{2a} = \frac{-4 \pm \sqrt{4^2 - 4 \cdot 1 \cdot (-3)}}{2 \cdot 1}$$
   $$x = \frac{-4 \pm \sqrt{16+12}}{2} = \frac{-4 \pm \sqrt{28}}{2}$$
   $$x = \frac{-4 \pm \sqrt{4 \cdot 7}}{2} = \frac{-4 \pm \sqrt{4}\sqrt{7}}{2}$$
   $$x = \frac{-4 \pm 2\sqrt{7}}{2} = \frac{2(-2 \pm \sqrt{7})}{2} = \frac{2}{2} \cdot \frac{-2 \pm \sqrt{7}}{1}$$
   $$x = -2 \pm \sqrt{7}$$

   The solutions are $-2 + \sqrt{7}$ and $-2 - \sqrt{7}$, or $-2 \pm \sqrt{7}$.

Practice this exercise:

1. Solve $2x^2 - 3x - 7 = 0$ using the quadratic formula.

Objective B Find approximate solutions of quadratic equations using a calculator.

Review this example for Objective B:

2. Use a calculator to approximate the solutions of $x^2+4x=3$ to the nearest tenth.

In the previous example, we used the quadratic formula to find that the solutions of this equation are $-2\pm\sqrt{7}$. Using a calculator and rounding to the nearest tenth, we have

$$-2+\sqrt{7}\approx 0.6457513111\approx 0.6$$
$$-2-\sqrt{7}\approx -4.645751331\approx -4.6$$

The approximate solutions are 0.6 and –4.6.

Practice this exercise:

2. Use a calculator to approximate the solutions of $x^2-5x+2=0$ to the nearest tenth.

ADDITIONAL EXERCISES

Objective A Solve quadratic equations using the quadratic formula.

For extra help, see Examples 1–5 on pages 1120–1122 of your text and the Section 15.3 lecture video.

Solve. Try factoring first. If factoring is not possible or is difficult, use the quadratic formula.

1. $x^2-8x=20$

2. $n^2+6n+3=8$

3. $\frac{1}{x}+\frac{1}{x-2}=\frac{1}{3}$

4. $3p^2+2p+7=0$

Objective B Find approximate solutions of quadratic equations using a calculator.

For extra help, see Example 6 on page 1122 of your text and the Section 15.3 lecture video.

Solve using the quadratic formula. Use a calculator to approximate the solutions to the nearest tenth.

5. $x^2-3x-2=0$

6. $w^2-4w+2=0$

7. $3y^2+12y+10=0$

8. $4x^2=5x+8$

Name: Date:
Instructor: Section:

Chapter 15 QUADRATIC EQUATIONS

15.4 Formulas

Learning Objectives
A Solve a formula for a specified letter.

GUIDED EXAMPLES AND PRACTICE

Objective A Solve a formula for a specified letter.

Review this example for Objective A:

1. Solve $m^2 + n^2 = r^2$ for n.

We assume that n is positive.

$$m^2 + n^2 = r^2$$
$$n^2 = r^2 - m^2$$
$$n = \sqrt{r^2 - m^2}$$

Practice this exercise:

1. Solve $A = cd^2$ for d. Assume that d is positive.

ADDITIONAL EXERCISES

Objective A Solve a formula for a specified letter.

For extra help, see Examples 1–7 on pages 1127–1129 of your text and the Section 15.4 lecture video.

Solve for the indicated letter.

1. $\frac{1}{x} + \frac{1}{y} = \frac{1}{z}$, for y

2. $f = \sqrt{\frac{2l}{g}}$, for l

3. $y = \frac{5x}{w^2}$, for w

4. $L = \frac{l^2 - l}{5}$, for l

5. $p = q(1 + ab)$, for a

Name: Date:
Instructor: Section:

Chapter 15 QUADRATIC EQUATIONS

15.5 Applications and Problem Solving

Learning Objectives
A Solve applied problems using quadratic equations.

GUIDED EXAMPLES AND PRACTICE

Objective A Solve applied problems using quadratic equations.

Review this example for Objective A:

1. The width of a rectangle is 5 m less than the length. The area is 66 m^2. Find the length and the width.

1. *Familiarize.* We first make a drawing. Let l = the length. Then $l - 5$ = the width.

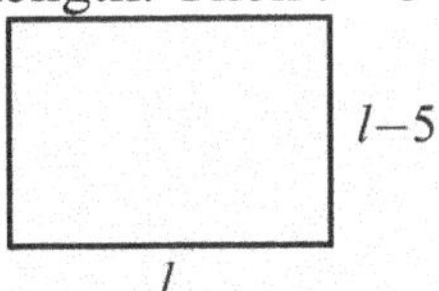

2. *Translate.* Recall that the area of a rectangle is $\text{length} \times \text{width}$. Thus, we have

$$l(l-5)=66.$$

3. *Solve.*

$$l(l-5)=66$$
$$l^2-5l=66$$
$$l^2-5l-66=0$$
$$(l-11)(l+6)=0$$
$$l-11=0 \quad or \quad l+6=0$$
$$l=11 \quad or \quad l=-6$$

4. *Check.* Length cannot be negative, so –6 is not a solution. If l = 11, then l – 5 = 11 – 5 = 6 and the area is $11 \cdot 6$, or 66 m^2. This checks.

5. *State.* The length is 11 m and the width is 6 m.

Practice this exercise:

1. The speed of a boat in still water is 10 km/h. The boat travels 24 km upstream and 24 km downstream in a total time of 5 hr. What is the speed of the stream?

ADDITIONAL EXERCISES

Objective A Solve applied problems using quadratic equations.

For extra help, see Examples 1–3 on pages 1133–1135 of your text and the Section 15.5 lecture video.

Solve.

1. The length of a TI-30X calculator is 9 cm greater than the width. The area is 136 cm^2. Find the length and the width.

2. Find the approximate answer. Round to the nearest tenth.
A picture frame, having a uniform thickness, measures 20 in. by 17 in. There is 108 in^2 of the picture showing. Find the thickness of the frame.

3. Complete the table to help with the familiarization.
An airplane flies 300 km against the wind and 450 km with the wind in a total time of 3 hr. The speed of the airplane in still air is 250 km/h. What is the speed of the wind?

	d	*r*	*t*
With wind			
Against wind			
Total time			

4. The speed of a boat in still water is 10 km/h. The boat travels 15 km upstream and 21 km downstream in a total time of 4 hours. What is the speed of the stream?

Name: Date:
Instructor: Section:

Chapter 15 QUADRATIC EQUATIONS

15.6 Graphs of Quadratic Equations

Learning Objectives
A Graph quadratic equations.
B Find the x-intercepts of a quadratic equation.

Key Terms
Use the vocabulary terms listed below to complete each statement in Exercises 1–4.

discriminant **line of symmetry** **parabola** **vertex**

1. The graph of a quadratic equation is called a(n) ________________ .

2. The top or bottom point where the curve of the graph of a quadratic equation changes is called the ________________ .

3. If the graph of a quadratic equation is folded on its ________________ , the two halves will match exactly.

4. In the quadratic formula, the radicand, $b^2 - 4ac$ is called the ________________ .

GUIDED EXAMPLES AND PRACTICE

Objective A Graph quadratic equations.

Review this example for Objective A:

1. Graph $y = -x^2 + 4x$.

First we find the vertex. The x-coordinate is

$$-\frac{b}{2a} = -\frac{4}{2\cdot(-1)} = 2.$$

We substitute 2 for x into the equation to find the y-coordinate of the vertex.

$y = -x^2 + 4x = -2^2 + 4\cdot 2 = 4$

The vertex is (2, 4). The line of symmetry is $x = 2$. Now we choose some x-values on either side of the vertex, find the corresponding y-values, plot points, and graph the parabola.

Practice this exercise:

1. Graph $y = x^2 + x - 4$.

For $x = -1$, $y = -(-1)^2 + 4(-1) = -5$
For $x = 0$, $y = -0^2 + 4(0) = 0$
For $x = 1$, $y = -1^2 + 4(1) = 3$
For $x = 3$, $y = -3^2 + 4(3) = 3$
For $x = 4$, $y = -4^2 + 4(4) = 0$

x	y	
-1	-5	
0	0	
1	3	
2	4	← Vertex
3	3	
4	0	

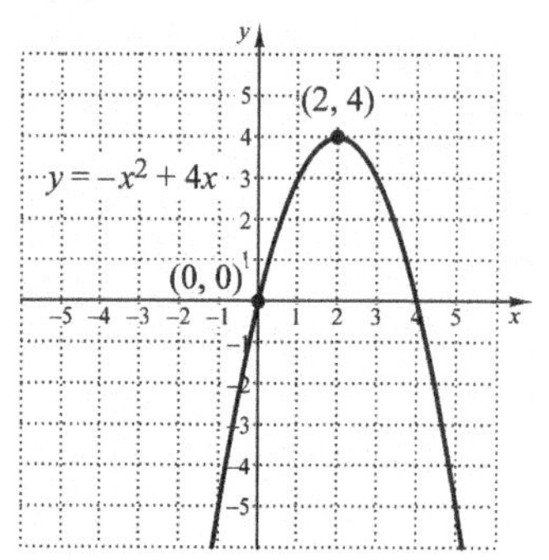

Objective B Find the x-intercepts of a quadratic equation.

Review this example for Objective B:

2. Find the x-intercepts of $y = x^2 - 4x - 3$.

We solve $x^2 - 4x - 3 = 0$. Using the quadratic formula, we get $x = 2 \pm \sqrt{7}$. Then the x-intercepts are $(2-\sqrt{7},\ 0)$ and $(2+\sqrt{7},\ 0)$.

Practice this exercise:

2. Find the x-intercepts of $y = x^2 - 2x - 5$.

ADDITIONAL EXERCISES

Objective A Graph quadratic equations.

For extra help, see Examples 1–3 on pages 1141–1143 of your text and the Section 15.6 lecture video.

Graph the quadratic equation.

1. $y = x^2 - 4x + 4$

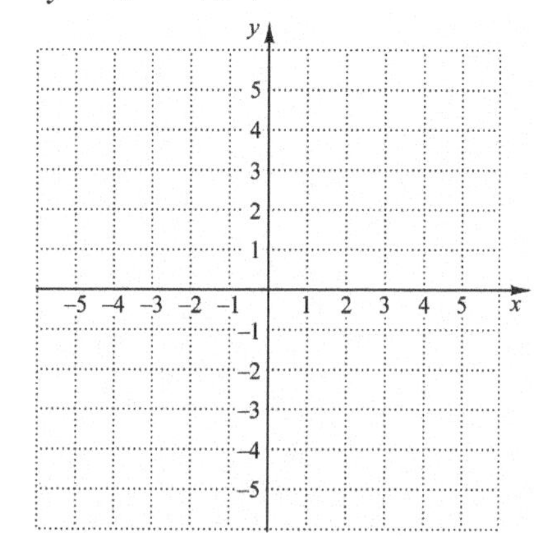

2. $y = -2x^2 + 6x + 1$

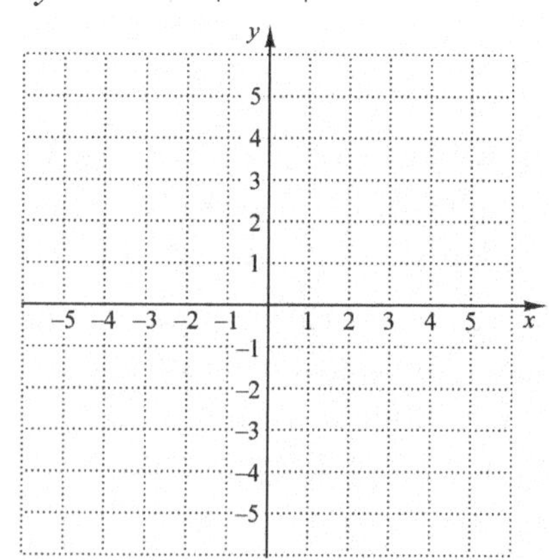

Name: Date:
Instructor: Section:

3. $y = 3 - x^2$

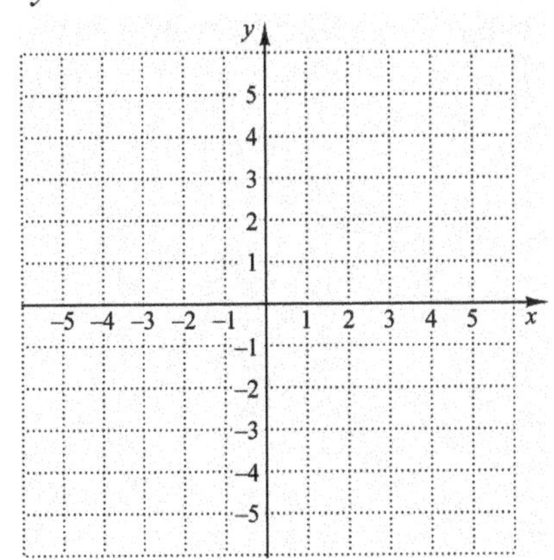

4. $y = x^2 + x$

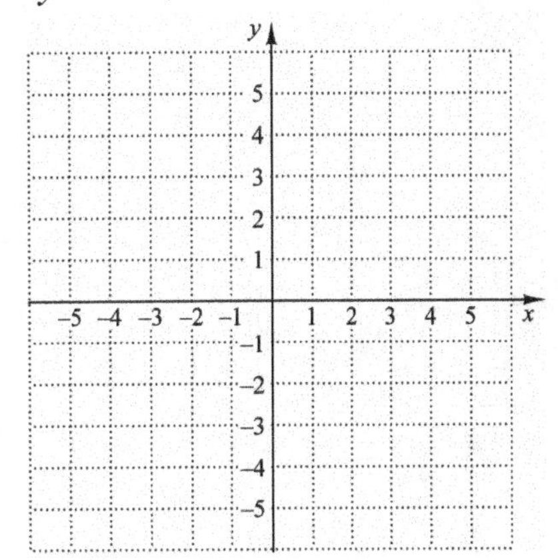

Objective B Find the x-intercepts of a quadratic equation.

For extra help, see Example 4 on page 1144 of your text and the Section 15.6 lecture video.

Find the x-intercepts.

5. $y = x^2 + 10x$

6. $y = x^2 + 14x + 49$

7. $y = -x^2 + 2x + 2$

8. $y = x^2 + 4$

Name: Date:
Instructor: Section:

Chapter 15 QUADRATIC EQUATIONS

15.7 Functions

Learning Objectives

A Determine whether a correspondence is a function.
B Given a function described by an equation, find function values (outputs) for specified values (inputs).
C Draw a graph of a function.
D Determine whether a graph is that of a function.
E Solve applied problems involving functions and their graphs.

Key Terms

Use the vocabulary terms listed below to complete each statement in Exercises 1–8.

domain	**function**	**input**	**linear**
output	**quadratic**	**range**	**relation**

1. A(n) ____________________ is a correspondence between two sets such that each member of the first set corresponds to exactly one member of the second set.

2. A(n) ____________________ is a correspondence between two sets such that each member of the first set corresponds to at least one member of the second set.

3. The first set in a correspondence is the ____________________ .

4. The second set in a correspondence is the ____________________

5. The function $f(x) = 2x - 5$ is an example of a(n) ____________________ function.

6. The function $f(x) = x^2 - x + 6$ is an example of a(n) ____________________ function.

7. In the statement $f(4) = 7$, 4 is a(n) ____________________ .

8. In the statement $f(4) = 7$, 7 is a(n) ____________________ .

GUIDED EXAMPLES AND PRACTICE

Objective A Determine whether a correspondence is a function.

Review this example for Objective A:

1. Determine whether each correspondence is a function.

a)

Domain Range

$f:$
$1 \to 3$
$2 \to -5$
$3 \to 8$
$4 \to -4$

b)

Domain Range

$g:$
$A \to m$
$B \to s$
$C \to t$
$C \to w$

a) f is a function because each member of the domain corresponds to exactly one member of the range.
b) g is not a function because one member of the domain, C, corresponds to more than one member of the range.

Practice this exercise:

1. Determine whether the correspondence is a function.

Domain Range

$h:$
$1 \to 7$
$2 \to 7$
$3 \to 5$
$4 \to 1$

Objective B Given a function described by an equation, find function values (outputs) for specified values (inputs).

Review this example for Objective B:

2. Find $f(-1)$ for $f(x) = 2x^2 - 1$.

$$f(-1) = 2(-1)^2 - 1 = 2 - 1 = 1$$

Practice this exercise:

2. Find $g(2)$ for $g(x) = 3x - 5$.

Objective C Draw a graph of a function.

Review this example for Objective C:

3. Graph $g(x) = -0.2x^2$.

We find some function values, plot points, and draw the graph.

x	y
−5	−5
−3	−1.8
0	0
3	−1.8
5	−5

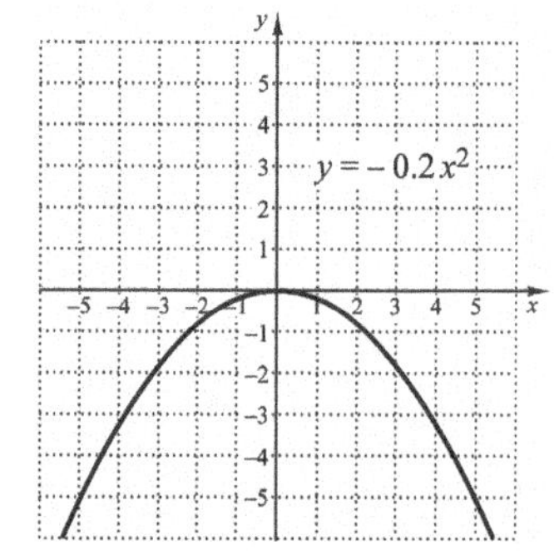

Practice this exercise:

3. Graph $f(x) = x - 1$.

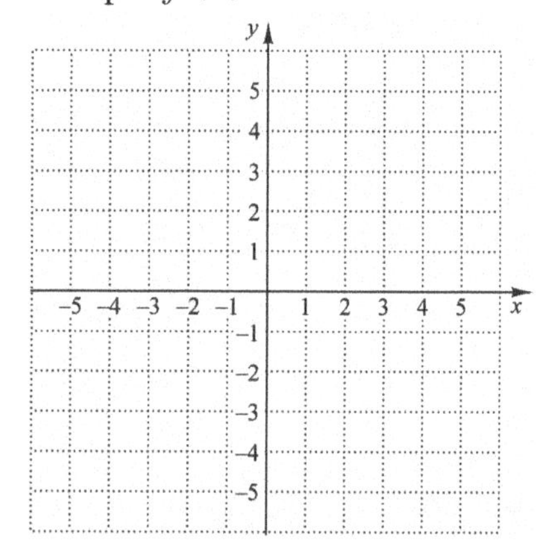

Name: Date:
Instructor: Section:

Objective D Determine whether a graph is that of a function.

Review this example for Objective D:

4. Determine whether each graph is the graph of a function.

a) b)

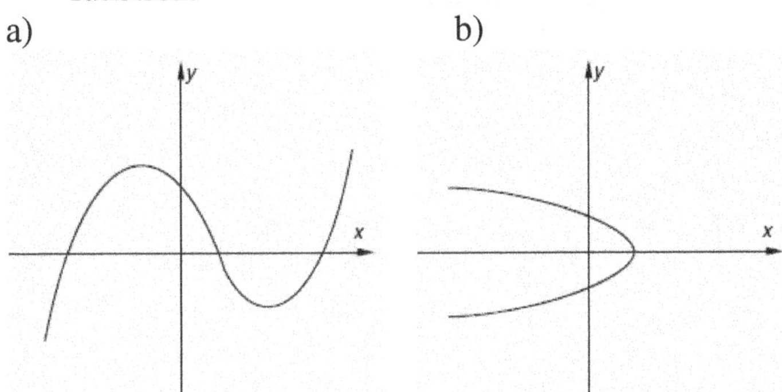

a) The graph is that of a function because no vertical line can cross the graph at more than one point. This can be confirmed with a straight edge.
b) The graph is not that of a function because a vertical line can be drawn that crosses the graph more than once.

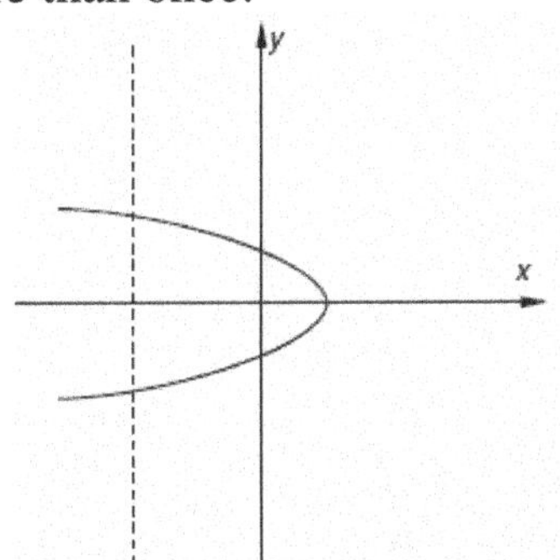

Practice this exercise:

4. Determine whether the graph is the graph of a function.

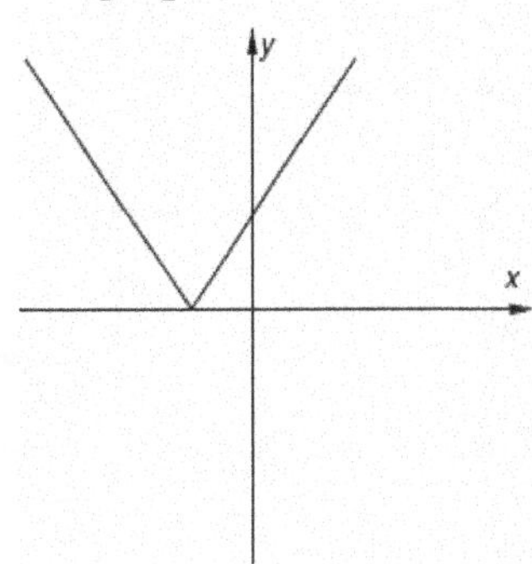

Objective E Solve applied problems involving functions and their graphs.

Review this example for Objective E:

5. The graph below shows the number of Americans over age 65 as a function of the year. (The data is projected for 2000-2030.)

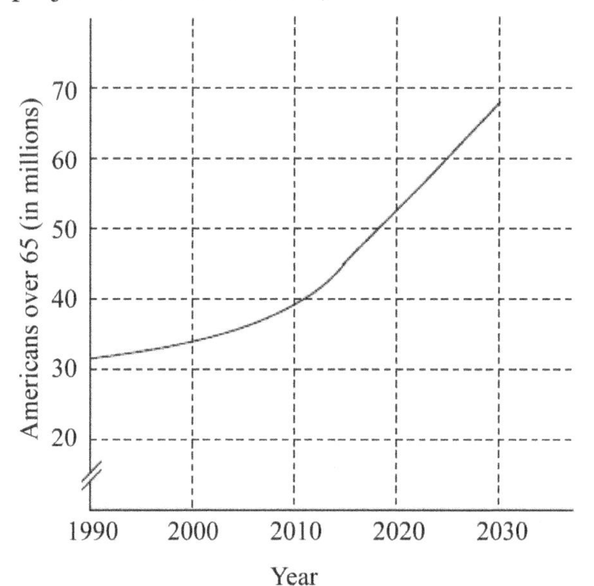

Use the graph to approximate the number of Americans over age 65 in 2000.

Locate 2000 on the horizontal axis, move directly up to the graph, and then across to the vertical axis. We see that the output that corresponds to the input 2000 is about 34, so there will be about 34 million Americans over age 65 in 2000.

Practice this exercise:

5. Use the graph at left to approximate the number of Americans over age 65 in 2020.

ADDITIONAL EXERCISES

Objective A Determine whether a correspondence is a function.

For extra help, see Examples 1–2 on pages 1149–1150 of your text and the Section 15.7 lecture video.

Determine whether each correspondence is a function.

1.

Domain	Range
3	10
4	11
5	12
6	13
7	

2.

Domain	Range
a	u
b	v
c	w
d	x
e	z

Name: Date:
Instructor: Section:

3.

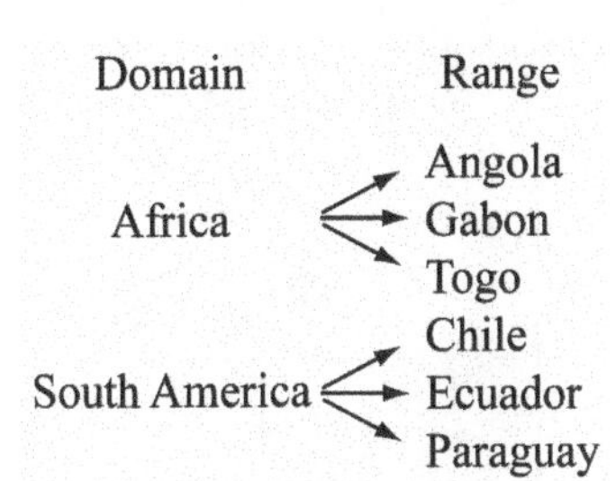

4.

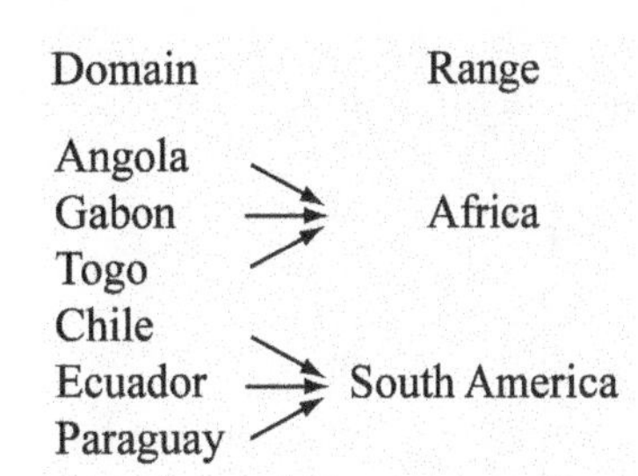

Objective B Given a function described by an equation, find function values (outputs) for specified values (inputs).

For extra help, see Example 3 on page 1151 of your text and the Section 15.7 lecture video.

Find the function values.

5. $h(a) = 4a - 1$

a) $h(2)$ **b)** $h(-2.5)$

c) $h(0)$ **d)** $h\left(\frac{1}{2}\right)$

6. $h(p) = p^2 - 2p + 3$

a) $h(0)$ **b)** $h(1)$

c) $h(-1)$ **d)** $h(4)$

7. $g(w) = |w| + 3$

a) $g(-15)$ **b)** $g(15)$

c) $g(0)$ **d)** $g(3)$

8. $f(x) = x^4$

a) $f(1)$ **b)** $f(10)$

c) $f(-1)$ **d)** $f(0)$

Objective C Draw a graph of a function.
For extra help, see Examples 4–6 on pages 1152–1153 of your text and the Section 15.7 lecture video.
Graph each function.

9. $f(x) = 3x - 4$

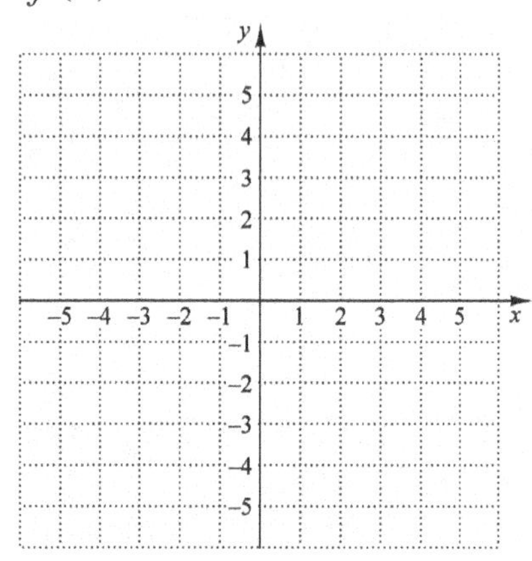

10. $g(x) = -\frac{1}{2}x + 3$

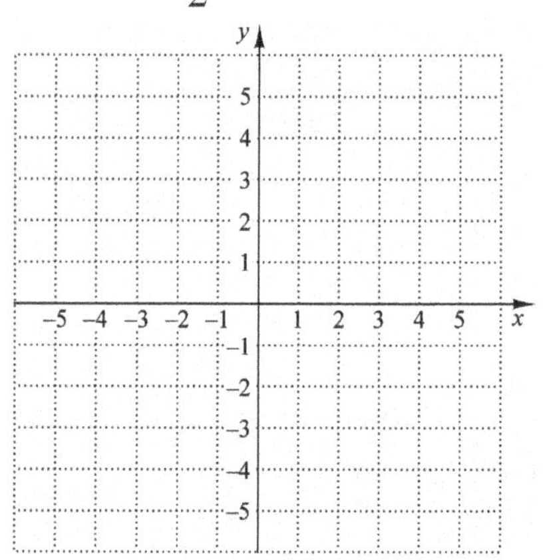

11. $g(x) = 4 - |x|$

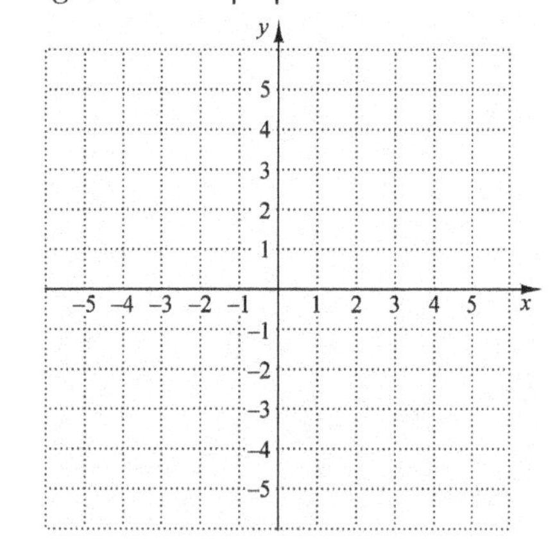

12. $f(x) = 2x^2 - 3$

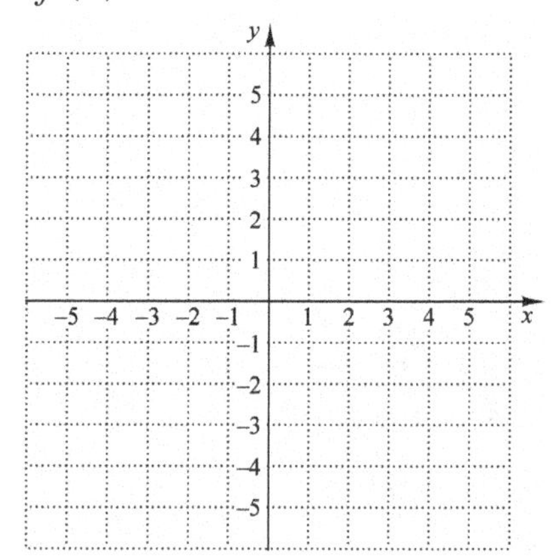

Name: Date:
Instructor: Section:

Objective D Determine whether a graph is that of a function.

For extra help, see Example 7 on page 1154 of your text and the Section 15.7 lecture video.

Determine whether each of the following is a graph of a function.

13.

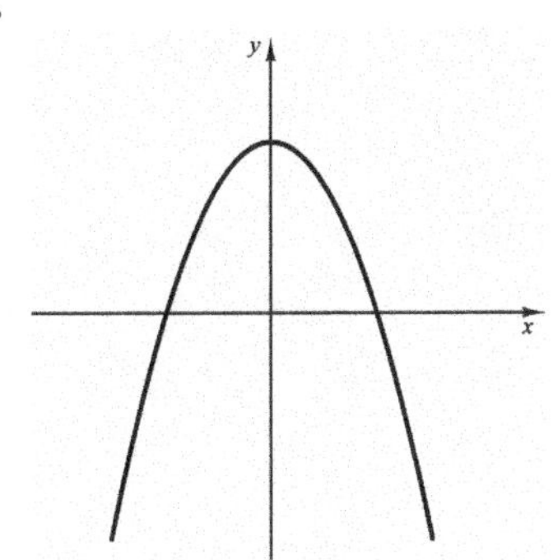

14.

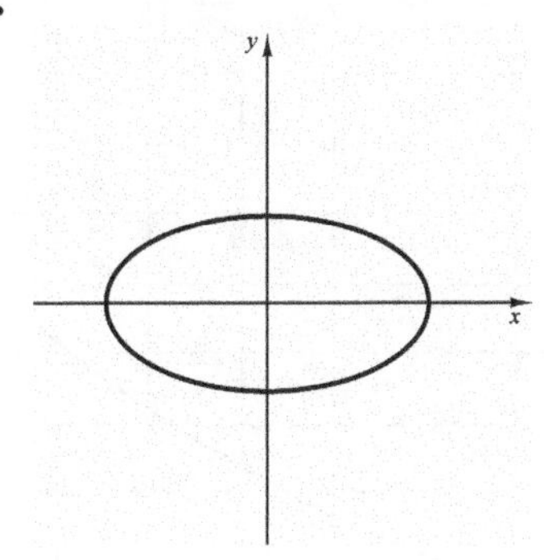

15.

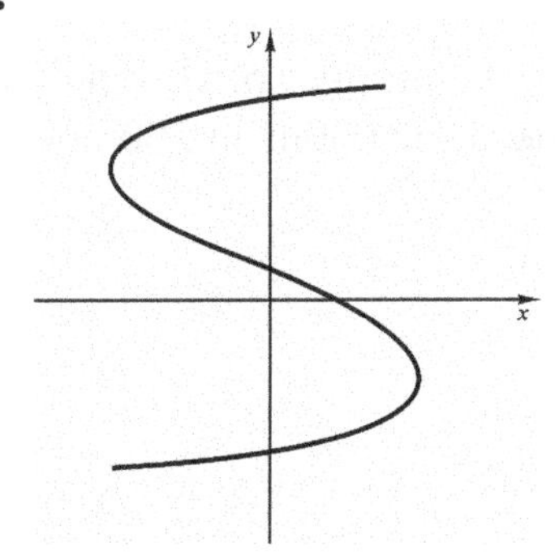

16.

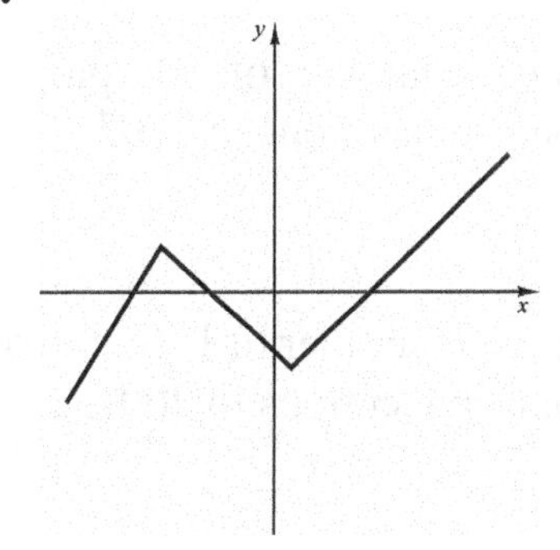

Objective E Solve applied problems involving functions and their graphs.
For extra help, see Example 8 on page 1155 of your text and the Section 15.7 lecture video.
The graph below shows the number of U.S. prepaid mobile phone connections as a function of the year.

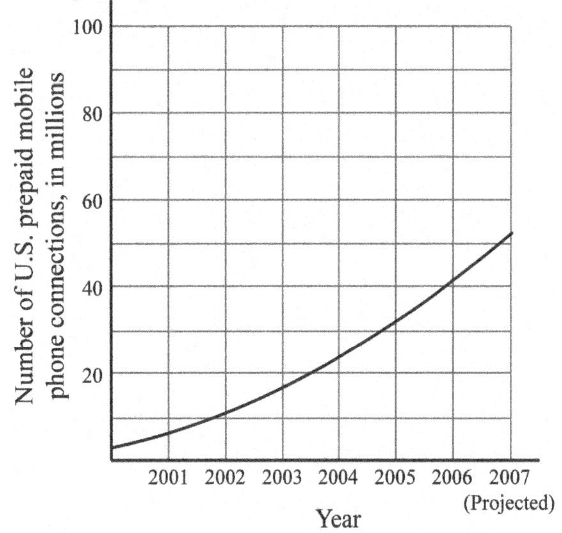

17. Approximate the number of U.S. prepaid mobile phone connections in 2005.

18. Approximate the number of U.S. prepaid mobile phone connections in 2007.

19. Approximate the number of U.S. prepaid mobile phone connections in 2002.

Odd Answers

Chapter 1 WHOLE NUMBERS

Section 1.1

Key Terms

1. expanded
2. whole
3. equation
4. digit
5. natural
6. standard
7. inequality

Practice

1. 2 hundreds
2. 0
3. 2 thousands + 0 hundreds + 8 tens + 7 ones, or 2 thousands + 8 tens + 7 ones
4. Five million, four hundred eighty-seven thousand, two hundred three
5. 465,813
6. <

Objective A

1. 8 ones
3. 9

Objective B

5. 1 thousand + 8 hundreds + 1 ten + 3 ones
7. 8 ten thousands + 5 thousands + 1 hundred + 2 tens + 6 ones
9. 4 millions + 3 hundred thousands + 0 ten thousands + 6 thousands + 7 hundreds + 4 tens + 9 ones, or 4 millions + 3 hundred thousands + 6 thousands + 7 hundreds + 4 tens + 9 ones

Objective C

11. Seventy-seven thousand, four hundred twenty-two
13. 414,963

Objective D

15. >
17. >

Section 1.2

Key Terms

1. commutative law of addition
2. subtrahend
3. associative law of addition
4. difference

Practice

1. 66,024
2. 156 in.
3. 2772

Objective A

1. 181
3. 187

Objective B

5. 127 in.
7. 155 yd

Objective C

9. 233
11. 2934
13. 4549

Section 1.3

Key Terms

1. zero
2. that same number
3. distributive law
4. commutative law of multiplication
5. not defined
6. one
7. associative law of multiplication

Practice

1. 181,832
2. 169 sq in.
3. 315 R 14
4. 27,000
5. 17,000
6. 2000
7. 1600
8. 5

Objective A

1. 258
3. 160,271

Objective B

5. 1232 sq in.
7. 225 sq ft

Objective C

9. not defined
11. 189 R 3

Objective D

13. 760
15. 66,000

Objective E

17. 2600
19. 4

Section 1.4

Key Terms

1. solution of an equation
2. equation
3. variable

Practice

1. 8
2. 1
3. 2806
4. 38

Objective A

1. 13
3. 8

Objective B

5. 12
7. 25

Section 1.5

Key Terms

1. familiarize yourself with the situation
2. translate the problem to an equation

Practice

1. 350 mi

Objective A

1. 4181 customers
3. $50,400

Section 1.6

Key Terms

1. exponent
2. base
3. average

Practice

1. 2^5
2. 125
3. 49
4. 30
5. 55

Objective A

1. 8^5
3. 5^4

Objective B

5. 128
7. 216

Objective C

9. 75
11. 31
13. 58

Objective D

15. 1232
17. 10

Section 1.7

Key Terms

1. multiple
2. prime
3. divisible
4. composite

Practice

1. Yes
2. 1, 2, 4, 5, 10, 20
3. 13, 26, 39, 52, 65, 78, 91, 104, 117, 130
4. No
5. composite
6. $3 \cdot 3 \cdot 7$

Objective A

1. No
3. 1, 2, 3, 4, 5, 6, 10, 12, 15, 20, 30, 60
5. 1, 2, 5, 10, 25, 50

Objective B

7. 9, 18, 27, 36, 45, 54, 63, 72, 81, 90
9. Yes

Objective C

11. composite
13. neither

Objective D

15. $2 \cdot 3 \cdot 3 \cdot 3$
17. $2 \cdot 3 \cdot 11$

Section 1.8

Key Terms

1. even
2. divisible by 5
3. divisible by 3
4. divisible

Practice

1. 18,225 is divisible by 3, 5, and 9

Objective A

1. 732 is divisible by 2, 3, and 6
3. 57, 171, 48, 7317, 5001

Section 1.9

Key Terms

1. prime factorization
2. factorization
3. least common multiple; multiple

Practice

1. 48
2. 40

Objective A

1. 70
3. 180

Chapter 2 FRACTION NOTATION

Section 2.1

Key Terms

1. denominator
2. canceling
3. equivalent
4. numerator
5. simplest

Practice

1. numerator: 5; denominator: 6
2. $\frac{4}{5}$
3. 5; 1; 0
4. $\frac{35}{48}$
5. $\frac{15}{20}$
6. $\frac{3}{8}$

Objective A

1. numerator: 2; denominator: 9
3. $\frac{3}{6}$

Objective B

5. 0
7. undefined

Objective C

9. $\frac{1}{21}$
11. $\frac{9}{32}$

Objective D

13. $\frac{8}{12}$
15. $\frac{77}{147}$

Objective E

17. $\frac{1}{3}$
19. $\frac{10}{13}$

Section 2.2

Key Terms

1. zero
2. reciprocal
3. one
4. one

Practice

1. $\frac{2}{9}$
2. $\frac{1}{13}$
3. $\frac{3}{4}$
4. $x = 36$

Objective A

1. $\frac{12}{35}$
3. $\frac{27}{7}$

Objective B

5. 9
7. $\frac{1}{19}$

Objective C

9. $\frac{6}{7}$
11. $\frac{12}{11}$

Objective D

13. $a = \frac{2}{3}$
15. $t = \frac{1}{6}$

Section 2.3

Key Terms

1. different denominators
2. least common denominator
3. like denominators

Practice

1. $\frac{2}{3}$
2. $\frac{21}{20}$
3. $\frac{1}{3}$
4. $\frac{17}{40}$
5. $\frac{2}{3} > \frac{5}{9}$
6. $\frac{1}{24}$

Objective A

1. $\frac{6}{7}$
3. $\frac{4}{9}$
5. $\frac{5}{4}$

Objective B

7. $\frac{1}{2}$
9. $\frac{5}{24}$

Objective C

11. $\frac{5}{4} < \frac{3}{2}$
13. $\frac{7}{10} > \frac{5}{9}$

Objective D

15. $t = \frac{1}{2}$

17. $x = \frac{3}{10}$

Section 2.4

Key Terms

1. mixed numeral(s); fraction notation
2. mixed numeral(s)
3. whole numbers; fractions

Practice

1. $\frac{19}{5}$
2. $1\frac{5}{6}$
3. $7\frac{5}{12}$
4. $5\frac{5}{8}$
5. $17\frac{3}{5}$
6. $2\frac{21}{26}$

Objective A

1. $\frac{23}{5}$

3. $2\frac{7}{15}$

Objective B

5. $11\frac{1}{7}$

7. $17\frac{11}{24}$

Objective C

9. $3\frac{3}{5}$

11. $3\frac{11}{18}$

Objective D

13. $19\frac{1}{2}$

15. $27\frac{43}{45}$

Objective E

17. $1\frac{3}{5}$

19. $\frac{63}{80}$

Section 2.5

Key Terms

1. solve
2. area
3. check

Practice

1. 24 mpg
2. $2\frac{3}{4}$ in.
3. $\frac{3}{8}$ mi

Objective A

1. $23\frac{3}{4}$ ft^2

3. $\frac{313}{120}$ mi

5. $14\frac{3}{8}$ ft

Section 2.6

Key Terms

1. order of operations; simplify
2. average

Practice

1. $\frac{17}{10}$
2. $\frac{33}{70}$
3. 0

Objective A

1. $\frac{17}{54}$
3. $22\frac{11}{18}$

Objective B

5. $\frac{1}{2}$
7. $\frac{1}{2}$

Chapter 3 DECIMAL NOTATION

Section 3.1

Key Terms

1. decimal notation
2. decimal point
3. arithmetic number(s); whole numbers; fraction(s)

Practice

1. fifty-nine and seven hundredths
2. $\frac{1609}{100}$
3. 2.59
4. 91.23
5. 2.11
6. 327.2

Objective A

1. one hundred nineteen and nine tenths
3. two and seven hundred eighty-nine thousandths

Objective B

5. $\frac{41,376}{10,000}$
7. 2.13

Objective C

9. 0.9
11. 147.18

Objective D

13. 234.1
15. 234.065

Section 3.2

Key Terms

1. credit
2. debit
3. balance forward; debit(s); credit(s)
4. place-value digits

Practice

1. 29.042
2. 207.848
3. $y = 114.92$

4. $423.92

Objective A

1. 41.17 3. 0.507

Objective B

5. 25.851 7. 1.8595

Objective C

9. $w = 19.9946$ 11. $x = 22{,}984.69$

Section 3.3

Key Terms

1. cents; dollar 2. ¢; cents 3. $; dollars

Practice

1. 11.575 2. 0.143 3. 85,043
4. 6,200,000,000 5. 12,549¢ 6. $2.45

Objective A

1. 36.8 3. 0.000675

Objective B

5. 4713¢ 7. $43.51 9. 32,800,000

Section 3.4

Key Terms

1. C 2. A 3. D
4. B

Practice

1. 51.3 2. 8.3 3. 0.039
4. 12,340 5. $y = 9.3$ 6. 26.799
7. 2.9 mi

Objective A

1. 3.35 3. 0.345

Objective B

5. $w = 0.05702$ 7. $h = 25$

Objective C

9. 997.94 11. $29.91

Section 3.5

Key Terms

1. repeating decimal 2. terminating decimal

Practice

1. 0.68 2. $0.\overline{5}$ 3. 0.15

4. 6.65

Objective A

1. 0.75   3. 0.125   5. $0.\overline{27}$

Objective B

7. 0.6; 0.57; 0.571   9. 32.8 mpg

Objective C

11. –64.32   13. –212.275

Section 3.6
Practice

1. b

Objective A

1. 109   3. c   5. $150

Section 3.7
Key Terms

1. variable   2. estimate

Practice

1. $324.65

Objective A

1. 2.8°F   3. Perimeter: 45.8 cm; area: 116.28 sq cm

Chapter 4 PERCENT NOTATION

Section 4.1
Key Terms

1. proportions   2. rate   3. cross products

Practice

1. $\frac{8}{3}$ 2. $\frac{12}{10}$; $\frac{10}{12}$ 3. $\frac{4}{9}$
4. $6.50 per hour   5. Yes   6. $x = 18$
7. $x = \frac{18}{5}$ 8. 180 mi

Objective A

1. $\frac{7}{9}$ 3. $\frac{5}{6}$; $\frac{6}{5}$ 5. $\frac{5}{3}$

Objective B

7. 3 gal/day   9. 5.3 cal/min

Objective C

11. No   13. No

Objective D

15. $n = 4$ 17. $y = 16.72$

Objective E

19. 780 mi 21. 200 deer

Section 4.2

Key Terms

1. left 2. right

Practice

1. $\frac{13}{100}$; $13\times\frac{1}{100}$; 13×0.01 2. 0.548
3. 150%

Objective A

1. $\frac{85}{100}$; $85\times\frac{1}{100}$; 85×0.01 3. $\frac{4.8}{100}$; $4.8\times\frac{1}{100}$; 4.8×0.01

Objective B

5. 1.50 7. 83%

Section 4.3

Key Terms

1. fraction equivalent 2. decimal equivalent

Practice

1. 87.5% 2. $\frac{13}{20}$

Objective A

1. 56% 3. 62.5%

Objective B

5. $\frac{3}{8}$ 7. $\frac{5}{6}$

Section 4.4

Key Terms

1. is 2. what 3. %
4. of

Practice

1. $16\%\cdot b = 224$ 2. 40% 3. \$96
4. 60

Objective A

1. $a = 18\%\times93$ 3. $65 = p\cdot150$

Objective B

5. 20.8
7. 80

Section 4.5

Key Terms

1. whole
2. part

Practice

1. $\frac{85}{100}=\frac{34}{b}$
2. $\frac{N}{100}=\frac{30}{45}$
3. $\frac{61}{100}=\frac{a}{320}$
4. 90

Objective A

1. $\frac{N}{100}=\frac{16}{80}$
3. $\frac{20}{b}=\frac{45}{100}$

Objective B

5. 114
7. 13.3

Section 4.6

Key Terms

1. amount of increase/decrease; original amount

Practice

1. $2150
2. 34%

Objective A

1. 48
3. 55%

Objective B

5. 22%
7. 275%

Section 4.7

Key Terms

1. Sales
2. Sales tax
3. Sale price

Practice

1. $378
2. $1500
3. $455

Objective A

1. 5%
3. $1598

Objective B

5. $300
7. $6500

Objective C

9. $125; $375
11. 12%; $1320
13. 6%; $235

Section 4.8

Key Terms

1. simple interest
2. compound interest

Practice

1. $18
2. $3896.76
3. interest: $20.06; amount applied to principal: $29.94

Objective A

1. $36
3. $335.94
5. $2435.51

Objective B

7. $3514.98
9. $5016.46

Objective C

11. interest: $59.65; amount applied to principal: $99.35
13. At 13.2%, the interest is $15.90 less than at 18%.
    The principal at 13.2% is reduced $15.90 more than at 18%.

Chapter 5 DATA, GRAPHS, AND STATISTICS

Section 5.1

Key Terms

1. median
2. average
3. statistic
4. mode

Practice

1. 5 mi
2. 56°
3. 85°
4. $56
5. Brand A

Objectives A, B, C

1. average: 13.5; median: 12; mode: 12
3. average: 43.8; median: 49; modes: 32 and 53
5. average: $7.78; median: $7.49; mode: $7.49

Objective D

7. Wynn's: average: 7.4375; Penn's: average: 7.5; Penn's tastes better.

Section 5.2

Key Terms

1. pictograph
2. table

Practice

1. Garden Veggie
2. 40 calories

Objective A

1. 198
3. Moderate walking

Objective B

5. 7500
7. 2002 and 2003

Section 5.3

Key Terms

1. bar graph
2. line graph

Practice

1. Office work
2.

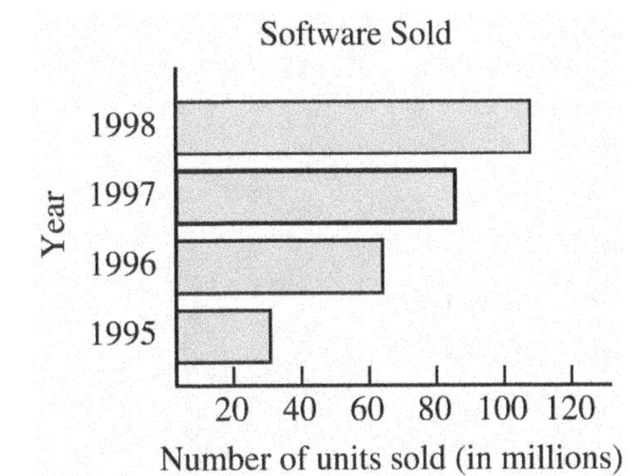

3. Increased
4.

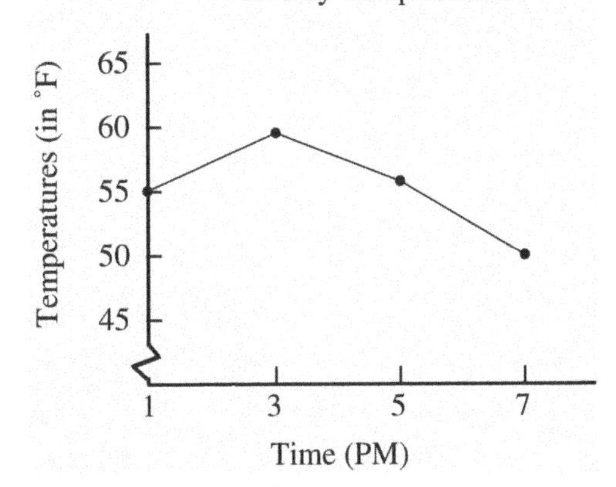

Objective A

1. Airline pilot
3. Airline pilot

Objective B

5.

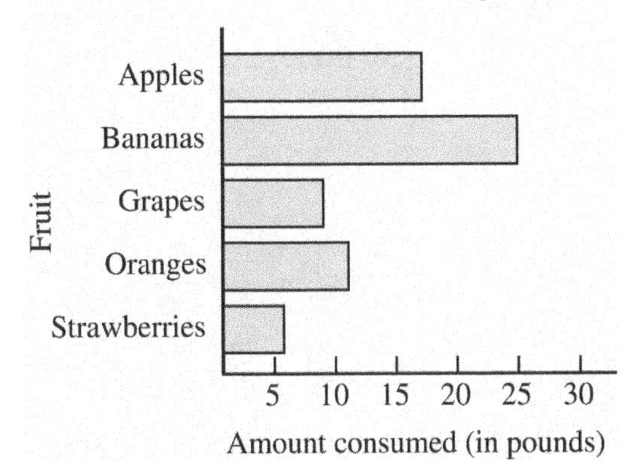

7. 6 pounds

Objective C

9. About 14 per hundred
11. 1995

Objective D

13.

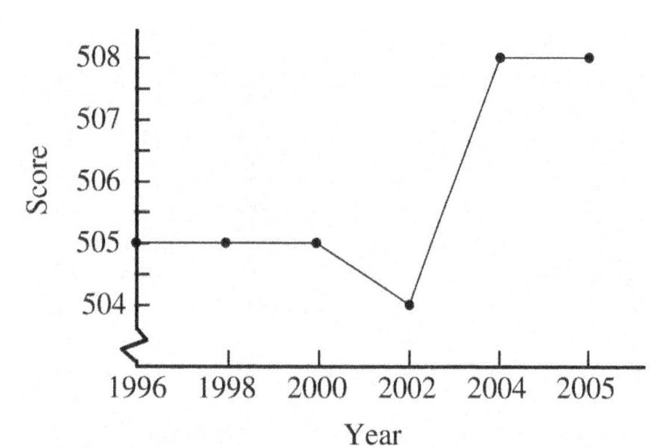

15. 1996 and 2000; 2004 and 2005

Section 5.4

Key Terms

1. pie chart
2. wedge; 100%

Practice

1. (a) 24%; (b) 10,120 homes
2.

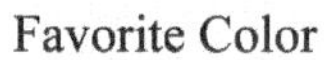

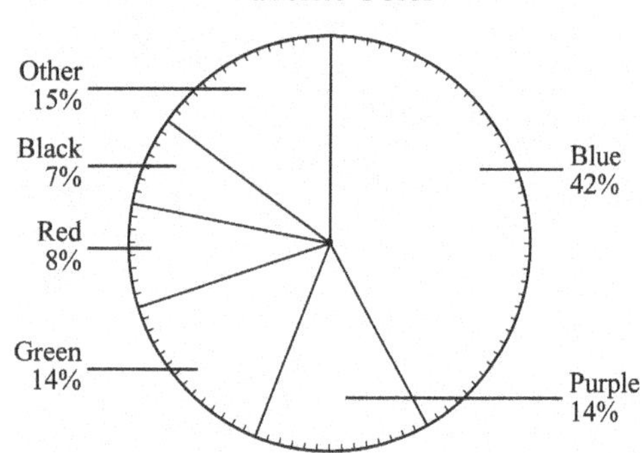

Objective A

1. 45%
3. 9%
5. 30 pints

Objective B

7.

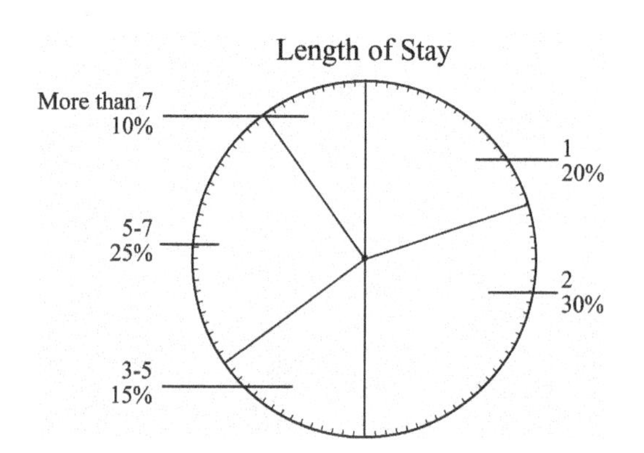

Chapter 6 GEOMETRY

Section 6.1

Key Terms

1. coplanar
2. perpendicular
3. ray
4. parallel
5. segment

Practice

1. $\overline{MN}$
2. $\overrightarrow{RP}$
3. m, $\overleftrightarrow{WX}$, $\overleftrightarrow{XW}$, $\overleftrightarrow{WY}$, $\overleftrightarrow{YW}$, $\overleftrightarrow{XY}$, or $\overleftrightarrow{YX}$
4. angle MNP, angle PNM, $\angle MNP$, $\angle PNM$, or $\angle N$
5. 70°
6. obtuse
7. not perpendicular
8. isosceles; acute
9. quadrilateral
10. 2520°
11. 40°

Objective A

1. E F (segment drawn) $\overline{EF}$, $\overline{FE}$
3. l, $\overleftrightarrow{ST}$, $\overleftrightarrow{TS}$, $\overleftrightarrow{SU}$, $\overleftrightarrow{US}$, $\overleftrightarrow{TU}$, $\overleftrightarrow{UT}$

Objective B

5. 25°

Objective C

7. obtuse
9. acute

Objective D

11. not perpendicular

Objective E

13. scalene; right
15. isosceles; obtuse

Objective F

17. 1080°
19. 20°
21. 115°

Section 6.2

Key Terms

1. polygon
2. rectangle
3. square

Practice

1. 20 cm
2. $1406

Objective A

1. 127 in.
3. 28.36 cm

Objective B

5. 110 ft; $240.90
7. 43.18 cm

Section 6.3
Key Terms

1. $A = b \cdot h$
2. $A = \frac{1}{2} \cdot h \cdot (a + b)$
3. $A = \frac{1}{2} \cdot b \cdot h$

Practice

1. 48 m^2
2. 40.96 yd^2
3. 58.5 cm^2
4. 6.75 m^2
5. 19.5 yd^2
6. 16,750 ft^2

Objective A

1. 5.76 ft^2
3. 96 ft^2

Objective B

5. 24 ft^2
7. 90 yd^2
9. 76 cm^2

Objective C

11. 404 ft^2
13. (a) $1379\frac{3}{4}$ ft^2; (b) $54.69

Section 6.4
Key Terms

1. diameter
2. circumference
3. radius; diameter

Practice

1. 10.5 in
2. 13 m
3. 65.94 in.
4. 132.665 m^2
5. 21.195 m^2

Objective A

1. 28 ft
3. 15.2 yd

Objective B

5. 88 ft
7. 95.456 yd

Objective C

9. 616 ft^2
11. 725.4656 yd^2

Objective D

13. 8 cm; 25.12 cm; 50.24 cm^2
15. 54.84 in.

Section 6.5
Key Terms

1. sphere
2. rectangular solid
3. circular cylinder
4. circular cone

Practice

1. 120 ft^3
2. 148 ft^2
3. 462 m^3
4. $\frac{1570}{3}$ in^3, or $523\frac{1}{3}$ in^3
5. 150.72 ft^3
6. 10,300.77 cm^3

Objective A

1. 48 m^3; 88 m^2
3. 192 m^3; 224 m^2

Objective B
5. 2034.72 m^3

Objective C
7. 14,130 cm^3
9. 57.87648 ft^3

Objective D
11. $837,333\frac{1}{3}$ m^3

Objective E
13. 113.04 cm^3
15. 96 in^3

Section 6.6

Key Terms
1. transversal
2. supplementary
3. congruent
4. vertical angles
5. complementary

Practice
1. 106°
2. Congruent
3. $m\angle 3 = 66°$
4. $m\angle 1 = 42°,\ m\angle 2 = 138°, m\angle 3 = 42°,\ m\angle 4 = 138°$

Objective A
1. 41°
3. 66°

Objective B
5. Not congruent
7. Not congruent

Objective C
9. $m\angle 3 = 60°$
11. $m\angle 6 = 60°$

Objective D
13. $m\angle 1 = m\angle 5 = m\angle 8 = 113°;\ m\angle 3 = m\angle 2 = m\angle 7 = m\angle 6 = 67°$
15. $\angle GHI \cong \angle FEI, 40°;\ \angle HGI \cong \angle EFI,\ \angle GIH \cong \angle FIE,\ \angle HIF \cong \angle EIG$

Section 6.7

Key Terms
1. diagonal
2. 180°
3. 360°
4. congruent; congruent; congruent

Practice
1. A
2. SAS
3. $m\angle T = 46°;\ m\angle U = 134°$

Objective A
1. $\angle N \cong \angle A,\ \angle F \cong \angle R,\ \angle S \cong \angle T,\ \overline{NF} \cong \overline{AR},\ \overline{FS} \cong \overline{RT},\ \overline{NS} \cong \overline{AT}$
3. SAS

Objective B
5. $m\angle D = 60°;\ m\angle E = m\angle G = 120°$
7. $ZW = 6\frac{1}{2},\ ZY = WX = 8\frac{1}{2}$

Section 6.8

Key Terms

1. congruent; proportional
2. congruent; congruent

Practice

1. $\frac{MN}{ED} = \frac{NP}{DF} = \frac{PM}{EF}$
2. 14

Objective A

1. $\angle R \leftrightarrow \angle U,\ \angle S \leftrightarrow \angle V,\ \angle T \leftrightarrow \angle W;\ \frac{RS}{UV} = \frac{RT}{UW} = \frac{ST}{VW}$

3. $\frac{YX}{QN} = \frac{XZ}{NP} = \frac{YZ}{QP}$

Objective B

5. $x = 9$
7. 16 ft

Chapter 7 INTRODUCTION TO REAL NUMBERS AND ALGEBRAIC EXPRESSIONS

Section 7.1

Key Terms

1. algebraic expression
2. variable
3. constant
4. substituting
5. evaluating
6. value

Practice

1. 8
2. Let x = the number; $6x - 9$

Objective A

1. 30
3. 4

Objective B

5. $w - 18$
7. Let n = the number; $40 + 2n$
9. $p - 0.25p$

Section 7.2

Key Terms

1. set
2. opposites
3. graph
4. absolute value
5. natural numbers
6. whole numbers
7. integers
8. rational numbers

Practice

1. –1200
2. Number line from -5 to 5 with a point at -2
3. –0.125
4. $-8 < 1$
5. $t > -3$
6. True
7. 59

Objective A

1. –3; 55
3. 525; –426

Objective B

5.

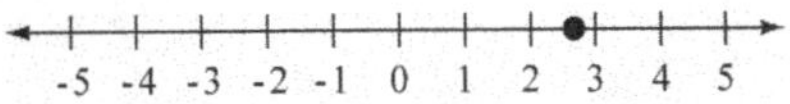

Objective C

7. –0.375

9. –1.25

Objective D

11. >

13. >

15. $x \leq -3$

Objective E

17. 12

19. 0

Section 7.3

Key Terms

1. positive
2. negative
3. zero
4. additive inverses

Practice

1. –4.6
2. 20
3. $124

Objective A

1. 0

3. –5.1

5. –45

Objective B

7. 54

9. $\frac{7}{3}$

Objective C

11. –1, or 1 under par

Section 7.4

Key Terms

1. difference
2. opposite

Practice

1. –10
2. 9
3. 116°

Objective A

1. 58

3. 3.37

5. 17.7

Objective B

7. 20°F

Section 7.5

Key Terms

1. negative
2. positive
3. positive
4. negative

Practice

1. 63
2. –$12

Objective A

1. –30
3. $-\frac{7}{10}$

Objective B

5. –12 lb
7. \$39.71

Section 7.6

Key Terms

1. not defined
2. positive
3. negative
4. reciprocals

Practice

1. –7
2. $\frac{1}{2}$
3. $\frac{33}{20}$
4. –\$2.50

Objective A

1. –13
3. Not defined

Objective B

5. $\frac{9}{4}$
7. $-3t$

Objective C

9. $-6.8\cdot\left(\frac{1}{1.3}\right)$
11. $\frac{7}{16}$

Objective D

13. 66.1%
15. –20.5

Section 7.7

Key Terms

1. equivalent
2. commutative
3. associative
4. distributive
5. terms
6. factor
7. like

Practice

1. $\frac{3t}{7t}$
2. $\frac{3}{4}$
3. $a+8$
4. $4(xy)$
5. $3x+12y-6z$
6. $9(4m-3n+p)$
7. $5a-2b$

Objective A

1. $\frac{4x}{14x}$
3. $-\frac{7}{5}$

Objective B

5. $xy+18$, or $18+yx$
7. $(4+n)+m$; $m+(n+4)$; $n+(4+m)$; answers may vary.

Objective C

9. $6x+24$
11. $24x+4y-36$
13. $2a, -1.4b, 10c$

Objective D

15. $-5(g-4)$
17. $\frac{1}{4}(3x-7y+1)$

Objective E

19. $-3g$
21. $8+44a-22b$

Section 7.8
Key Terms

1. grouping
2. exponential
3. divisions
4. additions

Practice

1. $3a-6b+c$
2. $m-6n$
3. –6
4. 0

Objective A

1. $-3x-y+10z$
3. $10m+3n+18$

Objective B

5. $4y-1$
7. $-3m+5n-18p$

Objective C

9. –18
11. $7x-1$

Objective D

13. 63
15. 44

Chapter 8 SOLVING EQUATIONS AND INEQUALITIES

Section 8.1
Key Terms

1. equation
2. solution
3. equivalent equations
4. addition principle

Practice

1. Yes
2. $x=-5$

Objective A

1. Yes
3. No
5. No

Objective B

7. $y=26$
9. $x=5.9$

Section 8.2
Key Terms

1. principle
2. inverse
3. identity
4. coefficient

Practice

1. $y = -6$
2. $v = \frac{5}{6}$

Objective A

1. $x = -6$
3. $a = \frac{4}{3}$

Section 8.3

Key Terms

1. distributive law
2. clear fractions
3. infinitely many solutions
4. no solution

Practice

1. $x = -3$
2. $x = -4$
3. $y = \frac{1}{4}$
4. $x = 5$
5. no solution
6. infinitely many solutions

Objective A

1. $x = 8$
3. $y = -6$

Objective B

5. $y = -4$
7. $x = 60$

Objective C

9. $y = 1$
11. no solution

Section 8.4

Key Terms

1. formula
2. evaluating

Practice

1. 220 mi
2. $x = \frac{y-5}{2}$

Objective A

1. 2 m^2
3. $53

Objective B

5. $q = 3A - p - r$
7. $x = 18 - y$

Section 8.5

Key Terms

1. of
2. is
3. what number
4. %

Practice

1. 15%
2. 160
3. 268

Objective A

1. 12
3. 25
5. a) 20%; b) $36

Section 8.6

Key Terms

1. familiarize
2. translate
3. solve
4. check
5. state

Practice

1. length: 30 ft; width: 14 ft

Objective A

1. 12 in. and 48 in.
3. 22, 23, 24

Section 8.7

Key Terms

1. inequality
2. graph
3. equivalent
4. set-builder notation

Practice

1. Yes
2. $x > 2$ 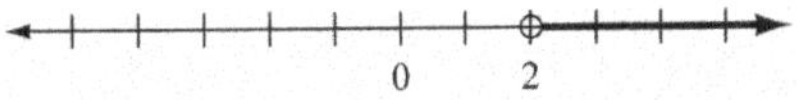
3. $\{x|x > -1\}$
4. $\{x|x < -4\}$
5. $\{y|y \geq -3\}$

Objective A

1. No
3. Yes
5. No

Objective B

7. $t \geq -2$ 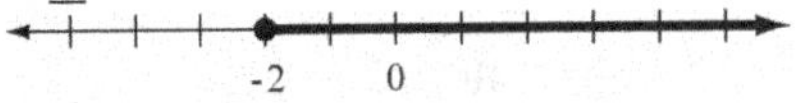
9. $-4 < x < 0$

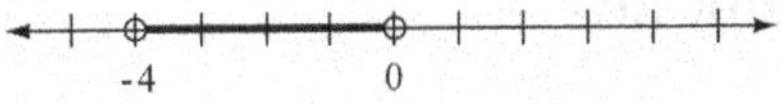

Objective C

11. $\{x|x \geq -5\}$ 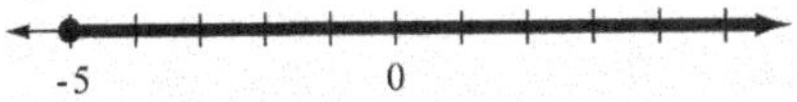
13. $\{y|y \geq 21\}$

Objective D

15. $\{x|x > 4\}$ 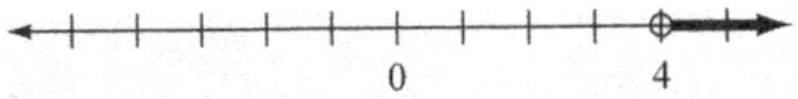
17. $\{p|p > -2\}$

Objective E

19. $\{x|x \geq 9\}$
21. $\left\{r \middle| r \leq -\frac{19}{2}\right\}$

Section 8.8

Key Terms

1. D
2. B
3. B
4. C
5. B
6. D

Practice

1. $x \leq 50$
2. 16 hr or more

Objective A

1. $n \leq 13$
3. 45 mph $< s <$ 65 mph

Objective B

5. $\{x|x \geq 66\}$
7. 300 or fewer cards

Chapter 9 GRAPHS OF LINEAR EQUATIONS

Section 9.1

Key Terms

1. axes
2. origin
3. coordinates
4. ordered pair
5. graph
6. y-intercept

Practice

1. A
2. IV
3. (0, 2)
4. No
5. 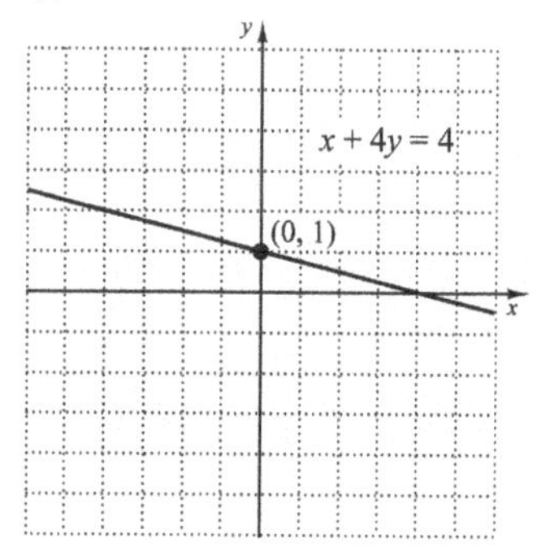

6. About 200 miles

Objective A

1. 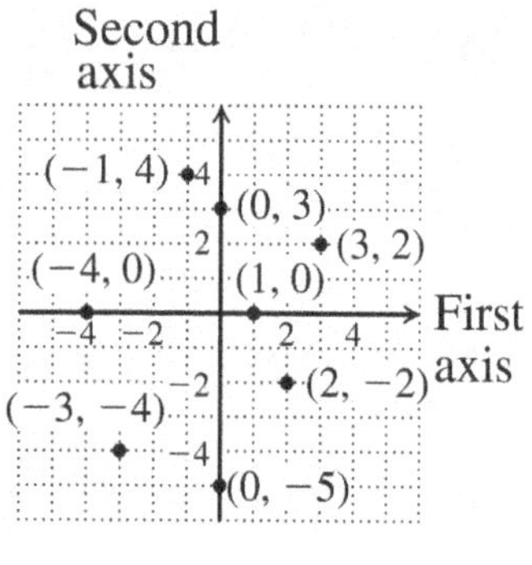

3. III
5. III; negative

Objective B

7. C: (4, 4); D: (3, 0)

Objective C

9. Yes
11. Yes

Objective D

13.

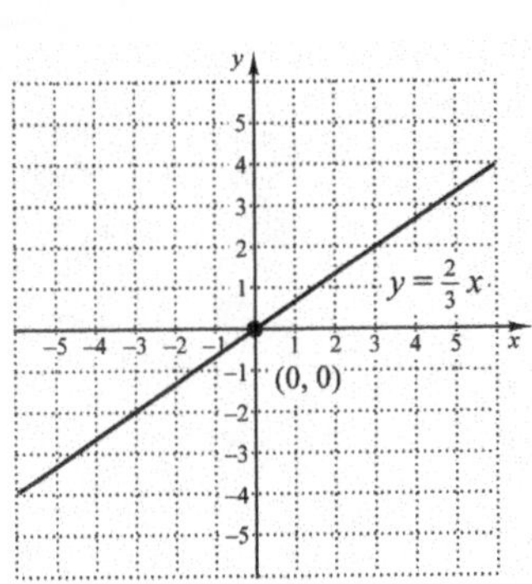

15.

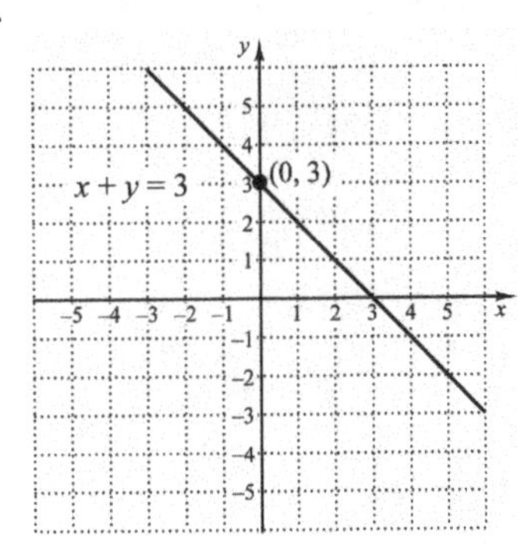

Objective E

17. \$435, \$477, \$603, \$666

19. \$450, \$225, \$0

Section 9.2

Key Terms

1. x-intercept
2. y-intercept
3. vertical line
4. horizontal line

Practice

1.

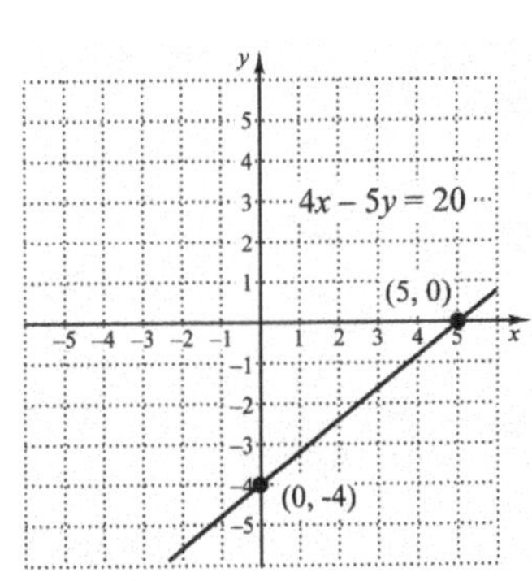

2.

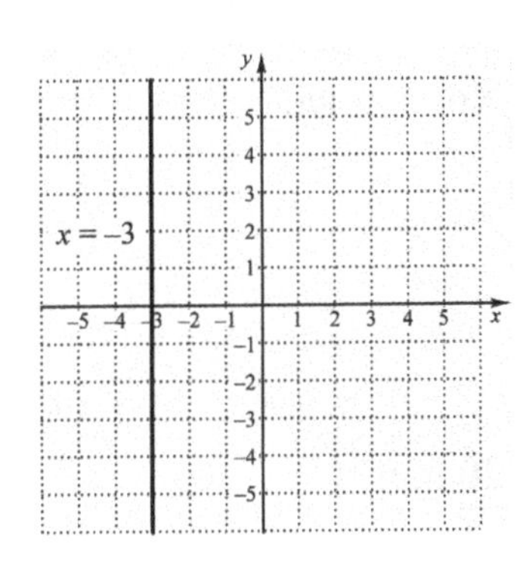

Objective A

1. x-intercept: (0, 3); y-intercept: (5, 0)

3.

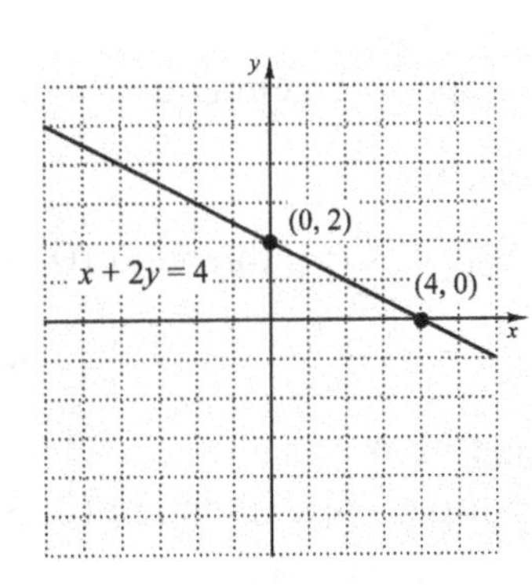

Objective B

5.

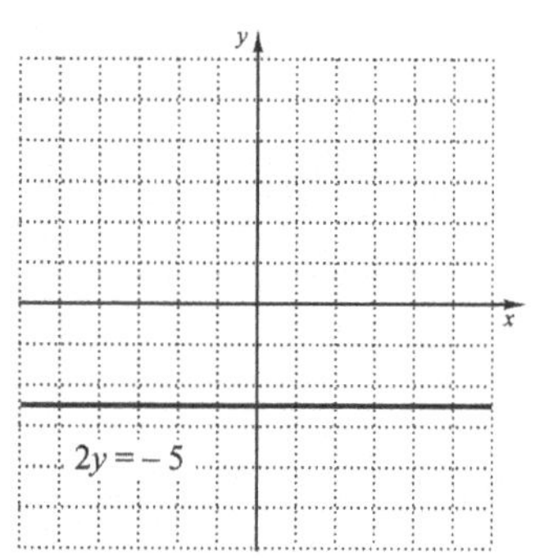

7. $x = -2$

Section 9.3

Key Terms

1. rise
2. run
3. slope
4. grade

Practice

1. $\frac{1}{2}$
2. $\frac{2}{3}$
3. 8%

Objective A

1. $-\frac{4}{9}$
3. Not defined

Objective B

5. –6
7. $-\frac{1}{2}$

Objective C

9. 2.5%
11. 20 mpg

Section 9.4

Key Terms

1. slope-intercept equation
2. slope
3. y-intercept

Practice

1. Slope: $\frac{4}{3}$; y-intercept (0, –4)
2. $y = 4x - 1$
3. $y = 4x + 3$
4. $y = x + 1$

Objective A

1. Slope: –5; y-intercept (0, 3)
3. Slope: 0; y-intercept (0, 56)
5. $y = 2.7x - 4.3$

Objective B

7. $y = \frac{2}{3}x + \frac{16}{3}$
9. $y = -5x + 7$

Objective C

11. $y=-\frac{1}{5}x+3$

13. $y=-\frac{1}{3}x+\frac{8}{3}$

Section 9.5
Practice

1. $y=\frac{3}{4}x+2$

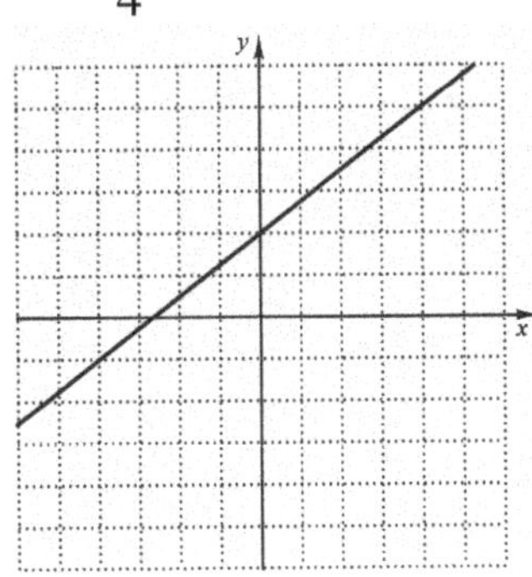

Objective A

1. Slope $=-2$; y-intercept $(0, 3)$

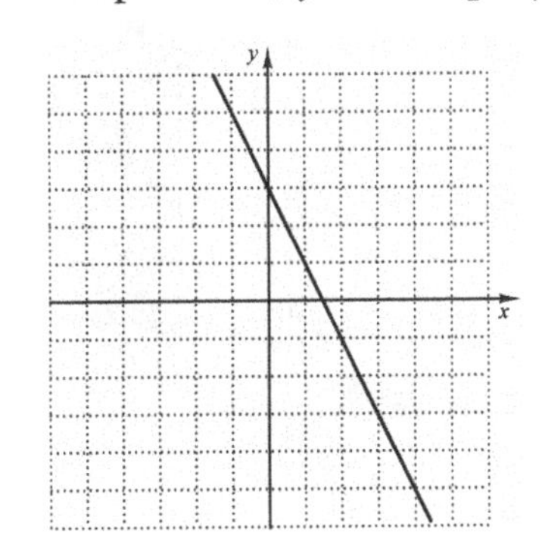

3. $y=-\frac{2}{5}x-2$

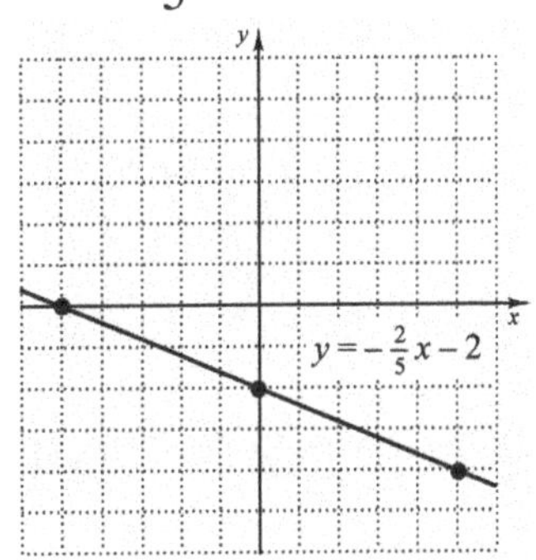

Section 9.6
Key Terms

1. parallel
2. perpendicular
3. vertical
4. horizontal

Practice

1. No
2. Yes

Objective A

1. Yes
3. No

Objective B

5. No
7. No
9. Neither

Section 9.7
Key Terms

1. solution
2. linear inequality
3. half-plane

4. test point

Practice

1. Yes

2. $x + y \leq 5$

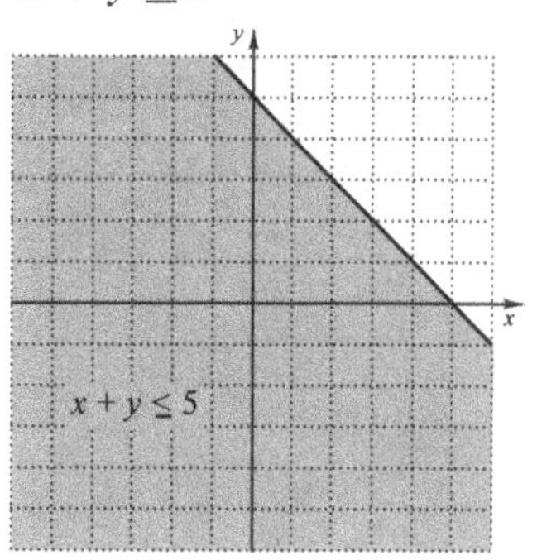

Objective A

1. No

3. Yes

Objective B

5.

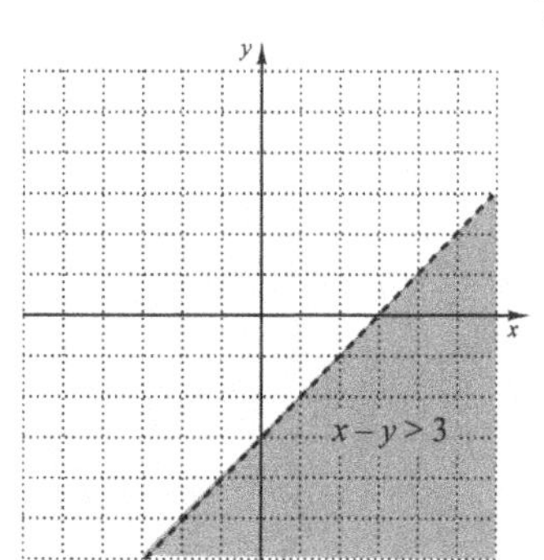

7.

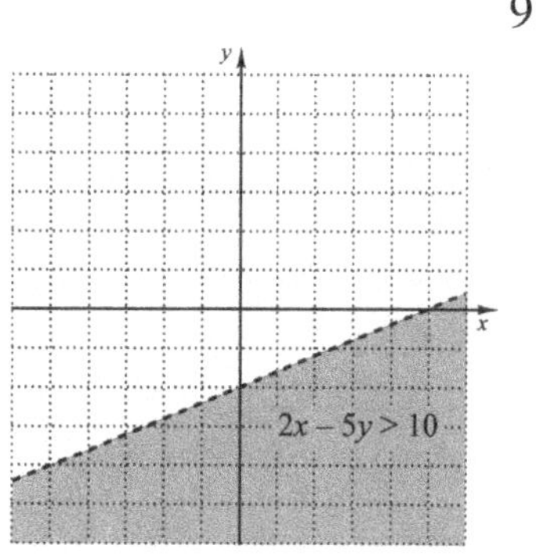

9.

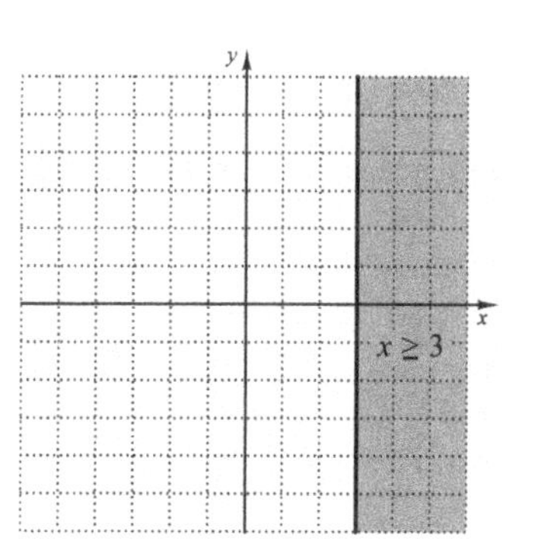

Chapter 10 POLYNOMIALS: OPERATIONS

Section 10.1

Key Terms

1. exponent
2. base
3. x-squared
4. x-cubed

Practice

1. $y \cdot y \cdot y \cdot y \cdot y$
2. 1
3. –32
4. x^7
5. xy^3
6. $\frac{2}{n^5}$

Objective A

1. $7 \cdot 7 \cdot 7$

3. $8x \cdot 8x \cdot 8x \cdot 8x$

Objective B

5. 1

7. 1

Objective C

9. 32

11. 13

Objective D

13. 7^{10}
15. x^{21}

Objective E

17. 11^5
19. t^9

Objective F

21. x^5
23. $\frac{1}{c}$

Section 10.2

Key Terms

1. exponential notation
2. scientific notation

Practice

1. b^{12}
2. $\frac{64a^6}{b^{10}}$
3. $\frac{y^8}{49}$
4. 0.000301
5. 5.67×10^5
6. 9.45×10^{-5}
7. 7.5×10^{11}
8. 7.8528×10^9 L

Objective A

1. 3^8
3. x^{14}

Objective B

5. $\frac{n^{20}}{m^8}$
7. $\frac{x^{10}}{9}$

Objective C

9. 6.4×10^8
11. 0.0000432

Objective D

13. 1.435×10^3
15. 4.0×10^7

Objective E

17. 3.6×10^{-7} m
19. 7.2×10^{32} neutrinos

Section 10.3

Key Terms

1. monomial
2. value
3. like terms
4. coefficient
5. degree
6. descending order
7. binomial
8. trinomial

Practice

1. –8
2. $-5y^4,\ 3y^2, -2$
3. $4y^5$ and $-3y^5$; -7 and 4
4. –8, 4, –7
5. $4x^3+x^2-5$
6. $-3x^2+9x-9$
7. 3
8. $x^3,\ x$
9. monomial

Objective A

1. -13
3. -5

Objective B

5. $5, -4x^2, x$

Objective C

7. $4x^3$ and $2x^3; -6x^2$ and $-x^2$
9. y^4 and $-y^4; 3y$ and $-5y$

Objective D

11. $-4, -5, 1.3, 8$

Objective E

13. $-9x$
15. $\frac{5}{4}x^4 - x^2 - 11$

Objective F

17. $x^5 + 5x^3 + 2x^2 - x - 6$
19. $9x^2 - 9x + 11$

Objective G

21. 4, 1, 0; 4
23. 0, 3, 5, 1; 5

Objective H

25. x^2
27. $4x^5 + 0x^4 - x^3 + x^2 + 0x - 1;$
    $4x^5 \quad - x^3 + x^2 \quad - 1$

Objective I

29. trinomial
31. binomial

Section 10.4

Key Terms

1. sign
2. opposite

Practice

1. $7x^3 - 8x^2 + 3$
2. $x^3 - 3x + 4$
3. $-4x^3 - x^2 - x + 10$
4. $22y$

Objective A

1. $3x^3 - 4$
3. $7t^4 + t^3 - 8t^2 - 6t + 7$

Objective B

5. $-(2y^2 - 4y);\ -2y^2 + 4y$
7. $3x^2 + 2x - 14$

Objective C

9. $-11t^2 - t$
11. $-6.9y^4 - 1.4y^2 - 3$

Objective D

13. $12x + 30$
15. $(y+2)(y+2);\ y^2 + 2y + 2y + 4$, or $y^2 + 4y + 4$

Section 10.5
Key Terms
1. term
2. collect

Practice
1. $-10n^6$
2. $2x^4 - 6x^3 - 10x^2$
3. $2x^2 - 5x - 7$
4. $x^5 + x^3 + x^2 - 12x + 4$

Objective A
1. $-16x^2$
3. $18x^4y^6$

Objective B
5. $12x^2 + 8x$
7. $18x^3 - 30x^2 + 24x$

Objective C
9. $x^2 + 9x + 8$
11. $4x^2 - 12x - 9$

Objective D
13. $x^3 - x^2 - 4$
15. $6n^4 + 13n^3 - 24n^2 - 22n + 15$

Section 10.6
Key Terms
1. binomials
2. FOIL
3. difference
4. square

Practice
1. $3y^2 + y - 10$
2. $x^2 - 36$
3. $4x^2 + 4x + 1$
4. $-9y^2 + 1$

Objective A
1. $a^2 - a - 42$
3. $1 - 3x - 40x^2$

Objective B
5. $9x^2 - 49$
7. $x^{10} - 4x^2$

Objective C
9. $x^2 + 10x + 25$
11. $36 - 24x^5 + 4x^{10}$

Objective D
13. $-6x^4 + 8x^3 + 14x^2$
15. $y^3 + 8$

Section 10.7
Key Terms
1. degree
2. like terms

Practice
1. -1
2. 7
3. $4xy^2 - 2x^2y + 10$
4. $2x^3y^2 + 5x^2y^2 + x^2y - 7xy^2$
5. $-2a^3b^2 - 5a^2b + ab^2 - 2ab$

6. $4x^2+20xy+25y^2$

Objective A

1. –59
3. 1423 calories

Objective B

5. Coefficients: 26, –10, 15; degrees: 7, 3, 0; 7

Objective C

7. $c-3d$
9. $22xy+13xz$

Objective D

11. $-x^2-2xy-5y^2$
13. $-u^2-4uv$

Objective E

15. $2x^3-3x^3y+xy^3$
17. $5a^5-2b+6a$

Objective F

19. $8x^2-2xy-3y^2$
21. $c^2+4cd+4d^2$

Section 10.8
Key Terms

1. quotient
2. divisor
3. dividend
4. remainder

Practice

1. $y^2-\frac{5}{4}y+3$
2. $x-6+\frac{-7}{x-2}$

Objective A

1. $-6x^3$
3. $-10x^2+8x+\frac{9}{2}$

Objective B

5. $x+5+\frac{8}{x-1}$
7. x^3+5

Chapter 11 POLYNOMIALS: FACTORING

Section 11.1
Key Terms

1. factor
2. factor
3. factorization
4. factoring by grouping

Practice

1. $4xy^2$
2. $3x^5(x^3-10x+5)$
3. $(x+3)(x^2-7)$

Objective A

1. x
3. $11x^2y$

Objective B

5. $5x^2(2x^3-1)$
7. $0.7x(2x^3-5x^2+6x+10)$

Objective C

9. $(x+1)(x^3+3)$
11. $(z+1)(2z^2+3)$

Section 11.2

Key Terms

1. leading coefficient
2. prime
3. positive
4. negative

Practice

1. $(x-1)(x-8)$

Objective A

1. $(p-7)(p+2)$
3. $(x+11)^2$
5. $(x-0.3)(x+0.2)$

Section 11.3

Key Terms

1. First
2. Last
3. Inside
4. Outside

Practice

1. $2(3z+1)(z+2)$

Objective A

1. $(3x+1)(x-1)$
3. $(7a^2+1)(2a^2+3)$

Section 11.4

Key Terms

1. common factor
2. leading coefficient
3. sum
4. split
5. grouping
6. multiplying

Practice

1. $(4x+1)(2x-1)$

Objective A

1. $(2x+5)(3x+2)$
3. $(7x-1)(2x-9)$
5. Prime

Section 11.5

Key Terms

1. trinomial square
2. difference of squares
3. sum of squares
4. factored completely

Practice

1. No
2. $(4x+1)^2$
3. Yes
4. $3(2+3x)(2-3x)$

Objective A

1. No
3. No

Objective B

5. $(x+6)^2$
7. $3(2a-7)^2$

Objective C

9. Yes
11. Yes

Objective D

13. $(8x+y)(8x-y)$
15. $3(x^2+4)(x+2)(x-2)$

Section 11.6

Key Terms

1. difference
2. perfect square
3. grouping
4. completely

Practice

1. $5(3x+4)(x-1)$

Objective A

1. $x(x-4)(x^2+5)$
3. $(4x-1)(u^2+3v)$
5. Prime

Section 11.7

Key Terms

1. quadratic equation
2. products
3. zero
4. factor

Practice

1. $-\frac{5}{6}, 0$
2. $-\frac{7}{4}, \frac{7}{4}$

Objective A

1. –3, 10
3. $\frac{1}{6}, \frac{1}{12}$

Objective B

5. –7, 3
7. –5, –3
9. –1, 2

Section 11.8

Key Terms

1. right
2. hypotenuse
3. legs
4. Pythagorean

Practice

1. Height: 10 cm; base: 6 cm

Objective A

1. Length: 26 in., width: 13 in.
3. 12 people

Chapter 12 RATIONAL EXPRESSIONS AND EQUATIONS

Section 12.1

Key Terms

1. rational
2. equivalent
3. multiply
4. simplify

Practice

1. −4, 4
2. $\frac{8x(x+1)}{7y(x+1)}$
3. $\frac{c-6}{c+6}$
4. $\frac{(t-5)(t-2)}{t^2(t+2)}$

Objective A

1. 0
3. $\frac{5}{4}$

Objective B

5. $\frac{(5a^2)(6c^2)}{(5a^2)(7d^4)}$
7. $\frac{(1-x)(-1)}{(2-x)(-1)}$

Objective C

9. $\frac{x+2}{3x^2+1}$
11. $-x-10$

Objective D

13. $\frac{x+1}{x-1}$
15. $\frac{2(y-3)(y+3)}{(y-8)(y-1)}$

Section 12.2

Key Terms

1. interchange
2. multiply

Practice

1. $\frac{a+3}{x+y}$
2. $\frac{(x+5)(x-3)}{(x-2)(x+2)}$

Objective A

1. $\frac{y}{3}$
3. $n+3$

Objective B

5. $\frac{(x-1)^2}{x^2+1}$
7. $\frac{(y+1)(y+2)^2}{9(y+3)}$

Section 12.3

Key Terms

1. least common multiple
2. least common denominator

Practice

1. 84
2. $\frac{11}{20}$
3. $m(m-3)(m+3)^2$

Objective A

1. 48
3. 600

Objective B

5. $\frac{53}{72}$

Objective C

7. $60x^2y^4$
9. $8a(a-2)(a+5)$

Section 12.4

Key Terms

1. least common denominator
2. equivalent expression
3. numerators
4. simplify

Practice

1. $\frac{8x-14}{(x+1)(x-4)(x+2)}$

Objective A

1. $\frac{6x-3}{x+5}$
3. $\frac{7x-41}{(x-7)(x-3)(x+1)}$

Section 12.5

Key Terms

1. denominator
2. least common denominator
3. numerators
4. multiply

Practice

1. $\frac{-x-2}{(x+6)(x-6)}$
2. $\frac{2y-23}{(y+5)(y-4)}$

Objective A

1. $\frac{1}{x-1}$
3. $\frac{2}{(x+4)(x-4)}$

Objective B

5. $\frac{19t-7}{t-2}$
7. $\frac{2x+1}{x-3}$

Section 12.6
Key Terms
1. rational
2. clear
3. LCM
4. check

Practice
1. –2, 4

Objective A
1. 1, 3
3. No solution

Section 12.7
Key Terms
1. ratio
2. rate
3. proportion
4. similar

Practice
1. car: 60 km/h; van: 50 km/h
2. 15 pages

Objective A
1. $1\frac{5}{7}$ hr
3. Maya: 65 km/h; Tara 45 km/h

Objective B
5. 24 mpg
7. 45 trout
9. $\frac{12}{5}$

Section 12.8
Practice
1. $\frac{6}{y^2}$

Objective A
1. $-\frac{1}{x}$
3. $\frac{x-5}{x-3}$

Section 12.9
Key Terms
1. direct
2. proportionality
3. inverse
4. variation

Practice
1. $y=\frac{3}{2}x$
2. \$231
3. $y=\frac{36}{x}$
4. $2\frac{2}{3}$ days

Objective A

1. $y = 7x$; 210
3. $y = \frac{6}{5}x$; 36

Objective B

5. (a) $t = \frac{3}{2}C$; (b) 12 teaspoons
7. 25 min

Objective C

9. $y = \frac{60}{x}$; 3
11. $y = \frac{360}{x}$; 18

Objective D

13. (a) $N = \frac{40}{P}$; (b) 1 gal
15. $\frac{8}{15}$ ampere

Chapter 13 SYSTEMS OF EQUATIONS

Section 13.1

Key Terms

1. system
2. solution
3. graph
4. intersect
5. are parallel
6. have the same graph

Practice

1. Yes
2. (4, 3)

Objective A

1. No
3. Yes

Objective B

5. No solution
7. (–3, 2)

Section 13.2

Key Terms

1. algebraic
2. substitute
3. solve
4. translate

Practice

1. (–2, –4)
2. (–1, 4)
3. Length: 4 ft; width: 2 ft

Objective A

1. (2, 10)
3. (–6, –1)

Objective B

5. (4, –1)
7. (–2, –3)

Objective C

9. Length: 144 m; width: 70 m
11. Length: 44 ft; width: 20 ft

Section 13.3

Key Terms

1. elimination
2. opposites
3. false
4. true

Practice

1. (–3, 5)
2. (3, –2)

Objective A

1. (7, 4)
3. (–7, 13)

Objective B

5. (–6, –4)
7. Infinite number of solutions

Section 13.4

Practice

1. Student: 115; non-student: 105

Objective A

1. Adults: 250; youth: 325
3. Small boxes: 3; large boxes: 7

Section 13.5

Key Terms

1. motion
2. distance
3. rate
4. time

Practice

1. 5 hr after the first train leaves

Objective A

1. 3 hr
3. In $\frac{5}{6}$ hr, or in 50 min

Chapter 14 RADICAL EXPRESSIONS AND EQUATIONS

Section 14.1

Key Terms

1. square
2. principal
3. radical
4. radicand

Practice

1. –8, 8
2. –10
3. 3.239
4. 19 spaces
5. $3a$
6. Yes
7. $x - 12$

Objective A

1. –20, 20
3. –9

Objective B

5. 22.023
7. 1.569

Objective C

9. 6 miles
11. 2.7 miles

Objective D

13. $17w$
15. x^2y-11

Objective E

17. yes
19. no

Objective F

21. $6a$
23. $h-10$

Section 14.2

Key Terms

1. nonnegative
2. radicands
3. perfect squares
4. half

Practice

1. $4w\sqrt{3}$
2. $16a^3\sqrt{2a}$
3. $4ab^2\sqrt{15ab}$

Objective A

1. $5\sqrt{t}$
3. $6d\sqrt{5}$

Objective B

5. $z^6\sqrt{z}$
7. $5(a+9)^8$

Objective C

9. $7\sqrt{6}$
11. $18s^5t^6\sqrt{st}$

Section 14.3

Key Terms

1. radicands
2. separately
3. rationalizing
4. perfect square

Practice

1. $\frac{5}{w}$
2. $\frac{4}{y^2}$
3. $\frac{3\sqrt{5}}{10}$

Objective A

1. 4
3. 2

Objective B

5. $\frac{7}{11}$
7. $\frac{2x}{15}$

Objective C

9. $\sqrt{5}$
11. $\frac{y^3\sqrt{x}}{x}$

Section 14.4
Key Terms
1. like radicals
2. conjugates

Practice
1. $\sqrt{5}$
2. 14
3. $\frac{28+7\sqrt{3}}{13}$

Objective A
1. $11\sqrt{7}$
3. 0

Objective B
5. $\sqrt{42}-3\sqrt{6}$
7. 9

Objective C
9. $\frac{\sqrt{30}-2\sqrt{3}}{3}$
11. $\frac{14+7\sqrt{x}}{4-x}$

Section 14.5
Key Terms
1. radical equation
2. principle of squaring

Practice
1. 6
2. 6
3. 7396 m

Objective A
1. 19
3. 10

Objective B
5. 3, 7
7. 23

Objective C
9. About 241 mi
11. 180 ft; 320 ft

Section 14.6
Key Terms
1. right triangle
2. hypotenuse
3. legs
4. Pythagorean equation

Practice
1. $a = 5$
2. About 12.207 ft

Objective A
1. $\sqrt{33} \approx 5.745$
3. 1

Objective B
5. $\sqrt{50{,}000}$ ft ≈ 223.607 ft
7. $\sqrt{189}$ ft ≈ 13.748 ft

Chapter 15 QUADRATIC EQUATIONS

Section 15.1

Key Terms

1. quadratic
2. standard
3. first
4. zero

Practice

1. $x^2 - 4x + 6 = 0$; $a = 1$, $b = -4$, $c = 6$
2. $-\frac{3}{2}$, 0
3. 4, 5
4. 6 sides

Objective A

1. $4x^2 - x + 3 = 0$; $a = 4$, $b = -1$, $c = 3$
3. $3x^2 + 6x + 2 = 0$; $a = 3$, $b = 6$, $c = 2$

Objective B

5. –10, 0
7. –2, 0

Objective C

9. –9, 6
11. –6, 4

Objective D

13. 54 diagonals

Section 15.2

Key Terms

1. two real
2. no real-number
3. only one
4. complete

Practice

1. $\frac{\sqrt{6}}{3}, -\frac{\sqrt{6}}{3}$
2. $-3 \pm \sqrt{7}$
3. $-1 \pm \sqrt{6}$
4. About 8.1 sec

Objective A

1. 12, –12
3. $\frac{4}{3}, -\frac{4}{3}$

Objective B

5. –17, 3
7. $6 \pm \sqrt{35}$

Objective C

9. 5, 11
11. $9 \pm \sqrt{57}$

Objective D

13. About 10.2 seconds

Section 15.3
Key Terms
1. principle of square roots
2. standard form
3. factoring
4. quadratic formula

Practice

1. $\dfrac{3 \pm \sqrt{65}}{4}$
2. 4.6, 0.4

Objective A

1. –2, 10
3. $4 \pm \sqrt{10}$

Objective B

5. –0.6, 3.6
7. –2.8, –1.2

Section 15.4
Practice

1. $d = \sqrt{\dfrac{A}{c}}$

Objective A

1. $y = \dfrac{xz}{x - z}$
3. $w = \dfrac{\sqrt{5xy}}{y}$
5. $a = \dfrac{p - q}{bq}$

Section 15.5
Practice
1. 2 km/h

Objective A
1. Length: 17 cm; width: 8 cm
3. 0 km/h (no wind), or 50 km/h

Section 15.6
Key Terms
1. parabola
2. vertex
3. line of symmetry
4. discriminant

Practice

1.

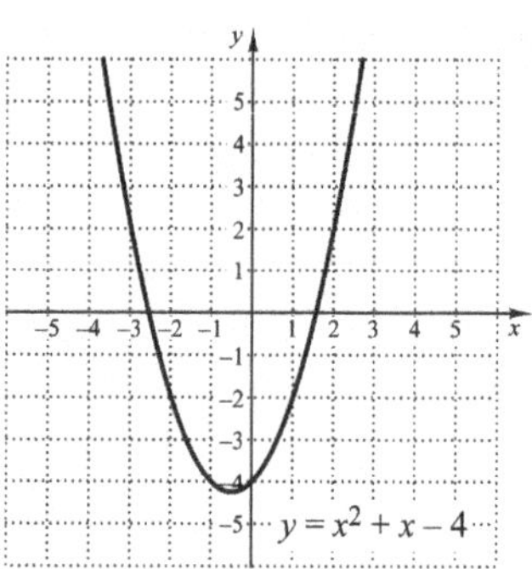

2. $(1-\sqrt{6}, 0)$; $(1+\sqrt{6}, 0)$

Objective A

1.

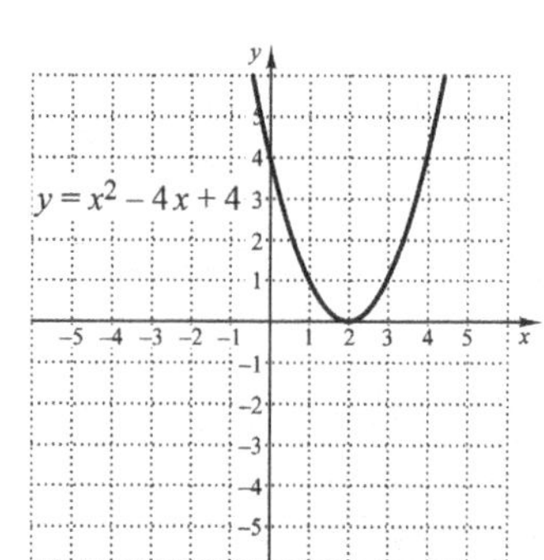

3.

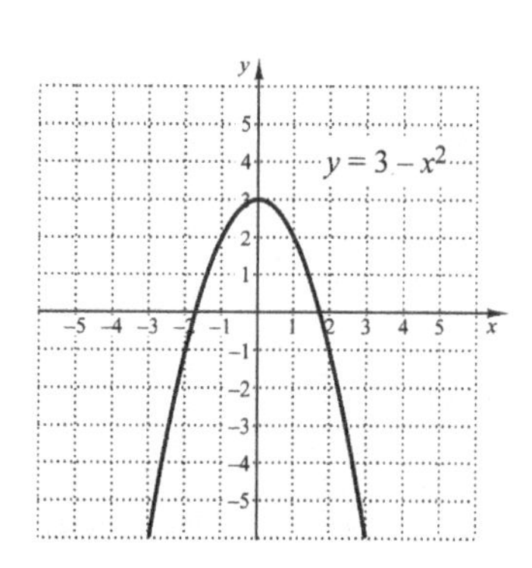

Objective B

5. $(-10, 0)$; $(0, 0)$

7. $(1-\sqrt{3}, 0)$; $(1+\sqrt{3}, 0)$

Section 15.7

Key Terms

1. function
2. relation
3. domain
4. range
5. linear
6. quadratic
7. input
8. output

Practice

1. Yes

2. $g(2) = 1$

3. $f(x) = x - 1$

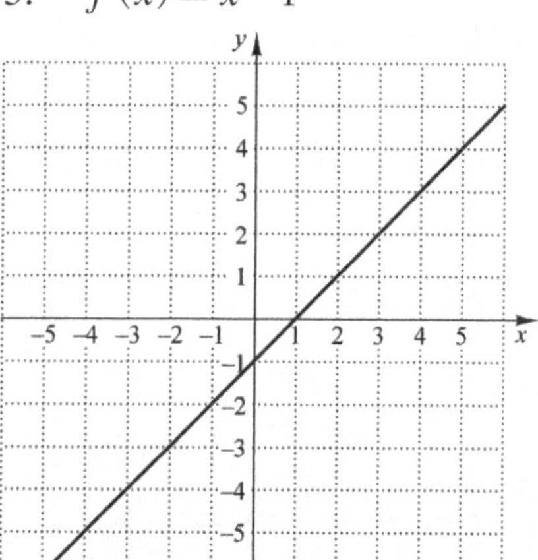

4. Yes

5. About 52 million Americans

Objective A

1. Yes
3. No

Objective B

5. (a) 7; (b) –11; (c) –1; (d) 1
7. (a) 18; (b) 18; (c) 3; (d) 6

Objective C

9.

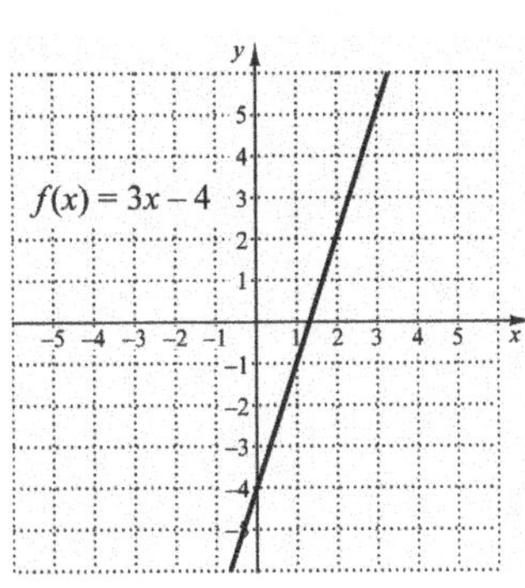

11.

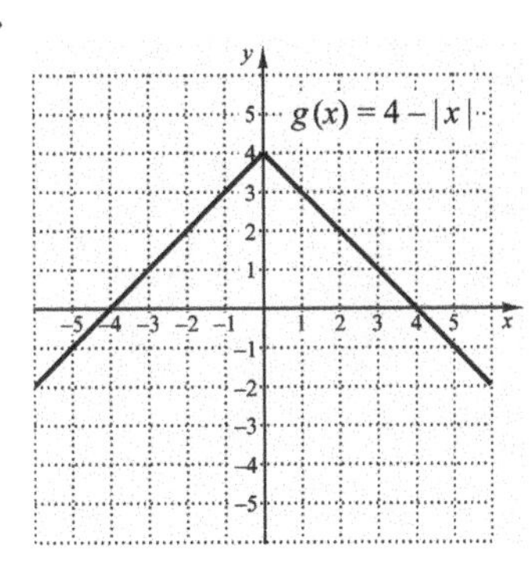

Objective D

13. Yes
15. No

Objective E

17. About 32 million connections
19. About 12 million connections